T0292753

Programmable Logic Controllers

Programmable Logic Controllers

Sixth Edition

W. Bolton

AMSTERDAM • BOSTON • HEIDELBERG • LONDON
NEW YORK • OXFORD • PARIS • SAN DIEGO
SAN FRANCISCO • SINGAPORE • SYDNEY • TOKYO
Newnes is an imprint of Elsevier

Newnes is an imprint of Elsevier
The Boulevard, Langford Lane, Kidlington, Oxford OX5 1GB
225 Wyman Street, Waltham MA 02451

Fifth edition 2009
Sixth edition 2015

Library of Congress Cataloging-in-Publication Data
A catalog record for this book is available from the Library of Congress

British Library Cataloguing-in-Publication Data
A catalogue record for this book is available from the British Library

ISBN 978-0-12-802929-9

For information on all publications visit our website at http://store.elsevier.com

Publisher: Jonathan Simpson
Acquisition Editor: Tim Pitts
Editorial Project Manager: Charlotte Kent
Production Project Manager: Melissa Read
Designer: Maria Ines Cruz
Printed and bound in the USA

Contents

Preface

Technological advances in recent years have resulted in the development of the programmable logic controller (PLC) and a consequential revolution of control engineering. This book, an introduction to PLCs, aims to ease the tasks of practicing engineers coming into contact with PLCs for the first time. It also provides a basic course for students in curricula such as the English technicians' courses for Nationals and Higher Nationals in Engineering, giving full syllabus coverage of the National and Higher National in Engineering units, company training programs, and serving as an introduction for first-year undergraduate courses in engineering.

The book addresses the problem of various programmable control manufacturers using different nomenclature and program forms by describing the principles involved and illustrating them with examples from a range of manufacturers. The text includes:

- The basic architecture of PLCs and the characteristics of commonly used input and outputs to such systems
- A discussion of the number systems: denary, binary, octal, hexadecimal, and BCD
- A painstaking methodical introduction, with many illustrations, describing how to program PLCs, whatever the manufacturer, and how to use internal relays, timers, counters, shift registers, sequencers, and data-handling facilities
- Consideration of the standards given by IEC 61131-3 and the programming methods of ladder, functional block diagram, instruction list, structured text, and sequential function chart
- Many worked examples, multiple-choice questions, and problems to assist the reader in developing the skills necessary to write programs for programmable logic controllers, with answers to all multiple-choice questions and problems given at the end of the book

Prerequisite Knowledge Assumed

This book assumes no background in computing. However, a basic knowledge of electrical and electronic principles is desirable.

Changes from the Fifth Edition

The fourth edition of this book was a complete restructuring and updating of the third edition and included a more detailed consideration of IEC 61131-3, including all the programming methods given in the standard, and the problems of safety, including a discussion of emergency stop relays and safety PLCs. The fifth edition built on this foundation by providing more explanatory text, more examples, and more problems and includes with each chapter a summary of its key points. The sixth edition has a new Chapter 1 with a comparison of relay, microprocessor and PLC controlled systems, an updated consideration of commercial PLCs, and more discussion of the merits and problems of the various PLC programming methods given by the IEC 61131 standard. Chapter 2 has had some new material on sensors included. The discussion of sequential function charts in Chapter 6 has been rewritten to give more detail of the method. In Chapter 10 the part concerned with the sequencer has been rewritten. The section of Chapter 13 concerned with forcing has been extended and Chapter 14 has had more case studies added.

Aims

This book aims to enable the reader to:

- Identify and explain the main design characteristics, internal architecture, and operating principles of programmable logic controllers.
- Use PLCs of different sizes and from different manufacturers.
- Use commonly used input and output devices with PLC systems, taking account of their characteristics.
- Explain the processing of inputs and outputs by PLCs so that input and output systems can be used correctly with PLCs.
- Use communication links involved with PLC systems, recognizing the protocols and networking methods involved.
- Use ladder programs involving internal relays, timers, counters, shift registers, sequencers, and data handling to tackle applications.
- Identify safety issues with PLC systems so they can be used safely.
- Use methods used for fault diagnosis, testing, and debugging.

Structure of the Book

The following figure outlines the structure of the book.

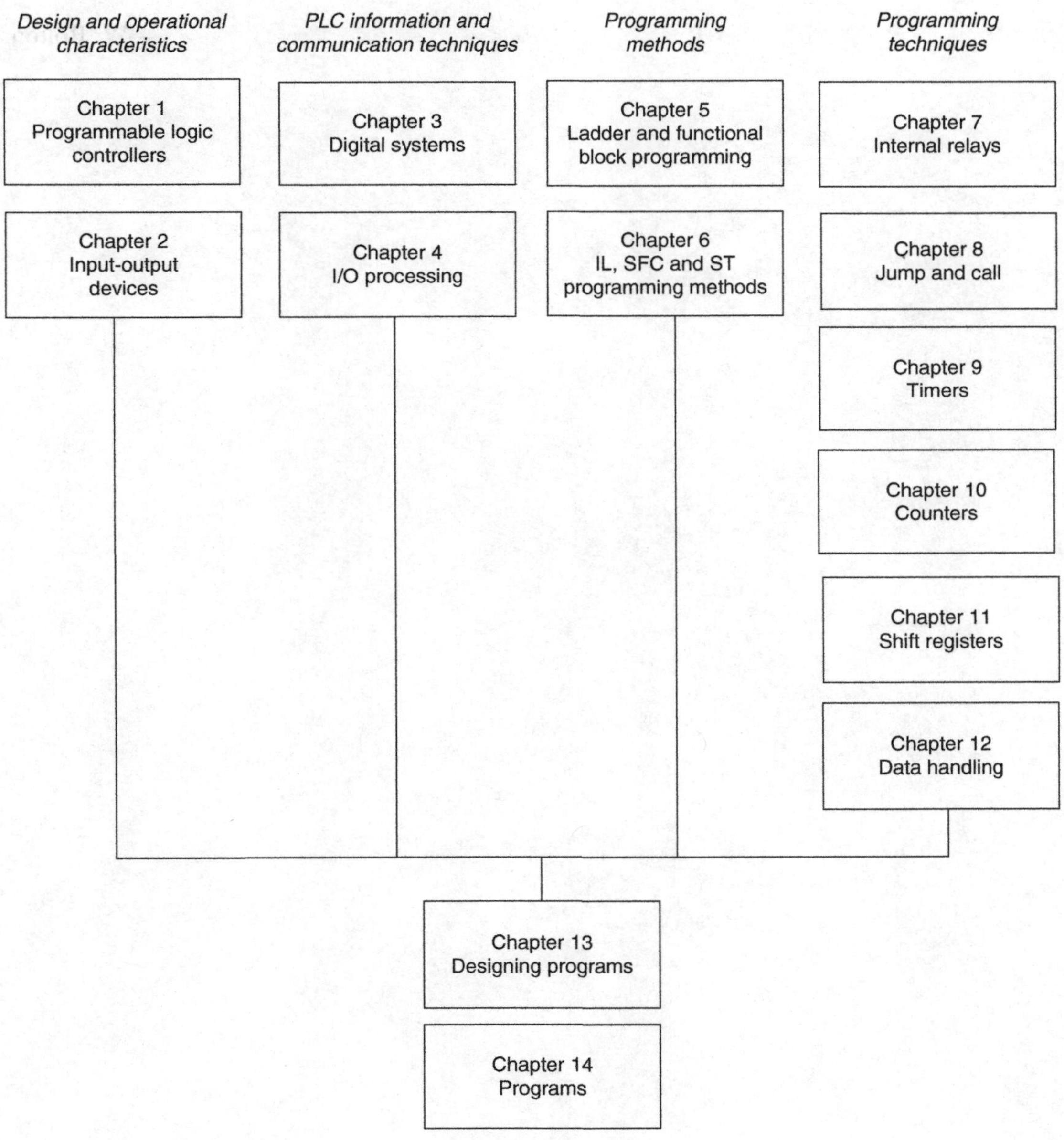

Acknowledgments

I am grateful to the many reviewers of the various editions of this book for their helpful feedback and comments.

—W. Bolton

CHAPTER 1

Programmable Logic Controllers

This chapter is an introduction to the programmable logic controller (PLC) and its general function, hardware forms, and internal architecture. PLCs are widely used for a range of automation tasks in areas such as industrial processes in manufacturing. This overview is followed by more detailed discussion in the following chapters. For a summary of the history, development, features, and comparison with other control systems, see the Wikipedia entry for Programmable logic controller.

1.1 Controllers

What type of task might a control system handle? It might be required to control a sequence of events, maintain some variable constant, or follow some prescribed change. For example, the control system for an automatic drilling machine (Figure 1.1a) might be required to start lowering the drill when the workpiece is in position, start drilling when the drill reaches the workpiece, stop drilling when the drill has produced the required depth of hole, retract the drill, and then switch off and wait for the next workpiece to be put in position before repeating the operation. Another control system (Figure 1.1b) might be used to control the number of items moving along a conveyor belt and direct them into a packing case. The inputs to such control systems might come from switches being closed or opened; for example, the presence of the workpiece might be indicated by it moving against a switch and closing it, or other sensors such as those used for temperature or flow rates. The controller might be required to run a motor to move an object to some position or to turn a valve, or perhaps a heater, on or off.

What form might a controller have? For the automatic drilling machine, we could wire up electrical circuits in which the closing or opening of switches would result in motors being switched on or valves being actuated. Thus, as a result, we might have a relay (Figure 1.2) closing or opening contacts which, in turn, switches on the current to a motor and causes the drill to rotate (Figure 1.3). Another switch might be used to activate a relay and switch on the current to a pneumatic or hydraulic valve, which results in pressure being switched to drive a piston in a cylinder and so results in the workpiece being pushed into the required position. Such electrical circuits would have to be specific to the automatic drilling machine. For controlling the number of items packed into a packing case, we could likewise wire up

W. Bolton: Programmable Logic Controllers, Sixth Edition. http://dx.doi.org/10.1016/B978-0-12-802929-9.00001-7

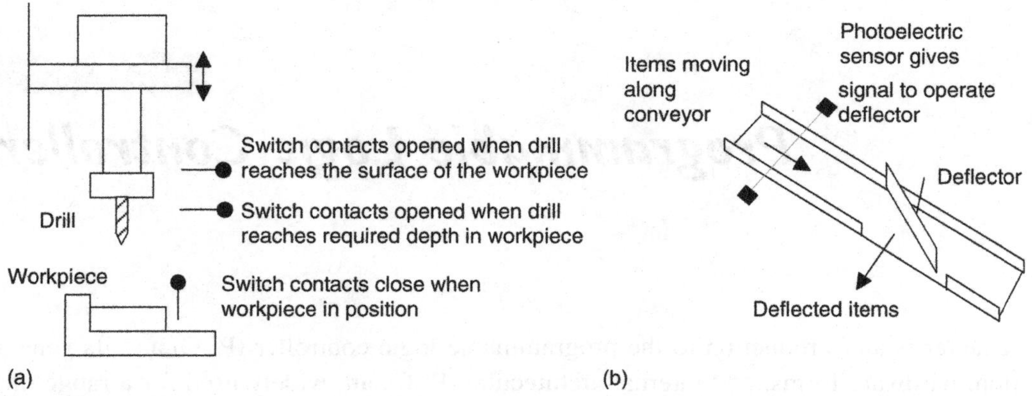

Figure 1.1: An example of a control task and some input sensors: (a) an automatic drilling machine; (b) a packing system.

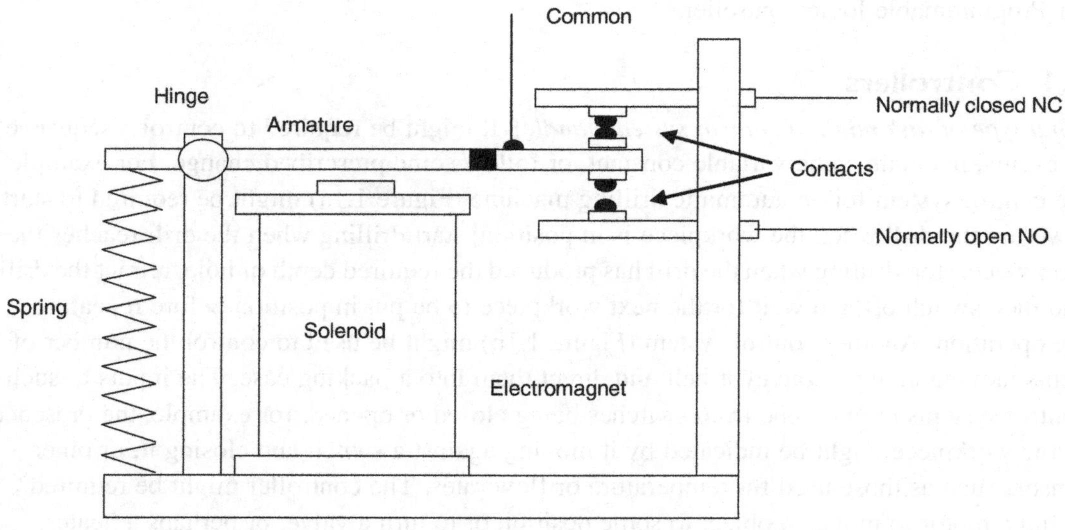

Figure 1.2: A basic relay.

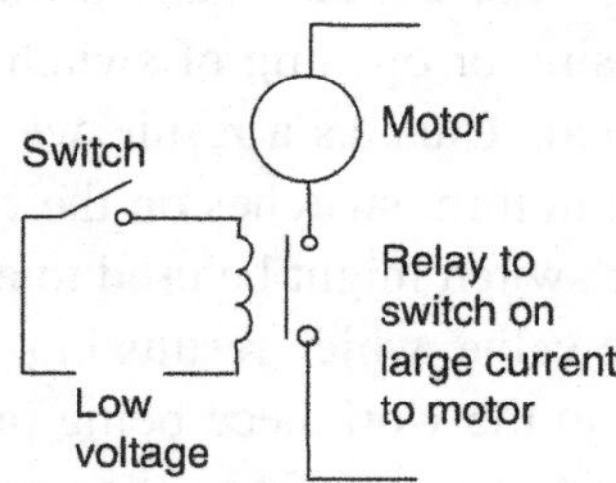

Figure 1.3: A control circuit.

electrical circuits involving sensors and motors. However, the controller circuits we devised for these two situations would be different. In the "traditional" form of control system, the rules governing the control system and when actions are initiated are determined by the wiring. When the rules used for the control actions are changed, the wiring has to be changed.

1.1.1 Relay-Controlled Systems

Relay-controlled systems are hard-wired systems. Figure 1.2 shows the basic elements of a simple relay. When a current is switched on to flow through the relay solenoid, normally-closed (NC) contacts open and normally-open (NO) contacts close. These contacts can be used to give control in a system. As an illustration consider a relay being used to operate a pneumatic or hydraulic valve, this then results in pressure being applied to drive a piston to move a workpiece. We can represent the situation by a control drawing. Figure 1.4 shows the standard symbols used for relays and Figure 1.5 shows the control drawing with the vertical lines representing the power rails and the horizontal lines to systems connected between them. The sequence of events is read from the top horizontal line downwards. Thus, in the top line of Figure 1.5(a), when the Off–On switch is closed, the relay is activated. This closes the contacts on the second line and so the solenoid valve is switched on. A more usual control drawing is shown in Figure 1.5(b) which has the relay switched

Figure 1.4: Relay symbols.

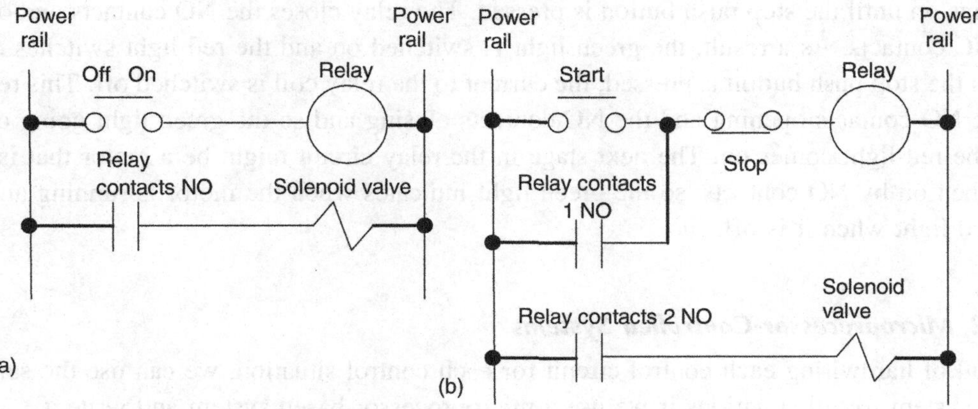

Figure 1.5: Relay-controlled system control drawings.

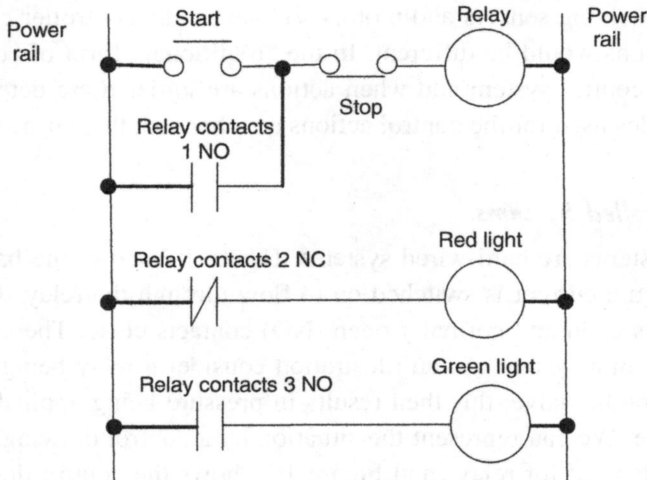

Figure 1.6: Relay circuit to control red and green lights.

on by a momentary NO push-button switch. This closes two sets of contacts. Contacts 1 latch the push button switch so that when the push stops there is still connection of power to the relay. Contacts 2 switch on the solenoid valve. The relay, and hence power to the solenoid valve, is switched off when the normally closed push-button switch is pressed. The control drawings are obviously only part of the control system as there will need to be further lines for when the solenoid valve has moved the workpiece the required distance so that it stops its action.

Figure 1.6 shows another example of a relay control system. When the start push button is closed, the relay coil is switched on and latches the push button switch so that the relay remains on until the stop push button is pressed. The relay closes the NO contacts and opens the NC contacts. As a result, the green light is switched on and the red light switches off. When the stop push button is pressed, the current to the relay coil is switched off. This results in the NO contacts opening and the NC contacts closing and so the green light going off and the red light comes on. The next stage in the relay circuit might be a motor that is switched on by NO contacts, so the green light indicates when the motor is running and the red light when it is off.

1.1.2 Microprocessor-Controlled Systems

Instead of hardwiring each control circuit for each control situation, we can use the same basic system for all situations if we use a microprocessor-based system and write a program to instruct the microprocessor how to react to each input signal from, say, switches and give the required outputs to, say, motors and valves. Thus we might have a program of the form:

```
If switch A closes

Output to motor circuit

If switch B closes

Output to valve circuit
```

By changing the instructions in the program, we can use the same microprocessor system to control a wide variety of situations.

As an illustration, the modern domestic washing machine uses a microprocessor system. Inputs to it arise from the dials used to select the required wash cycle, a switch to determine that the machine door is closed, a temperature sensor to determine the temperature of the water, and a switch to detect the level of the water. On the basis of these inputs the microprocessor is programmed to give outputs that switch on the drum motor and control its speed, open or close cold and hot water valves, switch on the drain pump, control the water heater, and control the door lock so that the machine cannot be opened until the washing cycle is completed.

1.1.3 The Programmable Logic Controller

A *programmable logic controller* (PLC) is a special form of microprocessor-based controller that uses programmable memory to store instructions and to implement functions such as logic, sequencing, timing, counting, and arithmetic in order to control machines and processes (Figure 1.7). It is designed to be operated by engineers with perhaps a limited knowledge of computers and computing languages. They are not designed so that only computer programmers can set up or change the programs. Thus, the designers of the PLC have preprogrammed it so that the control program can be entered using a simple, rather intuitive form of language (see Chapter 4). The term *logic* is used because programming is primarily concerned with implementing logic and switching operations; for example, if A *or* B occurs, switch on C; if A *and* B occurs, switch on D. Input devices (that is, sensors such as

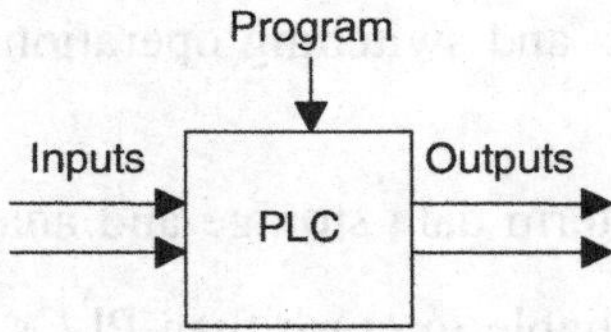

Figure 1.7: A programmable logic controller.

switches) and output devices (motors, valves, etc.) in the system being controlled are connected to the PLC. The operator then enters a sequence of instructions, a program, into the memory of the PLC. The controller then monitors the inputs and outputs according to this program and carries out the control rules for which it has been programmed.

PLCs have the great advantage that the same basic controller can be used with a wide range of control systems. To modify a control system and the rules that are to be used, all that is necessary is for an operator to key in a different set of instructions. There is no need to rewire. The result is a flexible, cost-effective system that can be used with control systems, which vary quite widely in their nature and complexity. When compared with relay systems, PLCs:

- Can easily implement changes as changes are implemented in software rather than more complex hardware modifications that would be the case with a relay system
- Can be readily expanded by adding new modules to the PLC whereas hardware changes are necessary with relay systems
- Are more robust and reliable than relay systems with their large number of mechanical components
- Are more compact than relay systems
- Require less maintenance than relay systems
- Can operate faster than relay systems.

PLCs are similar to computers, but whereas computers are optimized for calculation and display tasks, PLCs are optimized for control tasks and the industrial environment. Thus when compared to computers, PLCs:

- Are rugged and designed to withstand vibrations, temperature, humidity, and noise. The common personal computer is not designed for harsh environments.
- Have interfacing for inputs and outputs already inside the controller. PLCs in a rack format are easy to expand to tackle a larger number of inputs/outputs.
- Are easily programmed and have an easily understood programming language that is primarily concerned with logic and switching operations. As a consequence, they are more user friendly.
- They are not so good at long term data storage and analysis as personal computers.
- Personal computers are more liable to crash than PLCs that have greater reliability.

The first PLC was developed in 1969. PLCs are now widely used and extend from small, self-contained units for use with perhaps 20 digital inputs/outputs to modular systems that can be used for large numbers of inputs/outputs, handle digital or analog inputs/outputs, and carry out proportional-integral-derivative control modes. They are used in automation tasks for industrial processes in manufacturing such as machining, materials handling, automated assembly and packaging. However, for very simple automation tasks such as a household washing machine, a cheaper alternative is likely to be used. Where very demanding tasks are involved, for example aircraft flight control, a computer is likely to be used because of its ability to handle complex mathematics and its high speed of operation.

1.2 Hardware

Typically a PLC system has the basic functional components of processor unit, memory, power supply unit, input/output interface section, communications interface, and the programming device. Figure 1.8 shows the basic arrangement. The constituent elements are:

- The *processor unit* or *central processing unit* (CPU) is the unit containing the microprocessor. This unit interprets the input signals and carries out the control actions according to the program stored in its memory, communicating the decisions as action signals to the outputs.
- The *power supply unit* is needed to convert the mains AC voltage to the low DC voltage necessary for the processor and the circuits in the input and output interface modules.

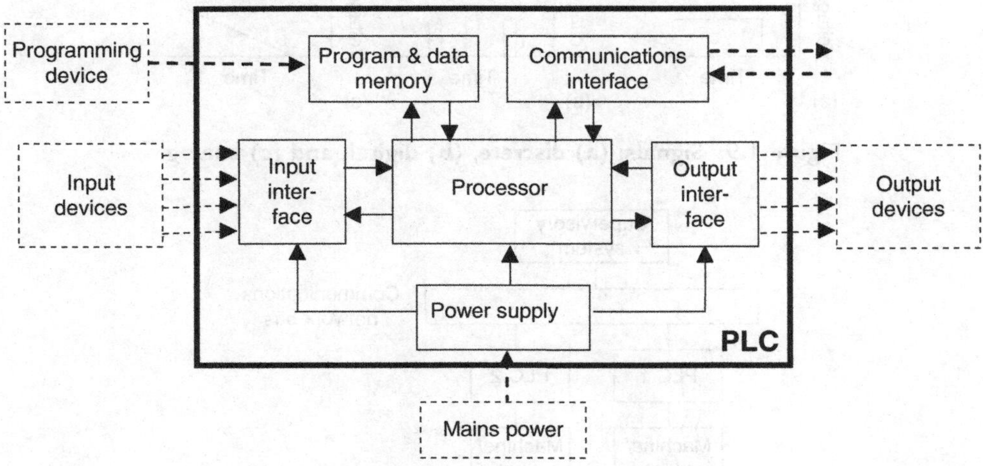

Figure 1.8: The PLC system.

- The *programming device* is used to enter the required program into the memory of the processor. The program is developed in the device and then transferred to the memory unit of the PLC.
- The *memory unit* is where the program containing the control actions to be exercised by the microprocessor is stored and where the data is stored from the input for processing and for the output.
- The *input and output sections* are where the processor receives information from external devices and communicates information to external devices. The inputs might thus be from switches, as illustrated in Figure 1.1a with the automatic drill, or other sensors such as photoelectric cells, as in the counter mechanism in Figure 1.1b, temperature sensors, flow sensors, or the like. The outputs might be to motor starter coils, solenoid valves, or similar things. (Input and output interfaces are discussed in Chapter 2.) Input and output devices can be classified as giving signals that are discrete, digital or analog (Figure 1.9). Devices giving *discrete* or *digital signals* are ones where the signals are either off or on. Thus a switch is a device giving a discrete signal, either no voltage or a voltage. *Digital* devices can be considered essentially as discrete devices that give a sequence of on/off signals. *Analog* devices give signals of which the size is proportional to the size of the variable being monitored. For example, a temperature sensor may give a voltage proportional to the temperature.
- The *communications interface* is used to receive and transmit data on communication networks from or to other remote PLCs (Figure 1.10). It is concerned with such actions as device verification, data acquisition, synchronization between user applications, and connection management.

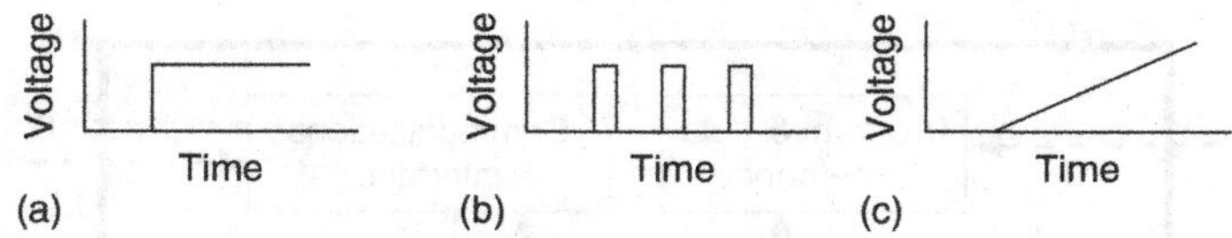

Figure 1.9: Signals: (a) discrete, (b) digital, and (c) analog.

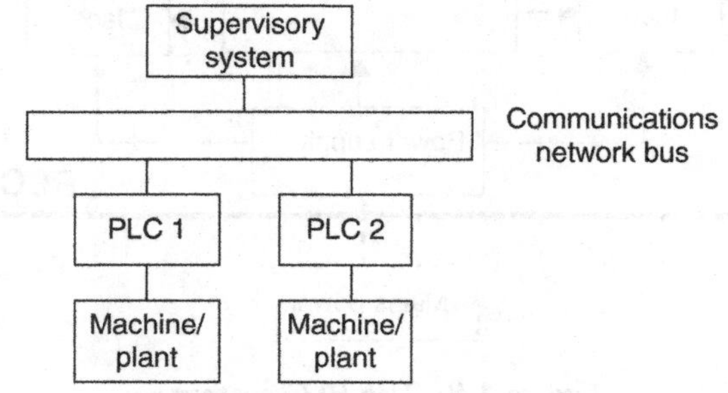

Figure 1.10: Basic communications model.

1.3 PLC Architecture

A PLC typically consists of a central processing unit (CPU) containing the system microprocessor, memory, and input/output circuitry. It can effectively be considered to be a unit containing vast numbers of separate relays, counters, timers and data storage units. These, however, do not exist physically in the PLC but are software-simulated.

The storage capacity of a memory unit is specified by the number of binary words that it can store. Thus, if a memory size is 256 words, it can store $256 \times 8 = 2048$ bits if 8-bit words are used and $256 \times 16 = 4096$ bits if 16-bit words are used. The term *byte* is used for a word of length 8 bits. Memory sizes are often specified in terms of the number of storage locations available, with 1K representing the number 2^{10}, that is, 1024. Thus a 4 Kbyte memory can store 4096 bytes, a 50 Kbyte memory 51 200 bytes.

1.3.1 Input/Output Unit

The input/output (I/O) unit in a PLC provides the circuitry for the interface between the system and the outside world, allowing for connections to be made through input/output channels to input devices such as sensors and output devices such as motors and solenoids. It is also through the input/output unit that programs are entered from a program panel. Every input/output point has a unique address that can be used by the CPU. It is like a row of houses along a road; number 10 might be the "house" used for an input from a particular sensor, whereas number 45 might be the "house" used for the output to a particular motor.

The input/output channels provide isolation and signal conditioning functions so that sensors and actuators can often be directly connected to them without the need for other circuitry (Figure 1.11). Electrical isolation from the external world is usually by means of *optoisolators* (the term *optocoupler* is also often used). Figure 1.12 shows the principle of an optoisolator. When a digital pulse passes through the light-emitting diode, a pulse of infrared radiation is produced. This pulse is detected by the phototransistor and gives rise to a voltage in that circuit. The gap between the light-emitting diode and the phototransistor gives electrical isolation, but the arrangement still allows for a digital pulse in one circuit to give rise to a digital pulse in another circuit.

Signal conditioning in the input channel, with isolation, enables a wide range of input signals to be supplied to it so that it is converted into a voltage compatible with the required for the microprocessor in the PLC (see Chapter 3 for more details). A range of inputs might be available with a larger PLC, such as 5 V, 24 V, 110 V, and 240 V digital/discrete, that is, on-off, signals. A small PLC is likely to have just one form of input, such as 24 V.

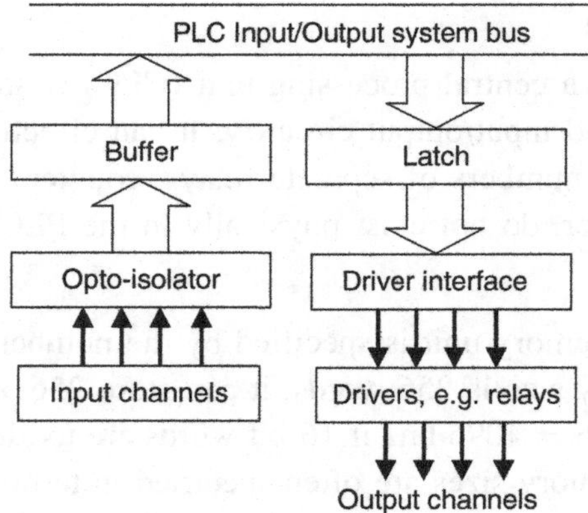

Figure 1.11: Architecture of PLC input/output channels.

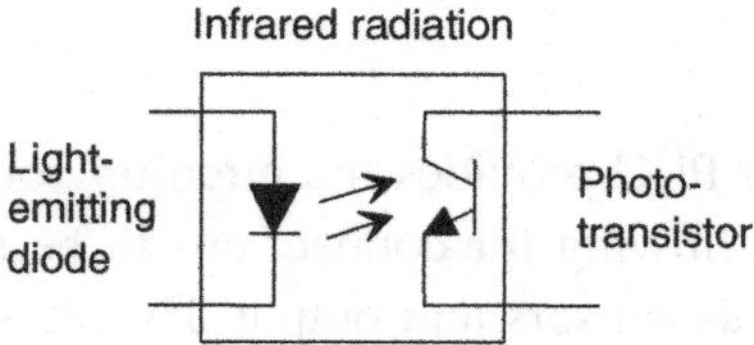

Figure 1.12: An optoisolator.

The output channels enable the PLC outputs to be available in a form suitable for direct connections to external circuits. Outputs are specified as being of relay type, transistor type, or triac type (see Chapter 3 for more details):

- With the *relay type*, the signal from the PLC output is used to operate a relay and is able to switch currents of the order of a few amperes in an external circuit. The relay not only allows small currents to switch much larger currents but also isolates the PLC from the external circuit. Relays are, however, relatively slow to operate. Relay outputs are suitable for AC and DC switching. They can withstand high surge currents and voltage transients.
- The *transistor type* of output uses a transistor to switch current through the external circuit. This gives a considerably faster switching action. It is, however, strictly for DC switching and is destroyed by overcurrent and high reverse voltage. For protection, either a fuse or built-in electronic protection is used. Optoisolators are used to provide isolation.
- *Triac* outputs, with optoisolators for isolation, can be used to control external loads that are connected to the AC power supply. It is strictly for AC operation and is very easily destroyed by overcurrent. Fuses are virtually always included to protect such outputs.

Thus, after signal conditioning with relays, transistors, or triacs, the output from the output channel might be a 24 V, 100 mA switching signal; a DC voltage of 110 V, 1 A, or perhaps 240 V; 1 A AC or 240 V, 2 A AC, from a triac output channel. With a small PLC, all the outputs might be of one type, such as 240 V AC, 1 A. With modular PLCs, however, a range of outputs can be accommodated by selection of the modules to be used.

1.3.2 Sourcing and Sinking

The terms *sourcing* and *sinking* are used to describe the way in which DC devices are connected to a PLC. With sourcing, using the conventional current flow direction as from positive to negative, an input device receives current from the input module, that is, the input module is the source of the current (Figure 1.13a). With sinking, using the conventional current flow direction, an input device supplies current to the input module, that is, the input module is the sink for the current (Figure 1.13b). If the current flows from the output module to an output load, the output module is referred to as *sourcing* (Figure 1.14a). If the current flows to the output module from an output load, the output module is referred to as *sinking* (Figure 1.14b).

It is important know the type of input or output concerned so that it can be correctly connected to the PLC. Thus, sensors with sourcing outputs should be connected to sinking PLC inputs and sensors with sinking outputs should be connected to sourcing PLC inputs. The interface with the PLC will not function and damage may occur if this guideline is not followed.

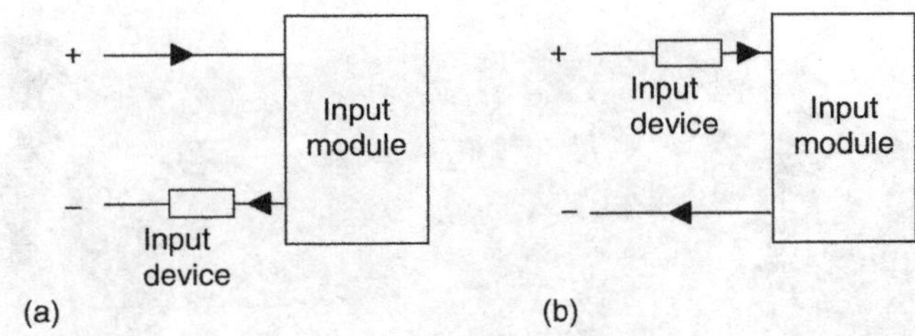

Figure 1.13: Inputs: (a) sourcing; (b) sinking.

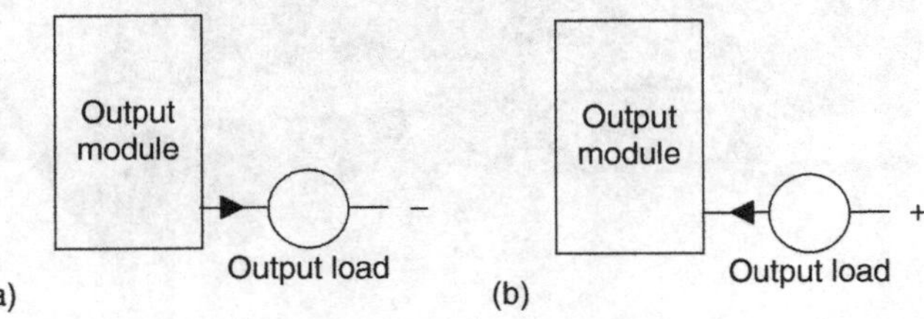

Figure 1.14: Outputs: (a) sourcing; (b) sinking.

1.4 PLC Systems

There are two common types of mechanical design for PLC systems—a *single box* and the *modular/rack types*. The single-box type (or, as it's sometimes called, compact or *brick*) is commonly used for small programmable controllers and is supplied as an integral compact package complete with power supply, processor, memory, and input/output units. Typically such a PLC might have 6, 8, 12, or 24 inputs and 4, 8, or 16 outputs and a memory that can store some 300 to 1000 instructions. For example, the Toshiba PLC brick TAR 116-6S has 8 inputs 120 V ac, 6 relay outputs, and 2 triac outputs while a bigger brick TDR140-6S has 24 inputs 24 V dc, 14 relay outputs and 2 triac outputs. Some compact systems have the capacity to be extended to cope with more inputs and outputs by linking input/output boxes to them.

Figure 1.15 shows an example of an Omron compact PLC for machine control, the CP1L. For this particular model, four CPU sizes are available, each with a choice of relay or transistor outputs. The combination of power supply, output, and the number of I/O points can be selected to meet the requirements. The base input/output brick, depending on the model concerned, has 10, 14, 20 or 30 inputs/outputs (I/O). The 10 I/O brick has 6 digital input points and four outputs, the 14 I/O brick has 8 digital input points and 6 outputs, the 20 I/O brick has 12 digital input points and 8 outputs and the 30 I/O brick has 18 digital input points and 12 outputs. The model can be selected to have outputs as relay or transistor sinking or sourcing. The 14, 20 and 30 I/O models can be extended to give more inputs/output, e.g. the 14 I/O model can be extended to give 54 input/outputs.

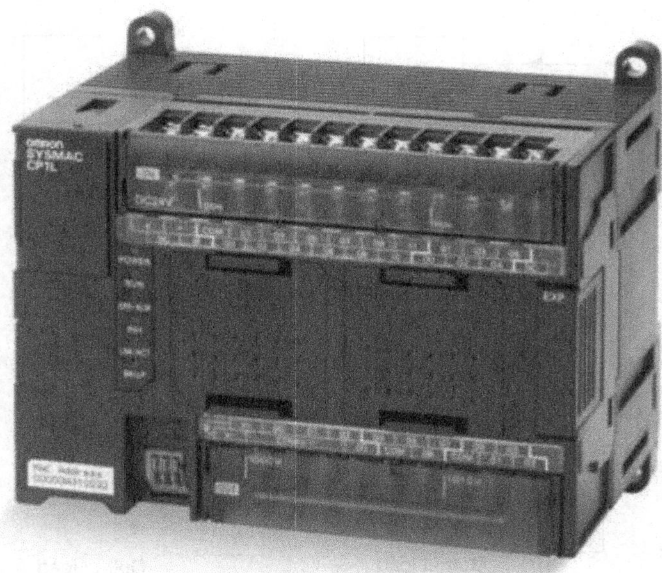

Figure 1.15: OMRON CP1L (By permission of Omron Industrial Automation).

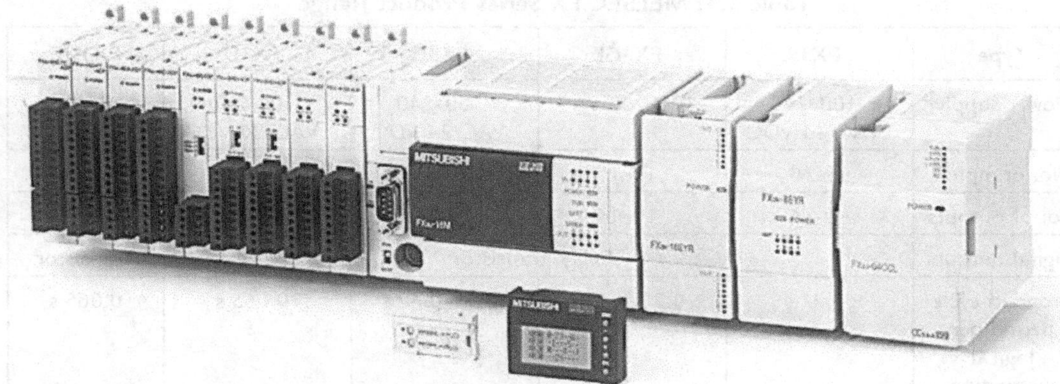

Figure 1.16: Mitsubishi Compact PLC – FX3U (By permission of Mitsubishi Electric Automation, Inc.)

Figure 1.16 shows the Mitsubishi FX3U compact PLC and Table 1.1 and Table 1.2 gives details of models in that Mitsubishi range.

Systems with larger numbers of inputs and outputs are likely to be modular and designed to fit in racks (Figure 1.17). The modular type consists of separate modules for power supply, processor, etc., which are often mounted on rails within a metal cabinet. The rack type can be used for all sizes of programmable controllers and has the various functional units packaged in individual modules that can be plugged into sockets in a base rack. The mix of modules required for a particular purpose is decided by the user and the appropriate ones then plugged into the rack. Thus it is comparatively easy to expand the number of input/output (I/O) connections by just adding more input/output modules or to expand the memory by adding more memory units. The power and data interfaces for modules in a rack are provided by copper conductors in the backplane of the rack. When modules are slid into a rack they engage with connectors in the backplane.

An example of such a modular system is provided by the Allen-Bradley PLC-5 PLC of Rockwell Automation which, at a minimum, will consist of the power supply, a programmable controller module and Input/Output (generally abbreviated to I/O) modules. There are a number of chassis available for mounting modules:

- *Chassis*. Some 1771 I/O chassis are built for back-panel mounting and some are built for rack mounting and are available in sizes of 4, 8, 12, or 16 I/O module slots.
- *Controller module*. PLC-5 controllers are available for a range of I/O capacity and memory size, e.g. PLC-5/11 with a maximum number of I/O of 512 and a maximum memory of 8000 words (see Chapter 3 for an explanation of the term 'word') and PLC-5/20 with a total number of I/O of 512 and a maximum memory of 16 000 words.

Table 1.1: MELSEC FX Series Product Range

Type	FX3S	FX3GE	FX3G	FX3U	FX3UC
Power supply	100–240 VAC/24 VDC	100–240 VAC	100–240 VAC/24 VDC	100–240 VAC/24 VDC	24 VDC
No. of inputs	6–16	14–16	8–36	8–64	8–48
No. of outputs	4–14	10–14	8–24	8–64	8–48
Digital outputs	Relay, transistor				Transistor
Program cycle period per logical instruction	0.21 s	0.21–0.42 s	0.21–0.42 s	0.065 s	0.065 s
User memory	4 k steps EEPROM (internal), EEPROM/ EPROM cassettes (optional)	32 k steps EEPROM (internal), EEPROM/ EPROM cassettes (optional)	32 k steps EEPROM (internal), EEPROM cassettes (optional)	64 k steps (standard), FLROM cassettes (optional)	64 k steps (standard), EEPROM cassettes (optional)
Dimensions in mm (WxHxD)	60–100x90x75	130–175x90x86	90–175x90x86	130–350x90x86	34–86x90x74

By permission of Mitsubishi Electric Automation, Inc.

They can be configured for a variety of communication networks, e.g. PLC-5/20C for use with ControlNet and PLC-5/20E for use with the Ethernet. They are single-slot modules that are placed in the left-most slot of a 1771 I/O chassis.

- *I/O modules*. The 1771 I/O modules are available in densities of 8, 16, or 32 I/O per module for signal interfaces to ac and dc sensors and actuators. Digital I/O modules have digital I/O circuits that interface to on/off sensors such as pushbutton and limit switches; and on/off actuators such as motor starters, pilot lights, and annunciators. Analog I/O modules perform the required A/D and D/A conversions using up to 16-bit resolution. Analog I/O can be user-configured for the desired fault-response state in the event that I/O communication is disrupted. This feature provides a safe reaction/response in case of a fault, limits the extent of faults, and provides a predictable fault response. 1771 I/O modules include optical coupling and filter circuitry for signal noise reduction. Digital I/O modules cover electrical ranges from 5 to 276 V AC or DC and relay contact output modules are available for ranges from 0 to 276V AC or 0 to 175 V DC. A range of analog signal levels can be accommodated, including standard analog inputs and outputs and direct thermocouple and RTD temperature inputs. As an illustration, there is the 1771-1B digital input module for 8 inputs with voltages in the range 10 to 27 V, the 1771-0VN for 32 digital outputs with voltages in the range 10 to 30 V, the analog input

Table 1.2: FX3U Main Units with 16 I/O

Model Number	FX3U-16MR/DS	FX3U-16MR/ES	FX3U-16MT/DSS	FX3U-16MT/DS	FX3U-16MT/ESS	FX3U-16MT/ES
Stocked Item	S	S	S	S	S	S
Rating	UL · cUL · CE (EMC)					
Integrated Inputs/ Outputs	16	16	16	16	16	16
Power Supply	24VDC	100-240VAC	24VDC	24VDC	100-240VAC	100-240VAC
Integrated Inputs	8	8	8	8	8	8
Integrated Outputs	8	8	8	8	8	8
Output Type	Relay	Relay	Transistor (Source)	Transistor (Sink)	Transistor (Source)	Transistor (Sink)
Power Consumption (W)	25	30	25	25	30	30
Weight (kg)	0.60	0.60	0.60	0.60	0.60	0.60
Dimensions (W x H x D) mm	130 x 90 x 86	130 x 90 x 86	130 x 90 x 86	130 x 90 x 86	130 x 90 x 86	130 x 90 x 86

Note: the rating row names the organisations for which conformity to certifications are given.
By permission of Mitsubishi Electric Automation, Inc.

module 1771-NIV for 8 inputs at ± 5 V dc, ± 20 mA and the analog output 1771-OFE2 for 4 outputs in the range 4 to 20 mA. A PLC-5 processor can communicate with I/O across a DeviceNet or Universal Remote I/O link.

- *Communication modules.* Communication modules can be used to add further communication ports to the PLC-5 controller beyond that provided by a controller module.

1.4.1 Security

Because PLCs can be connected to networks and contain real-time operating systems, there is the problem of security in that networks can be hacked and information fall into unauthorized hands or viruses inserted. PLCs can also be attacked when a computer they communicate with has been attacked.

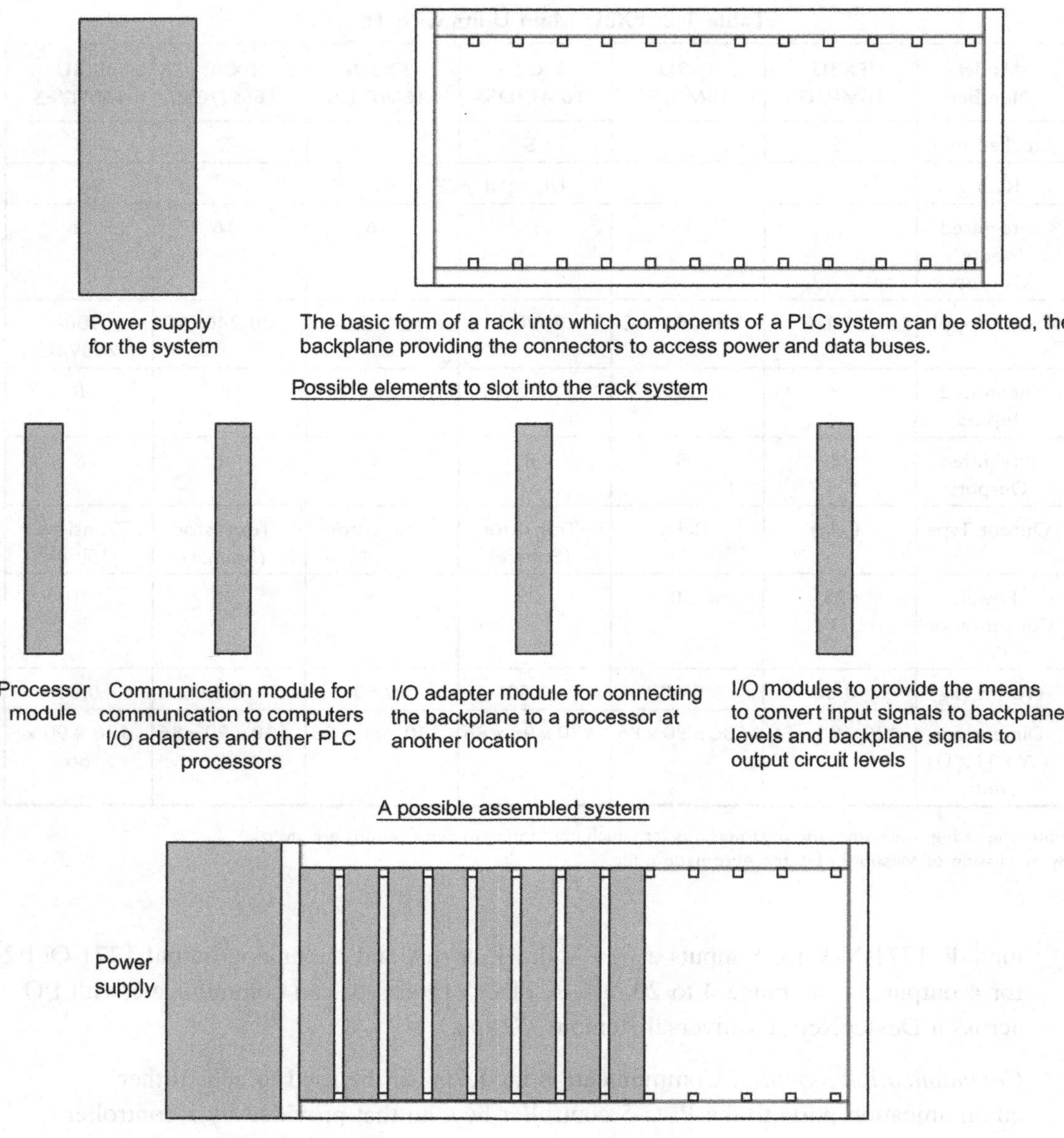
Power supply for the system
The basic form of a rack into which components of a PLC system can be slotted, the backplane providing the connectors to access power and data buses.
Possible elements to slot into the rack system
Processor module
Communication module for communication to computers I/O adapters and other PLC processors
I/O adapter module for connecting the backplane to a processor at another location
I/O modules to provide the means to convert input signals to backplane levels and backplane signals to output circuit levels
A possible assembled system
Power supply

Figure 1.17: A possible arrangement of a rack system.

1.5 Programs

Programs for use with PLCs can be written in a number of formats. To make it easier for engineers with no great knowledge of programming to write programs for PLCs, *ladder programming* was developed. Most PLC manufacturers adopted this method of writing programs; however, each tended to develop its own versions and so an international standard has been adopted for ladder programming and indeed all the methods used for programming PLCs. The standard, published in 1993, is International Electrotechnical Commission (IEC) 1131-3, now referred to as IEC 61131-3. The latest edition, dated 2013, is a compatible extension of the earlier version.

The IEC 61131-3 programming languages are ladder diagrams (LAD), instruction list (IL), sequential function charts (SFC), structured text (ST), and function block diagrams (FBD). The standard includes a library of pre-programmed functions and function blocks. Note that a function is the term used for a pre-programmed calculation, for example a function that gives the average value of two inputs, whereas the term function block is used when inputs are evaluated and give a value to an output, for example a counter function block which counts up the pulses to its input and gives an output signal when the count has reached a particular value. These are parts of a control program that is packaged so that it can be used in different parts of the same program or in different programs. The IEC standard gives formal definitions for each input and output parameter so that function blocks designed can different programmers can be readily interconnected. Any PLC that is IEC compliant supports these functions as a library with the code being written in a prom of flash ram on the device.

Two of the languages are graphical, i.e. structured text and instruction list, and so are entered into the programming device from a keyboard, one line at a time. The other languages, ladder diagrams, sequential function charts and function block diagrams, are graphical and so a program can be built up with graphical elements on the screen of the programming device.

1.5.1 The IEC Standard

The IEC 61131 standard covers the complete life cycle of PLCs:

Part 1: General definition of basic terminology and concepts.

Part 2: Electronic and mechanical equipment requirements and verification tests for PLCs and associated equipment.

Part 3: Programming languages. Five languages are defined: ladder diagram (LAD), sequential function charts (SFC), function block diagram (FBD), structured text (ST), and instruction list (IL).

1. Scope
2. Normative references
3. Terms and definitions
4. Architectural models
5. Compliance
6. Common elements
7. Textural languages (instruction list and structured text)

8. Graphic languages (ladder diagram and function block diagram)

Annex A: Formal specification of the language elements

Annex B: List of major changes and extensions of the third edition

Part 4: Guidance on selection, installation, and maintenance of PLCs.

Part 5: Software facilities needed for communication with other devices based on the Manufacturing Messaging Specification (MMS).

Part 6: Communications via fieldbus software facilities.

Part 7: Fuzzy control programming.

Part 8: Guidelines for the implementation of PLC programming languages defined in Part 3.

IEC 61131-6 covers the methods of programming for PLCs. Ladder programming (see Chapter 5) has evolved from electrical wiring diagrams for relay control systems, as in Figure 1.4, and has the advantage of being readily understood by those familiar with electrical wiring diagrams. It also has the advantage of enabling a maintenance engineer to readily trace faults as most programming stations tend to provide an animated display which shows the live state of contacts on the rungs of ladders. Ladder programming can be used to build quite large programs but is not so convenient when subroutines or program blocks are involved. Also programs that involve large numbers of sequences can prove unwieldy with the control of a sequence being mixed in with the application. While simple arithmetic operations can be carried out with ladder programs more complex calculations are rather cumbersome. Despite these issues, ladder programming is very widely used as it is so readily written and understood. Sequential function charts (see Chapter 6) have the merit of displaying all the operational states of a system, all the possible changes of the states and the conditions under which the changes can occur. They are better for showing sequences than ladder programs. Function block diagrams (see Chapter 5) have the advantage as a programming tool or making use of blocks of reusable software elements, logic gates being an example of such blocks. Structured text (see Chapter 6) is a programming language that strongly resembles the programming language Pascal. Instruction list (see Chapter 6) has a relatively simple structure and is useful for dealing with small programs where there only a few decision points and a limited number of changes in program execution flow. It is also harder to follow the program flow.

The IEC 61131-5 standard deals with PLC communications, as outlined in Figure 1.11, and so is concerned with the facilities that are relevant to allow PLCs that are connected by a communications network to exchange data and control information. It states the status information to be provided in a standard format at each subsystem in order to facilitate communication.

1.5.2 Programming PLCs

A programming device can be a handheld device, a desktop console, or a computer. Only when the program has been designed on the programming device and is ready is it transferred to the memory unit of the PLC.

- *Handheld programming devices* will normally contains enough memory to allow the unit to retain programs while being carried from one place to another.
- *Desktop consoles* are likely to have a visual display unit with a full keyboard and screen display.
- *Personal computers* are widely used for programming PLCs. A major advantage of using a computer is that the program can be stored on the hard disk or a CD and copies easily made. The computer is connected to the PLC by Ethernet, RS-232, RS-485 or RS-422 cabling.

PLC manufacturers have programming software for their PLCs. For example, Mitsubishi has MELSOFT. Mitsubishi's iQ Works software is a suite of four MELSOFT software packages that enable intuitive programming and setup of an iQ Platform system, including system/network configuration, Q and FX Series programming, Q Motion Controller and Servo setup, GOT1000 HMI screen design. Simulators and additional configuration software have been integrated into the base software, and Label programming across the entire system has been implemented. MELSOFT Navigator is the heart of iQ Works integrating the other MELSOFT programs included with iQ Works. Functions such as system configuration design, batch parameter setting, system labels, and batch read all help to reduce the total cost of ownership (TCO). MELSOFT GX Works 2 is the PLC maintenance and programming software. It supports all MELSEC controllers from the compact PLCs of the MELSEC FX series to the modular PLCs including MELSEC System Q and uses a Windows based environment. It supports the programming methods (see Chapter 4) of instruction list (IL), ladder diagram (LD) and sequential function chart (SFC) languages. You can switch back and forth between IL and LD at will while you are working. You can program your own function blocks, and a wide range of utilities is available for configuring special functions. The package includes powerful editors and diagnostics functions for configuring MELSEC networks and hardware, and extensive testing and monitoring functions to help get applications up and running quickly and efficiently. It offers offline simulation for all PLC types and thus enables simulation of all devices and application responses for realistic testing.

As another illustration, Siemens has SIMATIC STEP 7. This fully complies with the international standard IEC 61131-3 for PLC programming languages. With STEP 7, programmers can select from among various programming languages. Besides LAD and FBD, STEP 7 Basis also includes the IL programming language. Other additional options are available for IEC 61131-3 programming languages such as ST, called SIMATIC S7-SCL, or SFC, called SIMATIC S7-Graph, which provides an efficient way to describe sequential control systems graphically. Features of the whole engineering system include system diagnostic capabilities, process diagnostic tools, PLC simulation, remote maintenance, and plant documentation. S7-PLCSIM is an optional package for STEP 7 that allows simulation of a SIMATIC S7 control platform and testing of a user program on a PC, enabling testing and refining prior to physical hardware installation. By testing early in a project's development, overall project quality can be improved. Installation and commissioning can thus be quicker and less expensive because program faults can be detected and corrected early on during development.

Likewise, Rockell Automation have RSLogix for the Allen-Bradley PLC-5 family of PLCs. The RSLogix™ family of IEC-1131-compliant ladder logic programming packages have flexible, easy-to-use editors, common look-and-feel, diagnostics and troubleshooting tools and powerful, time-saving features and functionality. This family of products has been developed to operate on Microsoft® Windows® operating systems. RSLogix™ 5 supports the Allen-Bradley PLC-5® family of programmable controllers.

Summary

A *programmable logic controller* (PLC) is a special form of microprocessor-based controller that uses a programmable memory to store instructions and to implement functions such as logic, sequencing, timing, counting, and arithmetic to control machines and processes and is designed to be operated by engineers with perhaps a limited knowledge of computers and computing languages.

Typically, a PLC system has the basic functional components of processor unit, memory, power supply unit, input/output interface section, communications interface, and programming device. To operate the PLC system there is a need for it to access the data to be processed and the instructions, that is, the program, that informs it how the data is to be processed. Both are stored in the PLC memory for access during processing. The input/output channels provide isolation and signal conditioning functions so that sensors and actuators can often be directly connected to them without the need for other circuitry. Outputs are specified as being of relay type, transistor type, or triac type. The communications interface is used to receive and transmit data on communications networks from or to other remote PLCs. There are two common types of mechanical design for PLC systems—a single box and the modular/rack types.

The IEC 61131 defined the standards for PLCs, with 61131-3 defining the programming languages: ladder diagrams (LAD), instruction list (IL), sequential function charts (SFC), structured text (ST), and function block diagrams (FBD).

Problems

Questions 1 through 6 have four answer options: A, B, C or D. Choose the correct answer from the answer options.

1. The term *PLC* stands for:
   A. Personal logic computer
   B. Programmable local computer
   C. Personal logic controller
   D. Programmable logic controller

2. Decide whether each of these statements is true (T) or false (F): A transistor output channel from a PLC:
   (i) Is used for only DC switching.
   (ii) Is isolated from the output load by an optocoupler.

   Which option *best* describes the two statements?
   A. (i) T (ii) T
   B. (i) T (ii) F
   C. (i) F (ii) T
   D. (i) F (ii) F

3. Decide whether each of these statements is true (T) or false (F): A relay output channel from a PLC:
   (i) Is used for only DC switching.
   (ii) Can withstand transient overloads.

   Which option *best* describes the two statements?
   A. (i) T (ii) T
   B. (i) T (ii) F
   C. (i) F (ii) T
   D. (i) F (ii) F

4. Decide whether each of these statements is true (T) or false (F): A triac output channel from a PLC:
   (i) Is used for only AC output loads.
   (ii) Is isolated from the output load by an optocoupler.

   Which option *best* describes the two statements?
   A. (i) T (ii) T
   B. (i) T (ii) F
   C. (i) F (ii) T
   D. (i) F (ii) F

5. Decide whether each of these statements is true (T) or false (F): The term sourcing can be used for a device connected to a PLC when:
   (i) The input module of the PLC receive current from the input device.
   (ii) The output module of the PLC supplies current to the output load.

   Which option *best* describes the two statements?
   A. (i) T (ii) T
   B. (i) T (ii) F
   C. (i) F (ii) T
   D. (i) F (ii) F

6. Decide whether each of these statements is true (T) or false (F): The reason for including optocouplers on input/output units is to:
   (i) Provide a fuse mechanism that breaks the circuit if high voltages or currents occur.
   (ii) Isolate the CPU from high voltages or currents.

   Which option *best* describes the two statements?
   A. (i) T (ii) T
   B. (i) T (ii) F
   C. (i) F (ii) T
   D. (i) F (ii) F

7. Draw a block diagram showing in very general terms the main units in a PLC.

8. State the characteristics of the relay, transistor, and triac types of PLC output channels.

9. How many bits can a 2K memory unit store?

10. A PLC model has a number of different CPU units that can be ordered. One model has 10 I/O terminals of 6 DC outputs and 4 outputs and can be ordered for use with either AC or DC power supplies. The outputs can be selected as either relay output or transistor output with two forms of transistor output available –namely, sink or source type. Explain the capability of such a PLC and the significance of the various forms of output.

Lookup Tasks

11. Google "programmable logic controllers" on the Internet and look at the forms and specifications of PLCs available from various manufacturers. Then find a suitable PLC to meet a particular specification, such as one that would be suitable for six DC inputs and six relay outputs, or possibly six sinking transistor outputs, and a module system that would be suitable for five DC sourcing inputs, four DC sinking inputs, and 12 DC sinking transistor outputs.

12. Look up the IEC 61131-3 standard and find out what it covers.

CHAPTER 2

Input/Output Devices

This chapter is a brief consideration of typical input and output devices used with PLCs. The input devices considered include digital and analog devices such as mechanical switches for position detection, proximity switches, photoelectric switches, encoders, temperature and pressure switches, potentiometers, linear variable differential transformers, strain gauges, thermistors, thermotransistors, and thermocouples. Output devices considered include relays, contactors, solenoid valves, and motors.

2.1 Input Devices

The term *sensor* is used for an input device that provides a usable output in response to a specified physical input. For example, a thermocouple is a sensor that converts a temperature difference into an electrical output. The term *transducer* is generally used to refer to a device that converts a signal from one form to a different physical form. Thus sensors are often transducers, but also other devices can be transducers, such as a motor that converts an electrical input into rotation.

Sensors that give digital or discrete, that is, on/off, outputs can be easily connected to the input ports of PLCs. An analog sensor gives an output proportional to the measured variable. Such analog signals have to be converted to digital signals before they can be input to PLC ports.

The following are some of the more common terms used to define the performance of sensors:

- *Accuracy* is the extent to which the value indicated by a measurement system or element might be wrong. For example, a temperature sensor might have an accuracy of ±0.1°C. The *error* of a measurement is the difference between the result of the measurement and the true value of the quantity being measured. Errors can arise in a number of ways; the term *nonlinearity error* is used to describe the error that occurs as a result of assuming a linear relationship between the input and output over the working range, that is, a graph of output plotted against input is assumed to give a straight line. Few systems or elements, however, have a truly linear relationship and thus errors occur as a result of the assumption of linearity (Figure 2.1a). The term *hysteresis error*

W. Bolton: Programmable Logic Controllers, Sixth Edition. http://dx.doi.org/10.1016/B978-0-12-802929-9.00002-9

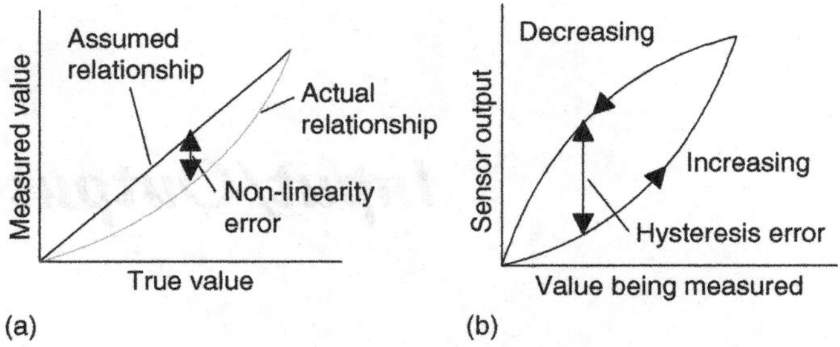

Figure 2.1: Some sources of error: (a) nonlinearity; (b) hysteresis.

(Figure 2.1b) is used for the difference in outputs given from the same value of quantity being measured according to whether that value has been reached by a continuously increasing change or a continuously decreasing change. Thus, you might obtain a different value from a thermometer used to measure the same temperature of a liquid if it is reached by the liquid warming up to the measured temperature or it is reached by the liquid cooling down to the measured temperature.

- The *range* of variable of a system is the limits between which the inputs can vary. For example, a resistance temperature sensor might be quoted as having a range of −200 to +800°C.
- When the input value to a sensor changes, it will take some time to reach and settle down to the steady-state value (Figure 2.2). The *response time* is the time that elapses after the input to a system or element is abruptly increased from zero to a constant value, up to the point at which the system or element gives an output corresponding to some specified percentage, such as 95%, of the value of the input. The *rise time* is the time taken for the output to rise to some specified percentage of the steady-state output. Often the rise time refers to the time taken for the output to rise from 10% of the steady-state

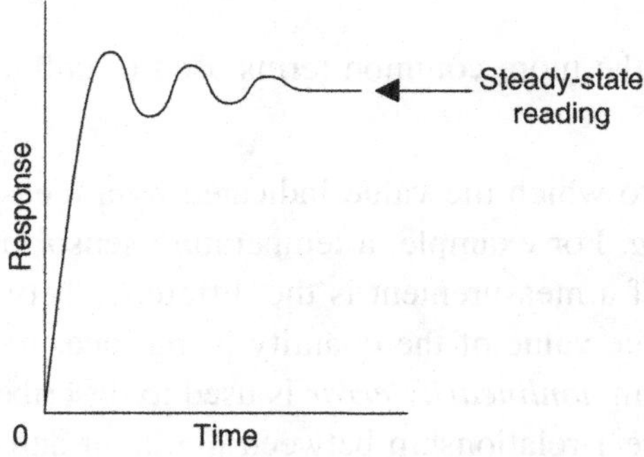

Figure 2.2: Response of a sensor or measurement system to a sudden input. You can see such a response when the current in an electrical circuit is suddenly switched on and an ammeter reading is observed.

value to 90% or 95% of the steady-state value. The *settling time* is the time taken for the output to settle to within some percentage, such as 2%, of the steady-state value.

- The *sensitivity* indicates how much the output of an instrument system or system element changes when the quantity being measured changes by a given amount, that is, the ratio ouput/input. For example, a thermocouple might have a sensitivity of 20 μV/°C and so give an output of 20 μV for each 1°C change in temperature.

- The *stability* of a system is its ability to give the same output when used to measure a constant input over a period of time. The term *drift* is often used to describe the change in output that occurs over time. The drift may be expressed as a percentage of the full range output. The term *zero drift* refers to the changes that occur in output when there is zero input.

- The term *repeatability* refers to the ability of a measurement system to give the same value for repeated measurements of the same value of a variable. Common causes of lack of repeatability are random fluctuations in the environment, such as changes in temperature and humidity. The error arising from repeatability is usually expressed as a percentage of the full range output. For example, a pressure sensor might be quoted as having a repeatability of ±0.1% of full range. With a range of 20 kPa this would be an error of ±20 Pa.

- The *reliability* of a measurement system, or the element in such a system, is defined as being the probability that it will operate to an agreed level of performance for a specified period, subject to specified environmental conditions. The agreed level of performance might be that the measurement system gives a particular accuracy.

As an illustration of the use of these terms in specification, the following were included in the specification of a MX100AP pressure sensor (see later in this chapter, Section 2.1.8, for an explanation of this sensor):

Supply voltage: 3 V (6 V max)

Supply current: 6 mA

Full-scale span: 60 mV

Range: 0 to 100 kPa

Sensitivity: 0.6 mV/kPa

Nonlinearity error: ±0.05% of full range

Temperature hysteresis: ±0.5% of full scale

Input resistance: 400 to 550 Ω

Response time: 1 ms (10% to 90%)

The following are examples of some of the commonly used PLC input devices and their sensors.

2.1.1 Mechanical Switches

A mechanical switch generates an on/off signal or signals as a result of some mechanical input causing the switch to open or close. Such a switch might be used to indicate the presence of a workpiece on a machining table, the workpiece pressing against the switch and so closing it. The absence of the workpiece is indicated by the switch being open and its presence by it being closed. Thus, with the arrangement shown in Figure 2.3a, the input signals to a single input channel of the PLC are thus the logic levels:

Workpiece not present: 0

Workpiece present: 1

The 1 level might correspond to a 24 V DC input, the 0 to a 0 V input.

With the arrangement shown in Figure 2.3b, when the switch is open the supply voltage is applied to the PLC input; when the switch is closed the input voltage drops to a low value. The logic levels are thus:

Workpiece not present: 1
Workpiece present: 0

Switches are available with *normally open (NO)* or *normally closed (NC)* contacts or can be configured as either by choice of the relevant contacts. An NO switch has its contacts open in the absence of a mechanical input and the mechanical input is used to close the switch. An NC switch has its contacts closed in the absence of a mechanical input and the mechanical input is used to open the switch. Mechanical switches are specified in terms of number of poles, that is, the number of separate circuits that can be completed by the same switching action, and number of throws, that is, the number of individual contacts for each pole.

A problem with mechanical switches is that when a switch is closed or opened, *bounce* can occur and the contacts do not make or open cleanly. Because they involve an elastic member,

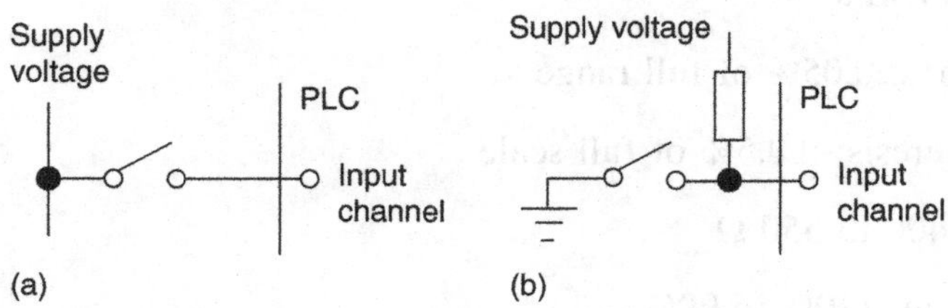

Figure 2.3: Switch sensors.

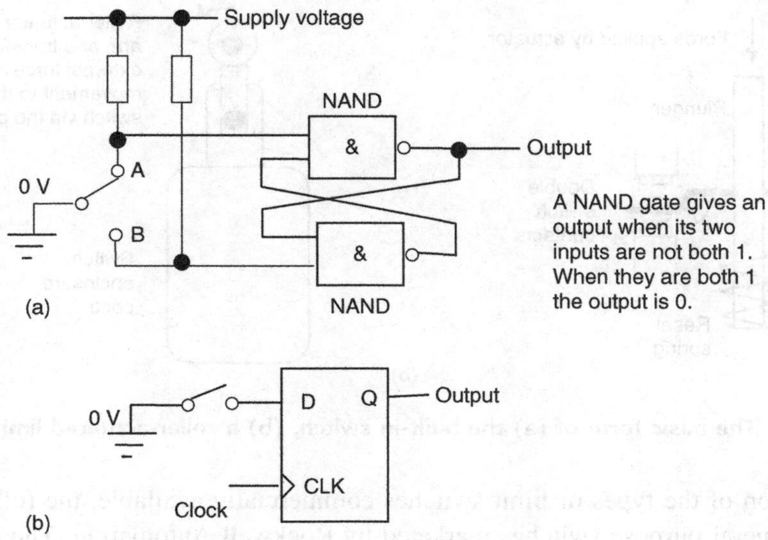

Figure 2.4: (a) A NAND gate circuit to debounce an SPDT switch; (b) a D flip-flop to debounce an SPST switch.

they bounce back and forth like an oscillating spring. This "bounce" may produce amplitudes that change logic levels over perhaps 20 ms, and so a single switch change may give rise to a number of signals rather than just the required single one. There are a number of ways of eliminating these spurious signals. One way is to include in the software program a delay of approximately 20 ms after the first detected signal transition before any further signals are read. A possibility for a single pole/double throw (SPDT) switch is to use two NAND logic gates (see Chapters 3 and 5), as illustrated in Figure 2.4a. When the switch is in position A, the output is a logic 1. When the switch moves to position B, the output becomes logic 0 and remains latched at this spot, even when the switch bounces. Figure 2.4b shows how a D flip-flop (see Chapter 3 for a discussion) can be used to debounce a single pole/single throw (SPST) switch. The output of the D flip-flop does not change until a position-edged clock signal is imposed, and if this is greater than the bounce time, the output is debounced.

The term *limit switch* applies to a switch that is used to detect the presence or absence of an object, the passage of a moving part and when an object has reached its end of travel. Because they were first used to determine the limit of travel of an object, they became known as limit switches. They are widely used as they are rugged, reliable and easily installed. The basic part of such a switch is a built-in electrical switch which is switched on or off by means of a plunger; the movement of this plunger being controlled by the actuator head of the limit switch which transfers the external force and movement to the built-in switch (Figure 2.5a). Depending on the object and movement being detected, the actuator head can take a number of forms. A common form is a roller rotating an arm to activate the switch contacts and having a spring return (Figure 2.5b).

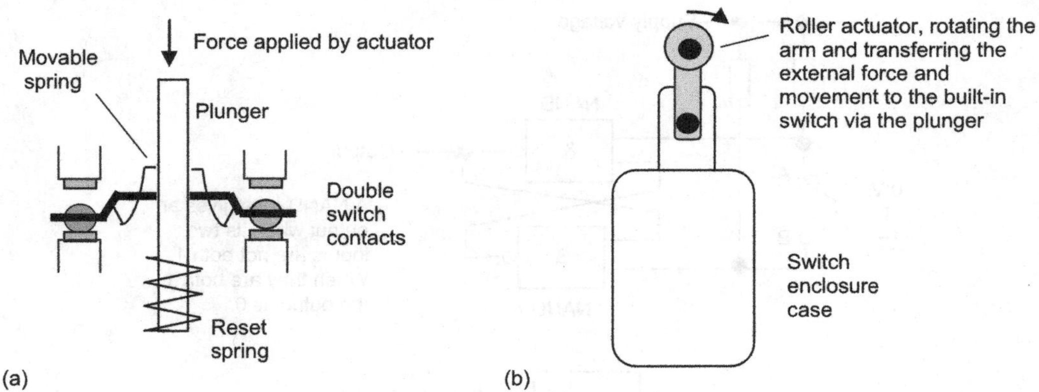

Figure 2.5: The basic form of (a) the built-in switch, (b) a roller-actuated limit switch.

As an illustration of the types of limit switches commercially available, the following are some of the general purpose switches marketed by Rockwell Automation. The 801 line of limit switches are general purpose for use in a wide variety of applications and a range of different contact arrangements are available. For a roller lever the contact operation can be slow action with spring return, snap action with spring return, ratchet type maintained or snap action maintained. With snap action, movement of the actuator creates a fast change in contact state once the actuator has reached the operating position, with the slow action relay the contacts are operated at a speed proportional to the speed of operation of the actuator. With the ratchet type of relay, when the lever is moved to the right, contacts are operated. The lever is spring return but the contacts remain in the operated position until the next movement of the roller lever. The snap action maintained type has contact operation when the lever is moved in one direction and restored when the lever is moved in the opposite direction. With the switches the angle through which the lever has to be rotated to activate the switch can be selected and can range from just a few degrees to about 25°.

Omron Industrial Automation also has a range of limit switches. For example, the D4CC miniature limit switches are available in a number of forms: with pin plunger, roller plunger, cross roller plunger, a high sensitive roller plunger, a sealed pin plunger, a sealed roller plunger, sealed cross roller plunger, a plastic rod, and a center roller lever.

As an illustration of the type of task that limit switches are used for, consider the movement of a lift between floors. A limit switch can be used at the floor level to detect the presence of the lift by the actuator head of the switch being actuated by the presence of the lift and so providing a signal which can be used to switch the lift motor on or off.

Liquid-level switches are used to control the level of liquids in tanks. Essentially, these are vertical floats that move with the liquid level, and this movement is used to operate switch contacts.

2.1.2 Proximity Switches

Proximity switches are used to detect the presence of an item without making contact with it. There are a number of forms of such switches, some being suitable only for metallic objects.

The *eddy current* type of proximity switch has a coil that is energized by a constant alternating current and produces a constant alternating magnetic field. When a metallic object is close to it, eddy currents are induced in it (Figure 2.6a). The magnetic field due to these eddy currents induces an EMF back in the coil with the result that the voltage amplitude needed to maintain the constant coil current changes. The voltage amplitude is thus a measure of the proximity of metallic objects. The voltage can be used to activate an electronic switch circuit, basically a transistor that has its output switched from low to high by the voltage change, creating an on/off device. The range over which such objects can be detected is typically about 0.5 to 20 mm.

Another switch type is the *reed switch*. This consists of two overlapping, but not touching, strips of a springy ferromagnetic material sealed in a glass or plastic envelope (Figure 2.6b). When a magnet or current-carrying coil is brought close to the switch, the strips become magnetized and attract each other. The contacts then close. The magnet closes the contacts when it is typically about 1 mm from the switch. Such a switch is widely used with burglar alarms to detect when a door is opened, with the magnet being in the door and the reed switch in the frame of the door. When the door opens, the switch opens.

A proximity switch that can be used with metallic and nonmetallic objects is the *capacitive proximity switch*. The capacitance of a pair of plates separated by some distance depends on the separation; the smaller the separation, the higher the capacitance. The sensor of the capacitive proximity switch is just one of the plates of the capacitor, the other plate being the metal object for which the proximity is to be detected (Figure 2.6c). Thus the proximity of the object is detected by a change in capacitance. The sensor can also be used to detect nonmetallic objects, since the capacitance of a capacitor depends on the dielectric between its plates. In this case the plates are the sensor and the earth and the nonmetallic object is the dielectric. The change in capacitance can be used to activate an electronic switch circuit

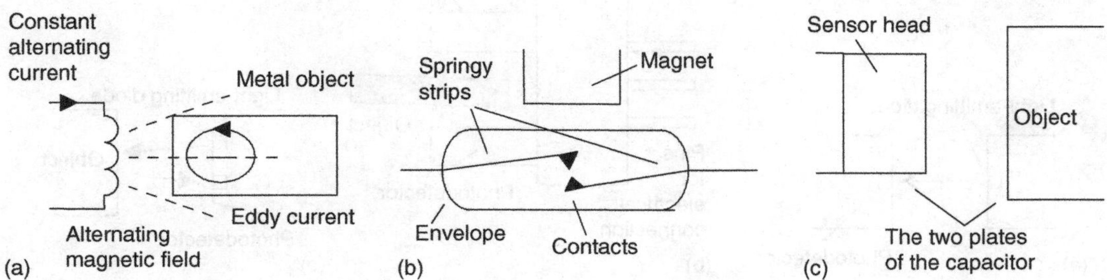

Figure 2.6: Proximity switches: (a) eddy current, (b) reed switch, and (c) capacitive.

and so create an on/off device. Capacitive proximity switches can be used to detect objects when they are typically between 4 mm and 60 mm from the sensor head. An example of the use of such a sensor might be to determine whether a cake is present inside a cardboard box, when such boxes move along a conveyor belt.

As an example of such a sensor, the Omron E2K-X capacitive sensor can be used with a wide range of metallic and non-metallic objects, e.g. glass, wood, and plastic, at distances between 3 and 30 mm. Capacitive proximity sensors also find applications as touch sensors in user interfaces such as computer touch pads and mobile phone touch screens. Such capacitive touch screens consist of an insulator such as glass which is coated with a transparent conductor. As the human body is an electrical conductor, when the surface of the screen is touched there is in a change in capacitance (see Wikipedia for more information).

Another type, the *inductive proximity switch*, consists of a coil wound a round a ferrous metallic core. When one end of this core is placed near a ferrous metal object, there is effectively a change in the amount of metallic core associated with the coil and so a change in its inductance. This change can be monitored using a resonant circuit, the presence of the ferrous metal object thus changing the current in that circuit. The current can be used to activate an electronic switch circuit and so create an on/off device. The range over which such objects can be detected is typically about 2 mm to 15 mm. An example of the use of such a sensor is to detect whether bottles passing along a conveyor belt have metal caps on. As an example, the Omron E2F sensor can be used to detect metallic objects up to 8 mm away.

2.1.3 Photoelectric Sensors and Switches

Photoelectric switch devices can either operate as *transmissive types*, in which the object being detected breaks a beam of light, usually infrared radiation, and stops it reaching the detector (Figure 2.7a), as in Figure 2.7b, which shows a U-shaped form in which the

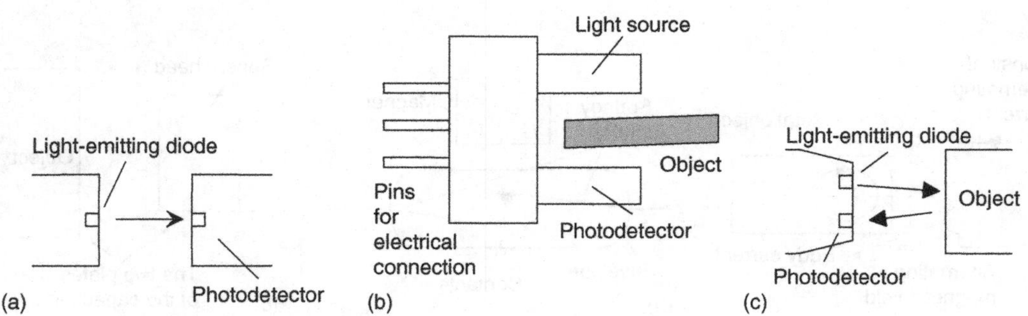

Figure 2.7: Photoelectric sensors.

object breaks the light beam; or *reflective types*, in which the object being detected reflects a beam of light onto the detector (Figure 2.7c). The transmissive form of sensor is typically used in applications involving the counting of parts moving along conveyor belts and breaking the light beam; the reflective form is used to detect whether transparent containers contain liquids to the required level.

The radiation emitter is generally a *light-emitting diode* (LED). The radiation detector might be a *phototransistor*, often a pair of transistors, known as a *Darlington pair*, to increase the sensitivity. Depending on the circuit used, the output can be made to switch to either high or low when light strikes the transistor. Such sensors are supplied as packages for sensing the presence of objects at close range, typically less than about 5 mm. Another possible detector is a *photodiode*. Depending on the circuit used, the output can be made to switch to either high or low when light strikes the diode. Yet another possibility is a *photoconductive cell*. The resistance of the photoconductive cell, often cadmium sulfide, depends on the intensity of the light falling on it.

With these sensors, light is converted to a current, voltage, or resistance change. If the output is to be used as a measure of the intensity of the light, rather than just the presence or absence of some object in the light path, the signal will need amplification and then conversion from analog to digital by an analog-to-digital converter. An alternative is to use a light-to-frequency converter, the light then being converted to a sequence of pulses, with the frequency of the pulses being a measure of the light intensity. Integrated circuit sensors, such as the Texas Instrument TSL220, incorporate the light sensor and the voltage-to-frequency converter (Figure 2.8).

2.1.4 Encoders

The term *encoder* is used for a device that provides a digital output as a result of angular or linear displacement. An *incremental encoder* detects changes in angular or linear displacement from some datum position; an *absolute encoder* gives the actual angular or linear position.

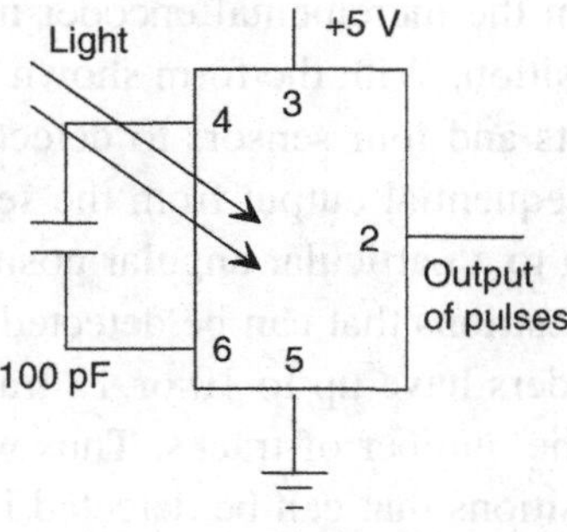

Figure 2.8: TSL220.

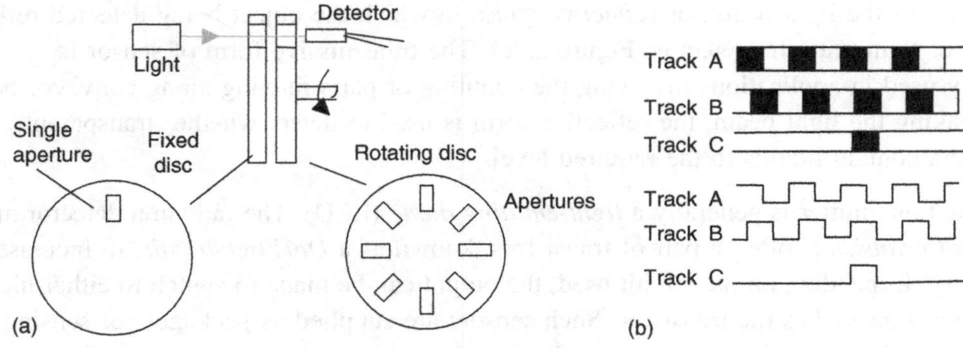

Figure 2.9: (a) Basic form of an incremental encoder, and (b) a three-track arrangement.

Figure 2.9 shows the basic form of an incremental encoder for the measurement of angular displacement. A beam of light, perhaps from an LED, passes through slots in a disc and is detected by a light sensor, such as a photodiode or phototransistor. When the disc rotates, the light beam is alternately transmitted and stopped, and so a pulsed output is produced from the light sensor. The number of pulses is proportional to the angle through which the disc has rotated, the resolution being proportional to the number of slots on a disc. With 60 slots, then, since one revolution is a rotation of 360°, a movement from one slot to the next is a rotation of 6°. By using offset slots it is possible to have over a thousand slots for one revolution and thus a much higher resolution.

This setup with just one track is a very basic form of incremental encoder with no way of determining the direction of rotation. With a single track, the output is the same for both directions of rotation. Thus, generally such encoders have two or three tracks with sensors (Figure 2.9b). With two tracks, one track is one-quarter of a cycle displaced from the other track. As a consequence, the output from one track will lead or lag that from the other track, depending on the direction of rotation. A third track of just a single aperture is also included; this gives one pulse per revolution and so can be used for counting the number of full revolutions.

The absolute encoder differs from the incremental encoder in having a pattern of slots that uniquely defines each angular position. With the form shown in Figure 2.10, the rotating disc has four concentric circles of slots and four sensors to detect the light pulses. The slots are arranged in such a way that the sequential output from the sensors is a number in the binary code, each number corresponding to a particular angular position. With four tracks there will be 4 bits, and so the number of positions that can be detected is $2^4 = 16$, that is, a resolution of $360/16 = 22.5°$. Typical encoders have up to 10 or 12 tracks. The number of bits in the binary number will be equal to the number of tracks. Thus with 10 tracks there will be 10 bits, and so the number of positions that can be detected is 2^{10}, that is, 1024, a resolution of $360/1024 = 0.35°$.

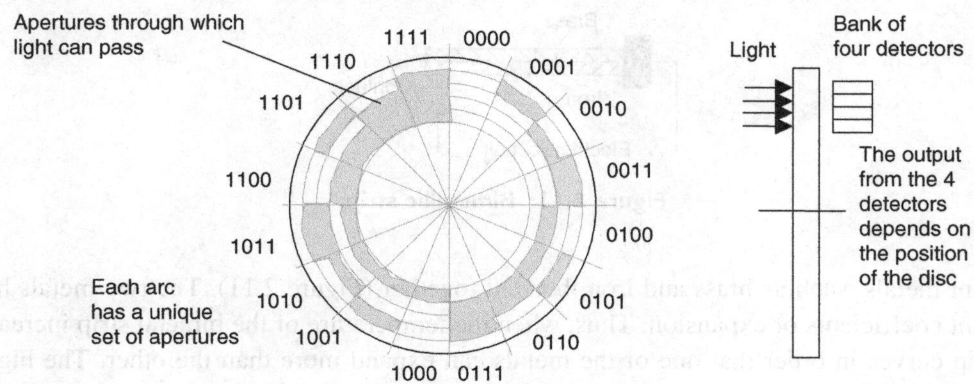

Figure 2.10: Basic form of the absolute encoder.

Table 2.1: Binary and Gray Codes

Binary				Gray			
0	0	0	0	0	0	0	0
0	0	0	1	0	0	0	1
0	0	1	0	0	0	1	1
0	0	1	1	0	0	1	0
0	1	0	0	0	1	1	0
0	1	0	1	0	1	1	1
0	1	1	0	0	1	0	1
0	1	1	1	0	1	0	0
1	0	0	0	1	1	0	0
1	0	0	1	1	1	0	1
1	0	1	0	1	1	1	1
1	0	1	1	1	1	1	0
1	1	0	0	1	0	1	0
1	1	0	1	1	0	1	1
1	1	1	0	1	0	0	1
1	1	1	1	1	0	0	0

Though the normal form of binary code is shown in the figure, in practice a modified form of binary code called the *Gray code* is generally used. This, unlike normal binary, has only 1 bit that changes in moving from one number to the next (see Table 2.1). This code provides data with the least uncertainty, but since we are likely to need to run systems with binary code, a circuit to convert Gray to binary code has to be used.

2.1.5 Temperature Sensors

A simple form of temperature sensor that can be used to provide an on/off signal when a particular temperature is reached is the *bimetal element*. This consists of two strips of

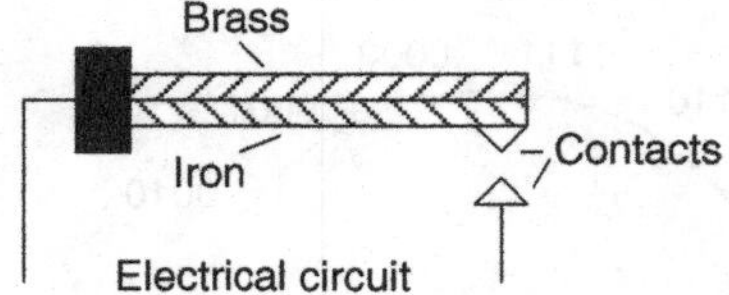

Figure 2.11: Bimetallic strip.

different metals, such as brass and iron, bonded together (Figure 2.11). The two metals have different coefficients of expansion. Thus, when the temperature of the bimetal strip increases, the strip curves in order that one of the metals can expand more than the other. The higher expansion metal is on the outside of the curve. As the strip cools, the bending effect is reversed. This movement of the strip can be used to make or break electrical contacts and hence, at some particular temperature, give an on/off current in an electrical circuit. The device is not very accurate but is commonly used in domestic central heating thermostats because it is a very simple, robust device.

Another form of temperature sensor is the *resistive temperature detector* (RTD). The electrical resistance of metals or semiconductors changes with temperature. In the case of a metal, the ones most commonly used are platinum, nickel, or nickel alloys. Such detectors can be used as one arm of a Wheatstone bridge and the output of the bridge taken as a measure of the temperature (Figure 2.12a). For such a bridge, there is no output when the resistors in the bridge arms are such that $P/Q = R/S$. Any departure of a resistance from this balance value results in an output. The resistance varies in a linear manner with temperature over a wide range of temperatures, though the actual change in resistance per degree is fairly small. A problem with a resistance thermometer is that the leads connecting it to the bridge can be quite long and themselves have significant resistance, which changes with temperature. One way of overcoming this problem is to use a three-wire circuit, as shown in Figure 2.12b. Then changes in lead resistance affect two arms of the bridge and balance out. Such detectors are very stable and very accurate, though expensive. They are available in the

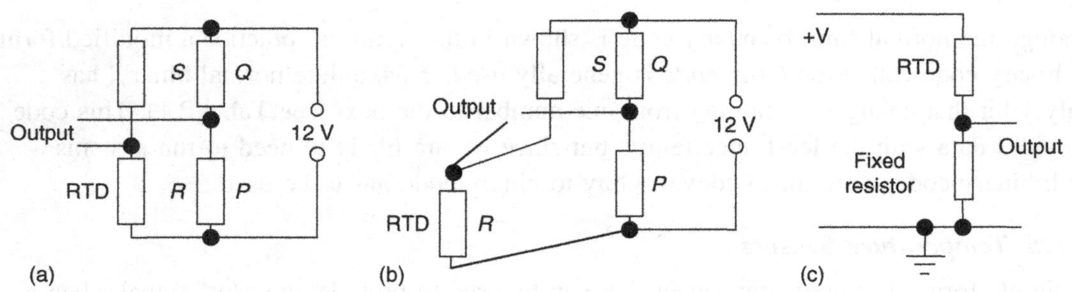

Figure 2.12: (a) A Wheatstone bridge, (b) a three-wire circuit, and (c) potential divider circuits.

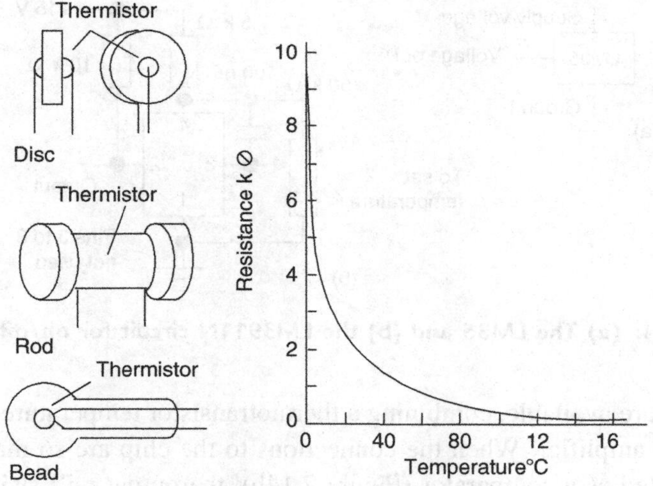

Figure 2.13: Common forms of thermistors and the typical variation of resistance with temperature for an NTC thermistor.

form of wire-wound elements inside ceramic tubes or as thin film elements deposited on a suitable substrate.

Semiconductors, such as thermistors (Figure 2.13), show very large changes in resistance with temperature. The change, however, is nonlinear. Those specified as NTC have negative temperature coefficients, that is, the resistance decreases with increasing temperature, and those specified as PTC have positive temperature coefficients, that is, the resistance increases with increasing temperature. They can be used with a Wheatstone bridge, but another possibility that is widely used is to employ a potential divider circuit with the change in resistance of the thermistor changing the voltage drop across a resistor (Figure 2.12c). The output from either type of circuit is an analog signal that is a measure of the temperature. Thermistors have the advantages of being cheap and small, giving large changes in resistance, and having fast reaction to temperature changes, though they have the disadvantage of being nonlinear, with limited temperature ranges.

Thermodiodes and *thermotransistors* are used as temperature sensors since the rate at which electrons and holes diffuse across semiconductor junctions is affected by the temperature. Integrated circuits can combine such a temperature-sensitive element with the relevant circuitry to give an output voltage related to temperature. A widely used integrated package is the LM35, which gives an output of 10 mV/°C when the supply voltage is +5 V (Figure 2.14a). A digital temperature switch can be produced with an analog sensor by feeding the analog output into a comparator amplifier, which compares it with some set value, producing an output that gives a logic 1 signal when the temperature voltage input is equal to or greater than the set point and otherwise gives a logic 0 signal. Integrated circuits,

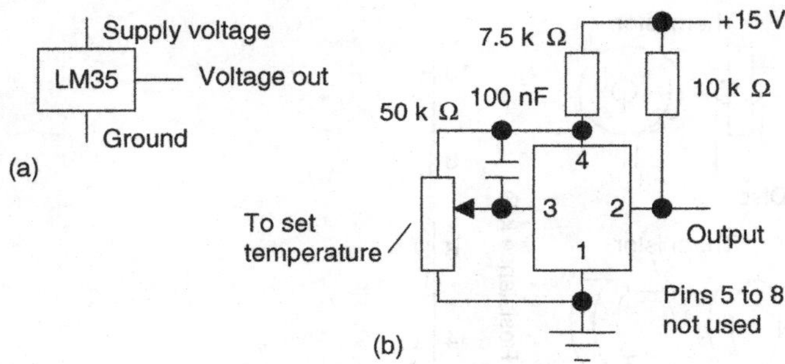

Figure 2.14: (a) The LM35 and (b) the LM3911N circuit for on/off control.

such as LM3911N, are available, combining a thermotransistor temperature-sensitive element with an operational amplifier. When the connections to the chip are so made that the amplifier is connected as a comparator (Figure 2.14b), the output will switch as the temperature traverses the set point and so directly give an on/off temperature controller. Such temperature sensors have the advantages of being cheap and giving a reasonably linear response. However, they have the disadvantage of a limited temperature range.

Another commonly used temperature sensor is the *thermocouple*. The thermocouple consists essentially of two dissimilar wires, A and B, forming a junction (Figure 2.15). When the junction is heated so that it is at a higher temperature than the other junctions in the circuit, which remain at a constant cold temperature, an EMF is produced that is related to the hot junction temperature. The EMF values for a thermocouple are given in Table 2.2, assuming that the cold junction is at 0°C. The thermocouple voltage is small and needs amplification before it can be fed to the analog channel input of a PLC. There is also circuitry required to compensate for the temperature of the cold junction, since often it will not be at 0°C, but room temperature and its temperature affects the value of the EMF. The amplification and compensation, together with filters to reduce the effect of interference from the mains supply, are often combined in a signal processing unit. Thermocouples have the advantages of being able to sense the temperature at almost any point, ruggedness, and being able to operate over a large temperature range. They have the disadvantages of giving a nonlinear response,

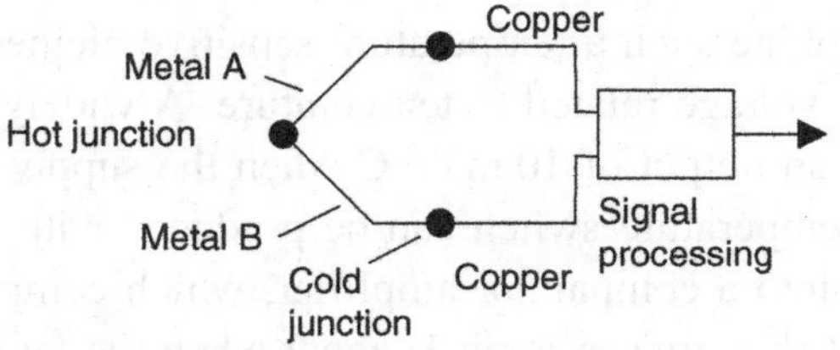

Figure 2.15: Thermocouple.

Table 2.2: Thermocouples

Ref.	Materials	Range (°C)	μV/°C
B	Platinum, 30% rhodium/platinum, 6% rhodium	0 to 1800	3
E	Chromel/constantan	−200 to 1000	63
J	Iron/constantan	−200 to 900	53
K	Chromel/alumel	−200 to 1300	41
N	Nirosil/nisil	−200 to 1300	28
R	Platinum/platinum, 13% rhodium	0 to 1400	6
S	Platinum/platinum, 10% rhodium	0 to 1400	6
T	Copper/constantan	−200 to 400	43

giving only small changes in EMF per degree change in temperature, and requiring temperature compensation for the cold junction.

2.1.6 Position/Displacement Sensors

The term *position sensor* is used for a sensor that gives a measure of the distance between a reference point and the current location of the target, while a *displacement sensor* gives a measure of the distance between the present position of the target and the previously recorded position.

Resistive linear and angular position sensors are widely used and relatively inexpensive. These are also called *linear and rotary potentiometers*. A DC voltage is provided across the full length of the track and the voltage signal between a contact that slides over the resistance track and one end of the track is related to the position of the sliding contact between the ends of the potentiometer resistance track (Figure 2.16). The potentiometer thus provides an analog linear or angular position sensor.

Another form of displacement sensor is the *linear variable differential transformer* (LVDT), which gives a voltage output related to the position of a ferrous rod. The LVDT consists of three symmetrically placed coils through which the ferrous rod moves (Figure 2.17). When

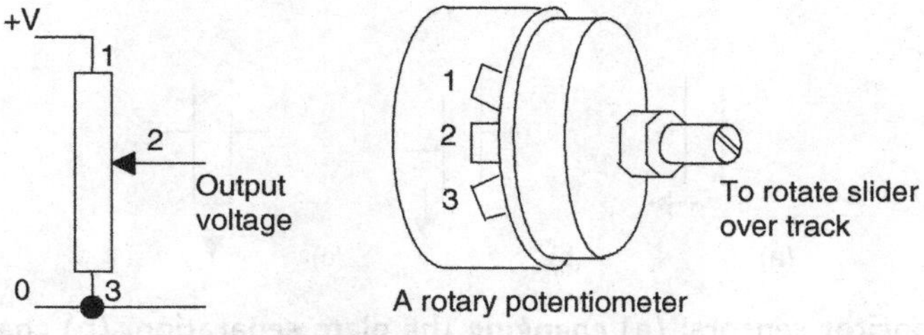

Figure 2.16: Potentiometer.

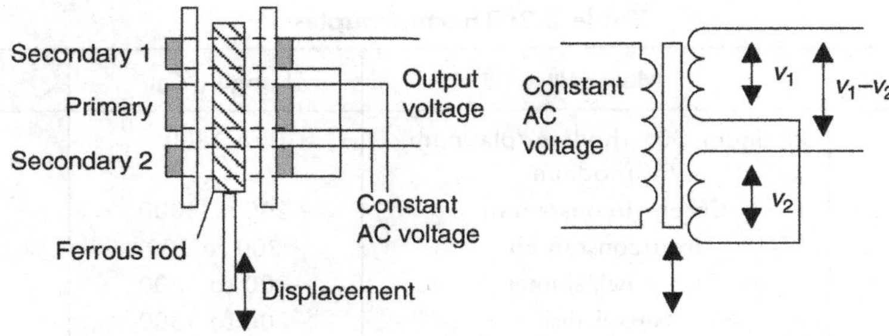

Figure 2.17: LVDT.

an alternating current is applied to the primary coil, alternating voltages, v_1 and v_2, are induced in the two secondary coils. When the ferrous rod core is centered between the two secondary coils, the voltages induced in them are equal. The outputs from the two secondary coils are connected so that their combined output is the difference between the two voltages, that is, $v_1 - v_2$. With the rod central, the two alternating voltages are equal and so there is no output voltage. When the rod is displaced from its central position, there is more of the rod in one secondary coil than the other. As a result, the size of the alternating voltage induced in one coil is greater than that in the other. The difference between the two secondary coil voltages, that is, the output, thus depends on the position of the ferrous rod. The output from the LVDT is an alternating voltage. This is usually converted to an analog DC voltage and amplified before inputting to the analog channel of a PLC.

Capacitive displacement sensors are essentially just parallel plate capacitors. The capacitance will change if the plate separation changes, the area of overlap of the plates changes, or a slab of dielectric is moved into or out of the plates (Figure 2.18). All these methods can be used to give linear displacement sensors. The change in capacitance has to be converted into a suitable electrical signal by signal conditioning.

2.1.7 Strain Gauges

When a wire or strip of semiconductor is stretched, its resistance changes. The fractional change in resistance is proportional to the fractional change in length, that is, strain.

$$\frac{\Delta R}{R} = G \times \text{strain},$$

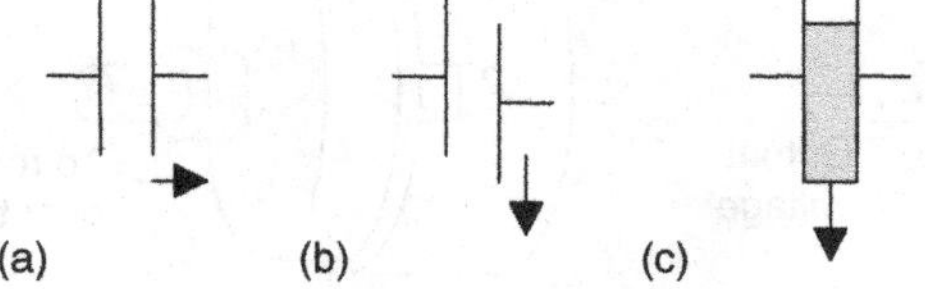

Figure 2.18: Capacitor sensors: (a) changing the plate separation, (b) changing the area of overlap, and (c) moving the dielectric.

where ΔR is the change in resistance for a wire of resistance R and G is a constant called the *gauge factor*. For metals, the gauge factor is about 2; for semiconductors, about 100. Metal resistance strain gauges are in the form of a flat coil so that they get a reasonable length of metal in a small area. Often they are etched from metal foil (Figure 2.19a) and attached to a backing of thin plastic film so that they can be stuck on surfaces, like postage stamps on an envelope. The change in resistance of the strain gauge, when subject to strain, is usually converted into a voltage signal by the use of a Wheatstone bridge. A problem that occurs is that the resistance of the strain gauge also changes with temperature, and thus some means of temperature compensation has to be used so that the output of the bridge is only a function of the strain. This can be achieved by placing a dummy strain gauge in an opposite arm of the bridge, that gauge not being subject to any strain but only the temperature (Figure 2.19b). A popular alternative is to use four active gauges as the arms of the bridge and arrange them so that one pair of opposite gauges is in tension and the other pair in compression. This not only gives temperature compensation; it also gives a much larger output change when strain is applied. The following paragraph illustrates systems employing such a form of compensation.

By attaching strain gauges to other devices, changes that result in strain of those devices can be transformed, by the strain gauges, to give voltage changes. They might, for example, be attached to a cantilever to which forces are applied at its free end (Figure 2.19c). The voltage change, resulting from the strain gauges and the Wheatstone bridge, then

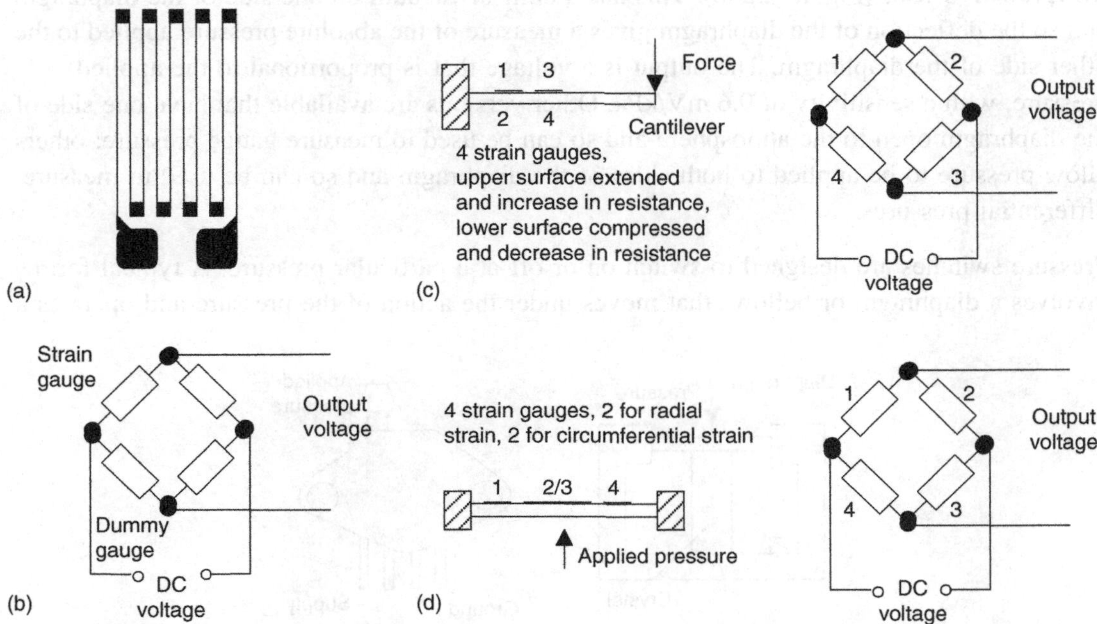

Figure 2.19: (a) Metal foil strain gauge, (b) a Wheatstone bridge circuit with compensation for temperature changes, (c) strain gauges used for a force sensor, and (d) a pressure sensor.

becomes a measure of the force. Another possibility is to attach strain gauges to a diaphragm, which deforms as a result of pressure (Figure 2.19d). The output from the gauges and associated Wheatstone bridge then becomes a measure of the pressure.

2.1.8 Pressure Sensors

Pressure sensors can be designed to give outputs that are proportional to the *difference in pressure* between two input ports. If one of the ports is left open to the atmosphere, the gauge measures pressure changes with respect to the atmosphere and the pressure measured is known as *gauge pressure*. The pressure is termed the *absolute pressure* if it is measured with respect to a vacuum. Commonly used pressure sensors that give responses related to the pressure are *diaphragm* and *bellows types*. The diaphragm type consists of a thin disc of metal or plastic, secured around its edges. When there is a pressure difference between the two sides of the diaphragm, its center deflects. The amount of deflection is related to the pressure difference. This deflection may be detected by strain gauges attached to the diaphragm (see Figure 2.19d), by a change in capacitance between it and a parallel fixed plate, or by using the deflection to squeeze a piezoelectric crystal (Figure 2.20a).

When a piezoelectric crystal is squeezed, there is a relative displacement of positive and negative charges within the crystal and the outer surfaces of the crystal become charged. Hence a potential difference appears across it. An example of such a sensor is the Motorola MPX100AP sensor (Figure 2.20b). This has a built-in vacuum on one side of the diaphragm and so the deflection of the diaphragm gives a measure of the absolute pressure applied to the other side of the diaphragm. The output is a voltage that is proportional to the applied pressure, with a sensitivity of 0.6 mV/kPa. Other versions are available that have one side of the diaphragm open to the atmosphere and so can be used to measure gauge pressure; others allow pressure to be applied to both sides of the diaphragm and so can be used to measure differential pressures.

Pressure switches are designed to switch on or off at a particular pressure. A typical form involves a diaphragm or bellows that moves under the action of the pressure and operates a

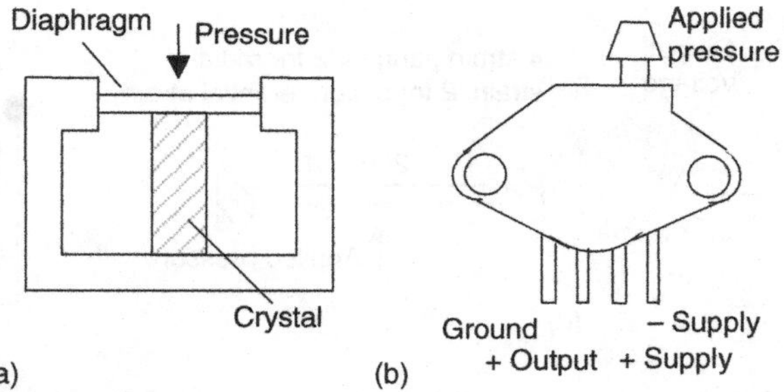

Figure 2.20: (a) A piezoelectric pressure sensor and (b) the MPX100AP.

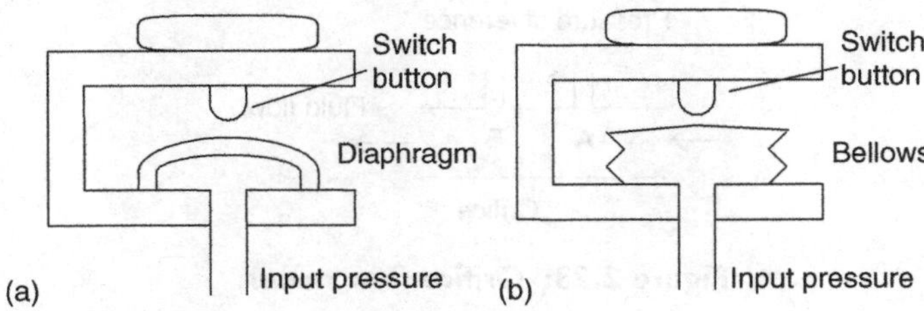

Figure 2.21: Examples of pressure switches.

mechanical switch. Figure 2.21 shows two possible forms. Diaphragms are less sensitive than bellows but can withstand greater pressures.

2.1.9 Liquid-Level Detectors

Pressure sensors may be used to monitor the depth of a liquid in a tank. The pressure due to a height of liquid h above some level is $h\rho g$, where ρ is the density of the liquid and g the acceleration due to gravity. Thus a commonly used method of determining the level of liquid in a tank is to measure the pressure due to the liquid above some datum level (Figure 2.22).

Often a sensor is just required to give a signal when the level in some container reaches a particular level. A float switch that is used for this purpose consists of a float containing a magnet that moves in a housing with a reed switch. As the float rises or falls, it turns the reed switch on or off, the reed switch being connected in a circuit that then switches a voltage on or off.

2.1.10 Fluid Flow Measurement

A common form of fluid flow meter is one based on measuring the difference in pressure that results when a fluid flows through a constriction. Figure 2.23 shows a commonly used form, the *orifice flow meter*. As a result of the fluid flowing through the orifice, the pressure at A is higher than that at B, the difference in pressure being a measure of the rate of flow. This pressure difference can be monitored by means of a diaphragm pressure gauge and thus becomes a measure of the rate of flow.

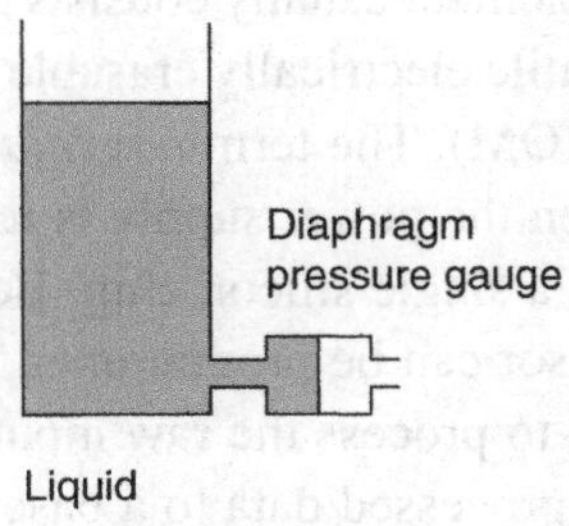

Figure 2.22: Liquid-level sensor.

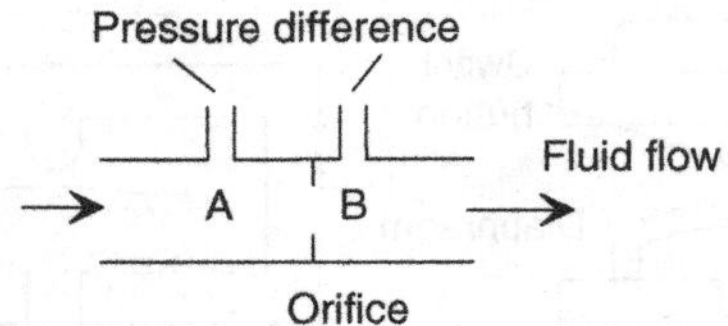

Figure 2.23: Orifice flow meter.

2.1.11 Ultrasonic Proximity Sensors

Ultrasonic proximity sensors direct ultrasonic sound waves (i.e. high frequency sound waves beyond the audible frequencies) to a target and measure the time taken for the sound waves to return; the further away the object the greater the time taken. Such sensors are used for distances to a target of the order of a few centimeters to a meter. The Omron E4C-DS30 ultrasonic proximity sensor has a range of 50–300 mm, the E4C-DS80 has 70–800 mm, and the E4C-DS100 sensor 90–1000 mm. Ideally, the target object should have a flat smooth surface to give a good reflection, as uneven or curved surfaces give a poorer reflection and so the ultrasonic proximity sensor must be positioned closer to such objects. Soft materials, e.g. foam, do not reflect sound waves well enough to be able to be detected. As an indicator of the variety of applications that the Omron sensors can be used for, they give detection of transparent trays, inspection of solvent tank levels, detection of sheet sag, detection of tires on a conveyor belt, detection of rubber sheet sag between supporting rollers and detection of the position of glass substrates in cassettes.

2.1.12 Smart Sensors

To use a sensor, we generally need to add signal conditioning circuitry, such as circuits which amplify and convert from analog to digital, to get the sensor signal in the right form, take account of any nonlinearities, and calibrate it. Additionally, we need to take account of drift, that is, a gradual change in the properties of a sensor over time. Some sensors have all these elements taken care of in a single package; they are called *smart sensors*.

The term *smart sensor* is thus used in discussing a sensor that is integrated with the required buffering and conditioning circuitry in a single element and provides functions beyond that of just a sensor. The circuitry with the element usually consists of data converters, a processor and firmware, and some form of nonvolatile electrically erasable programmable read only memory (EEPROM, which is similar to EPROM). The term *nonvolatile* is used because the memory has to retain certain parameters when the power supply is removed. Such smart sensors can have all their elements produced on a single silicon chip. Because the elements are processor-based devices, such a sensor can be programmed for specific requirements. For example, it can be programmed to process the raw input data, correcting for such things as nonlinearities, and then send the processed data to a base station. It can be programmed to send a warning signal when the measured parameter reaches some critical value.

The IEEE 1451.4 standard interface for smart sensors and actuators is based on an electronic data sheet (TEDS) format that is aimed at allowing installed analog transducers to be easily connected to digital measurement systems. The standard requires the nonvolatile EEPROM embedded memory to hold and communicate data, which will allow a plug-and-play capability. It thus would hold data for the identification and properties for the sensor and might also contain the calibration template, thus facilitating digital interrogation.

2.1.13 Sensors Ranges

To give some idea of the range of sensors that are used in control systems, the following is part of the extensive list of sensors available from Rockwell Automation for such applications:

- Condition sensors: give information to enable automatic sequencing of equipment. These include pressure sensors, temperature sensors, level sensors, flow switches and speed sensing switches.
- Presence sensing sensors: detect the distance, absence or presence of an object and include inductive proximity sensors, capacitive proximity sensors, ultrasonic sensors and photoelectric sensors. Photoelectric sensors are available for such applications as the detection of clear materials such as glass and plastic bottles, also with the use of filters to detect changes in color. Fibre optic sensors, consisting of a light sensor at the end of a fibre optic cable, can be used for packaging applications where the detection of very small objects is required.
- Limit sensors: electromechanical devices that consist of an actuator linked to a set of contacts so when an object comes into contact with the actuator, the device operates its contact to either make or break an electrical circuit.
- Safety interlock switches: used as a means for safeguarding plant by shutting off power, controlling personnel access and preventing a machine from starting when the safety guard is open.

2.2 Output Devices

The output ports of a PLC are relay or optoisolator with transistor or triac, depending on the devices that are to be switched on or off. Generally, the digital signal from an output channel of a PLC is used to control an actuator, which in turn controls some process. The term *actuator* is used for the device that transforms the electrical signal into some more powerful action, which then results in control of the process. The following are some examples.

2.2.1 Relay

When a current passes through a solenoid, a magnetic field is produced; this can then attract ferrous metal components in its vicinity. With the *relay*, this attraction is used to operate a

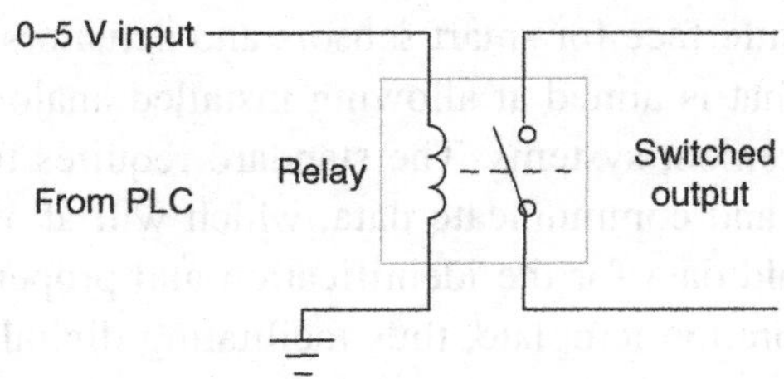

Figure 2.24: Relay used as an output device.

switch. Relays can thus be used to control a larger current or voltage and, additionally, to isolate the power used to initiate the switching action from that of the controlled power. For a relay connected to the output of a PLC, when the output switches on, the solenoid magnetic field is produced, and this pulls on the contacts and so closes a switch or switches (Figure 2.24). The result is that much larger currents can be switched on. Thus the relay might be used to switch on the current to a motor. The solenoid of a relay might be used to operate more than one set of contacts, the term *pole* being used for each set of contacts. Contacts can also be obtained as, in the absence of any input, either normally open (NO) or normally closed (NC). Thus, when selecting relays for a particular application, consideration has to be given to the number of poles required, the initial contact conditions, and the rated voltage and current.

The term *latching relay* is used for a relay whose contacts remain open or closed even after the power has been removed from the solenoid. The term *contactor* is used when large currents are being switched from large voltage sources.

2.2.2 Directional Control Valves

Another example of the use of a solenoid as an actuator is a *solenoid operated valve*. The valve may be used to control the directions of flow of pressurized air or oil and so used to operate other devices, such as a piston moving in a cylinder. Figure 2.25 shows one such form, a *spool valve*, used to control the movement of a piston in a cylinder. Pressurized air or hydraulic fluid is input from port P, which is connected to the pressure supply from a pump or compressor, and port T is connected to allow hydraulic fluid to return to the supply tank or, in the case of a pneumatic system, to vent the air to the atmosphere. With no current through the solenoid (Figure 2.25a), the hydraulic fluid or pressurized air is fed to the right of the piston and exhausted from the left, the result then being the movement of the piston to the left. When a current is passed through the solenoid, the spool valve switches the hydraulic fluid or pressurized air to the left of the piston and exhausts it from the right. The piston then moves to the right. The movement of the piston might be used to push a deflector to deflect items off a conveyor belt (refer back to Figure 1.1b) or implement some other form of displacement that requires power.

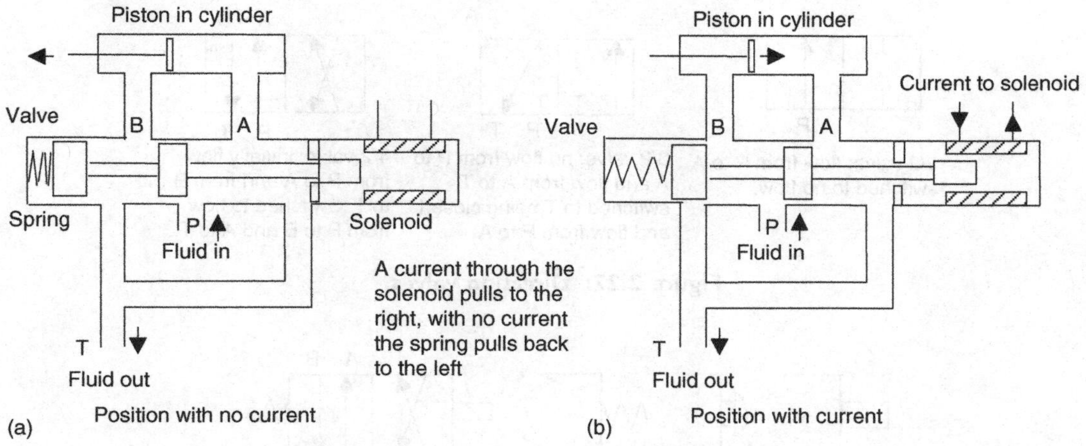

Figure 2.25: An example of a solenoid operated valve.

With the preceding valve the two control positions are shown in Figures 2.25a and 2.25b. Directional control valves are described by the number of ports and the number of control positions they contain. The valve shown in Figure 2.25 has four ports—A, B, P, and T—and two control positions. It is thus referred to as a *4/2 valve*. The basic symbol used on drawings for valves is a square, with one square used to describe each of the control positions. Thus the symbol for the valve in Figure 2.25 consists of two squares (Figure 2.26a). Within each square the switching positions are then described by arrows to indicate a flow direction, or a terminated line to indicate no flow path (Figure 2.26b). Pipe connections, that is, the inlet and output ports, for a valve are indicated by lines drawn outside the box and are drawn for just the box representing the unactuated or rest position for the valve. Figure 2.26c shows this for the valve shown in Figure 2.25. Figure 2.27 shows more examples of direction valves and their switching positions.

In diagrams, the actuation methods used with valves are added to the symbol; Figure 2.28 shows examples of such symbols. The valve shown in Figure 2.25 has a spring to give one position and a solenoid to give the other, so the symbol is as shown in Figure 2.28d.

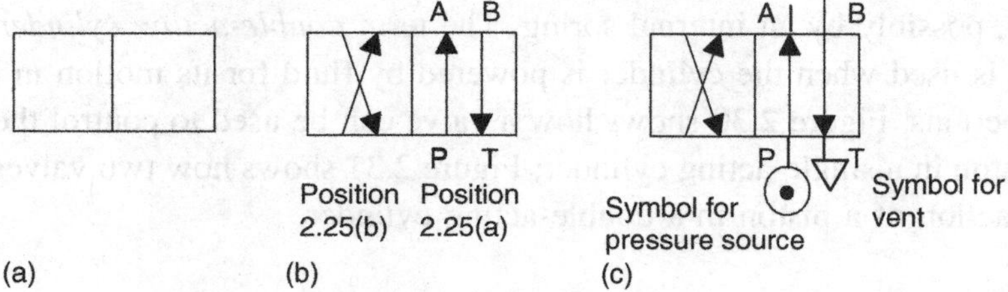

Figure 2.26: (a) The basic symbol for a two-position valve; (b) the 4/2 valve; and (c) external connections to the 4/2 valve. The P label is used to indicate a connection to a pressure supply and the T to an exhaust port.

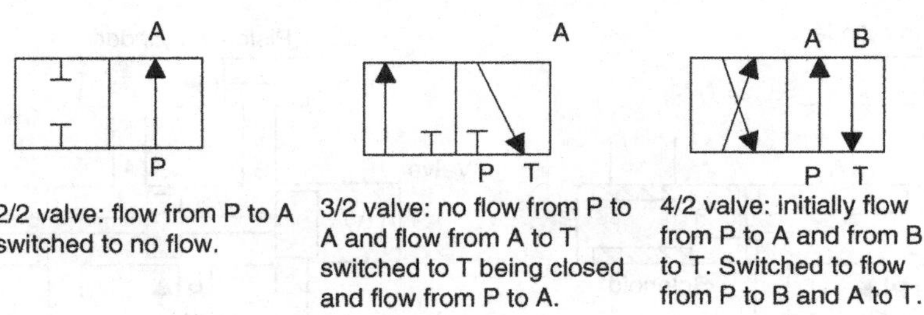
A
P
2/2 valve: flow from P to A switched to no flow.
A
P T
3/2 valve: no flow from P to A and flow from A to T switched to T being closed and flow from P to A.
A B
P T
4/2 valve: initially flow from P to A and from B to T. Switched to flow from P to B and A to T.

Figure 2.27: Direction valves.

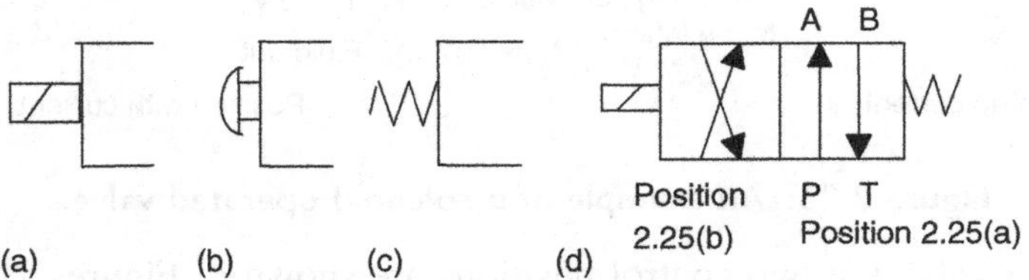
A B
Position 2.25(b)
P T
Position 2.25(a)
(a)
(b)
(c)
(d)

Figure 2.28: Actuation symbols: (a) solenoid, (b) push button, (c) spring operated, (d) a spring and solenoid operated 4/2 valve.

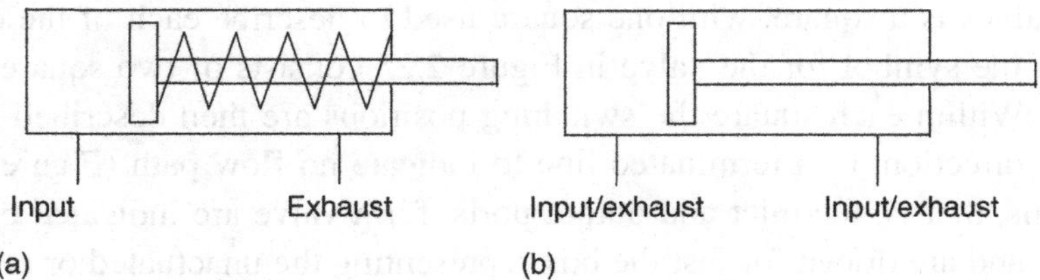
Input
Exhaust
Input/exhaust
Input/exhaust
(a)
(b)

Figure 2.29: Cylinders: (a) single acting, and (b) double acting.

Direction valves can be used to control the direction of motion of pistons in cylinders, the displacement of the pistons being used to implement the required actions. The term *single-acting cylinder* (Figure 2.29a) is used for one that is powered by the pressurized fluid being applied to one side of the piston to give motion in one direction, which is returned in the other direction, possibly by an internal spring. The term *double-acting cylinder* (Figure 2.29b) is used when the cylinder is powered by fluid for its motion in both piston movement directions. Figure 2.30 shows how a valve can be used to control the direction of motion of a piston in a single-acting cylinder; Figure 2.31 shows how two valves can be used to control the action of a piston in a double-acting cylinder.

2.2.3 Motors

A *DC motor* has coils of wire mounted in slots on a cylinder of ferromagnetic material, which is termed the *armature*. The armature is mounted on bearings and is free to rotate. It is

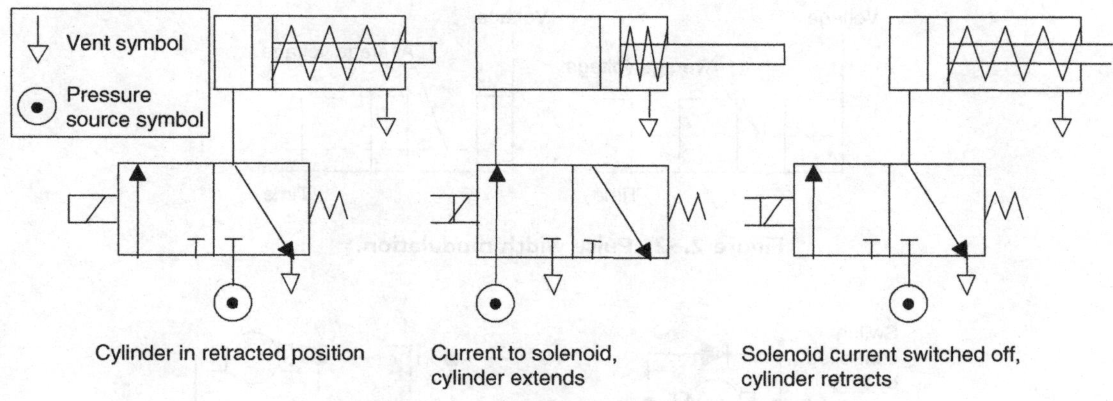

Figure 2.30: Control of a single-acting cylinder.

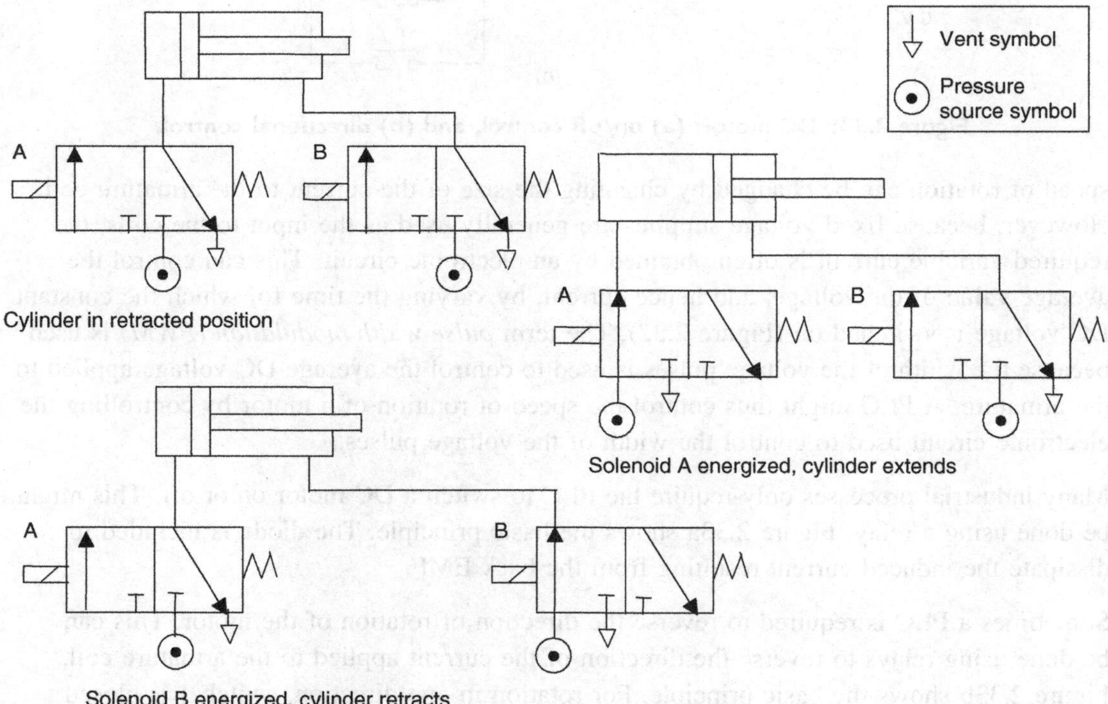

Figure 2.31: Control of a double-acting cylinder.

mounted in the magnetic field produced by permanent magnets or current passing through coils of wire, which are called the *field coils*. When a current passes through the armature coil, forces act on the coil and result in rotation. Brushes and a commutator are used to reverse the current through the coil every half rotation and so keep the coil rotating. The

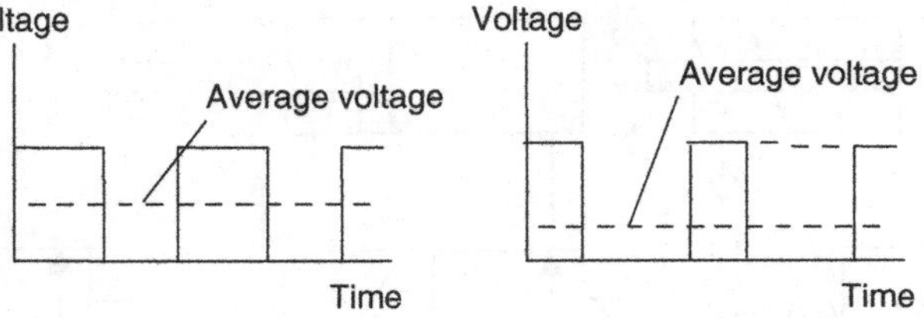

Figure 2.32: Pulse width modulation.

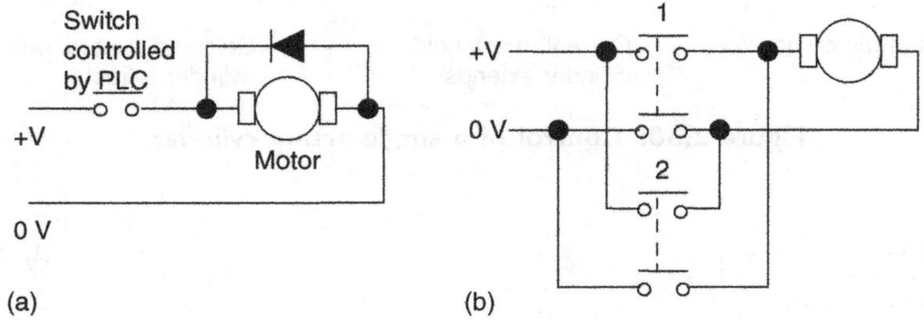

Figure 2.33: DC motor: (a) on/off control, and (b) directional control.

speed of rotation can be changed by changing the size of the current to the armature coil. However, because fixed voltage supplies are generally used as the input to the coils, the required variable current is often obtained by an electronic circuit. This can control the average value of the voltage, and hence current, by varying the time for which the constant DC voltage is switched on (Figure 2.32). The term *pulse width modulation (PWM)* is used because the width of the voltage pulses is used to control the average DC voltage applied to the armature. A PLC might thus control the speed of rotation of a motor by controlling the electronic circuit used to control the width of the voltage pulses.

Many industrial processes only require the PLC to switch a DC motor on or off. This might be done using a relay. Figure 2.33a shows the basic principle. The diode is included to dissipate the induced current resulting from the back EMF.

Sometimes a PLC is required to reverse the direction of rotation of the motor. This can be done using relays to reverse the direction of the current applied to the armature coil. Figure 2.33b shows the basic principle. For rotation in one direction, switch 1 is closed and switch 2 opened. For rotation in the other direction, switch 1 is opened and switch 2 closed.

Another form of DC motor is the *brushless DC motor*. This uses a permanent magnet for the magnetic field, but instead of the armature coil rotating as a result of the magnetic field of the magnet, the permanent magnet rotates within the stationary coil. With the conventional DC motor, a commutator has to be used to reverse the current through the coil every half rotation to keep the coil rotating in the same direction. With the brushless permanent magnet motor, electronic circuitry is used to reverse the current. The motor can be started and stopped by controlling the current to the stationary coil. Reversing the motor is more difficult, as reversing the current is not so easy, due to the electronic circuitry used for the commutator function. One method that is used is to incorporate sensors with the motor to detect the position of the north and south poles. These sensors can then cause the current to the coils to be switched at just the right moment to reverse the forces applied to the magnet. The speed of rotation can be controlled using pulse width modulation, that is, controlling the average value of pulses of a constant DC voltage.

Though AC motors are cheaper, more rugged, and more reliable than DC motors, maintaining constant speed and controlling that speed is generally more complex than with DC motors. As a consequence, DC motors, particularly brushless permanent magnet motors, tend to be more widely used for control purposes.

2.2.4 Stepper Motors

The *stepper* or *stepping motor* is a motor that produces rotation through equal angles, the so-called *steps*, for each digital pulse supplied to its input (Figure 2.34). Thus, if one input pulse produces a rotation of 1.8°, then 20 such pulses would give a rotation of 36.0°. To obtain one complete revolution through 360°, 200 digital pulses would be required. The motor can thus be used for accurate angular positioning.

If a stepping motor is used to drive a continuous belt (Figure 2.35), it can be used to give accurate linear positioning. Such a motor is used with computer printers, robots, machine tools, and a wide range of instruments for which accurate positioning is required.

There are two basic forms of stepper motor: the *permanent magnet* type, with a permanent magnet rotor, and the *variable reluctance* type, with a soft steel rotor. There is also a *hybrid*

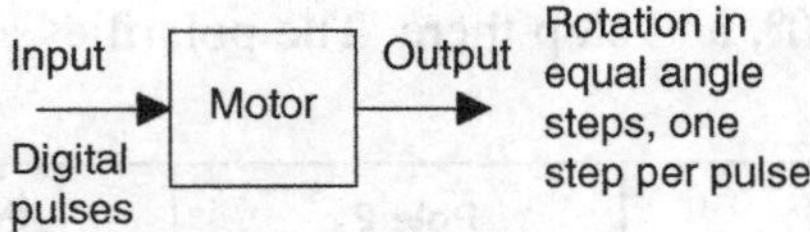

Figure 2.34: The stepping motor.

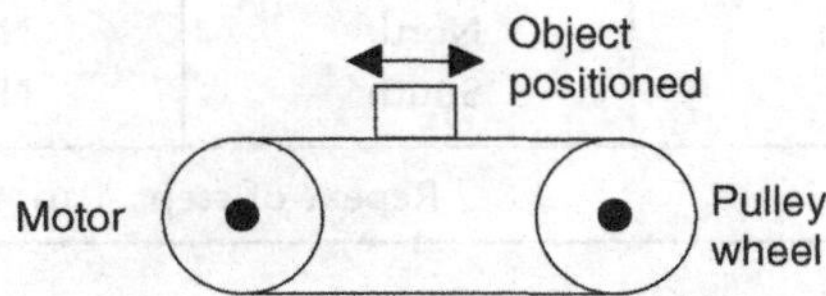

Figure 2.35: Linear positioning.

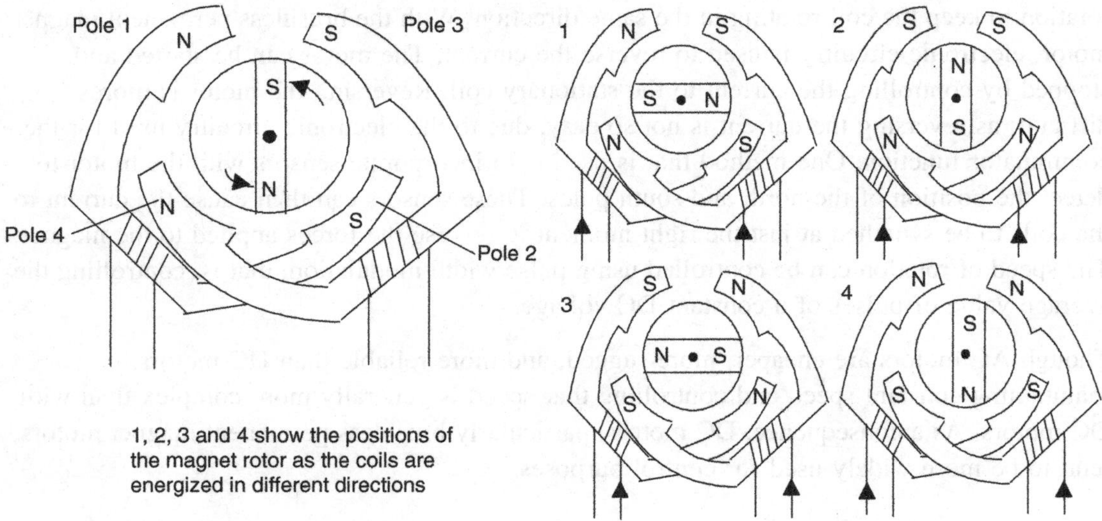

Figure 2.36: The basic principles of the permanent magnet stepper motor (2-phase) with 90° steps.

form combining both the permanent magnet and variable reluctance types. The most common type is the permanent magnet form.

Figure 2.36 shows the basic elements of the permanent magnet type with two pairs of stator poles. Each pole is activated by a current being passed through the appropriate field winding, the coils being such that opposite poles are produced on opposite coils. The current is supplied from a DC source to the windings through switches. With the currents switched through the coils such that the poles are as shown in Figure 2.36, the rotor will move to line up with the next pair of poles and stop there. This would be a rotation of 90°. If the current is then switched so that the polarities are reversed, the rotor will move a step to line up with the next pair of poles, at angle 180°, and stop there. The polarities associated with each step are as follows:

Step	Pole 1	Pole 2	Pole 3	Pole 4
1	North	South	South	North
2	South	North	South	North
3	South	North	North	South
4	North	South	North	South
5	Repeat of steps 1 to 4			

Thus in this case there are four possible rotor positions: 0°, 90°, 180°, and 270°.

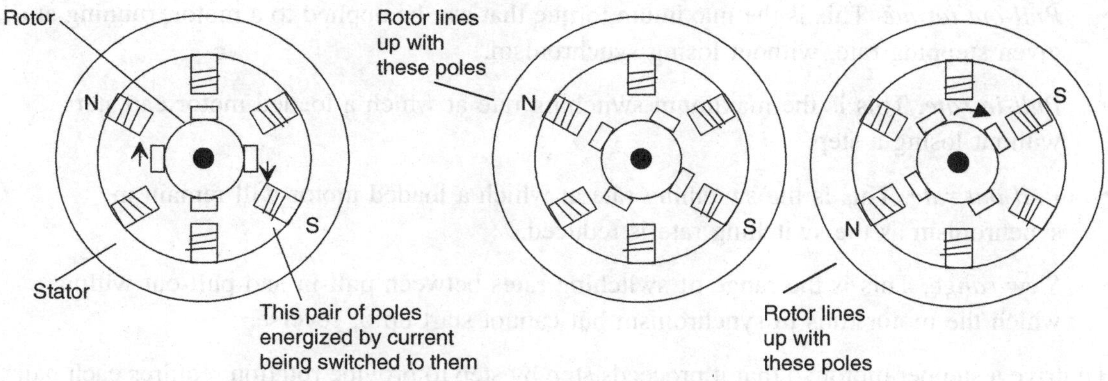

Figure 2.37: The principle of the variable reluctance stepper motor.

Figure 2.37 shows the basic principle of the *variable reluctance* type. The rotor is made of soft steel and has a number of teeth, the number being less than the number of poles on the stator. The stator has pairs of poles, each pair of which is activated and made into an electromagnet by a current being passed through the coils wrapped round it. When one pair of poles is activated, a magnetic field is produced that attracts the nearest pair of rotor teeth so that the teeth and poles line up. This is termed the position of *minimum reluctance*. By then switching the current to the next pair of poles, the rotor can be made to rotate to line up with those poles. Thus by sequentially switching the current from one pair of poles to the next, the rotor can be made to rotate in steps.

There is another version of the stepper motor—the *hybrid stepper*. This version combines features of both the permanent magnet and variable reluctance motors. Hybrid steppers have a permanent magnet rotor encased in iron caps that are cut to have teeth. The rotor sets itself in the minimum reluctance position when a pair of stator coils are energized.

The following are some of the terms commonly used in specifying stepper motors:

- *Phase*. This term refers to the number of independent windings on the stator. Two-phase motors tend to be used in light-duty applications, three-phase motors tend to be variable reluctance steppers, and four-phase motors tend to be used for higher-power applications.
- *Step angle*. This is the angle through which the rotor rotates for one switching change for the stator coils.
- *Holding torque*. This is the maximum torque that can be applied to a powered motor without moving it from its rest position and causing spindle rotation.
- *Pull-in torque*. This is the maximum torque against which a motor will start, for a given pulse rate, and reach synchronism without losing a step.

- *Pull-out torque*. This is the maximum torque that can be applied to a motor, running at a given stepping rate, without losing synchronism.
- *Pull-in rate*. This is the maximum switching rate at which a loaded motor can start without losing a step.
- *Pull-out rate*. This is the switching rate at which a loaded motor will remain in synchronism as the switching rate is reduced.
- *Slew range*. This is the range of switching rates between pull-in and pull-out within which the motor runs in synchronism but cannot start up or reverse.

To drive a stepper motor so that it proceeds step by step to provide rotation requires each pair of stator coils to be switched on and off in the required sequence when the input is a sequence of pulses (Figure 2.38). Driver circuits are available to give the correct sequencing. Figure 2.39 shows an example: the SAA 1027 for a four-phase unipolar stepper. Motors are termed *unipolar* if they are wired so that the current can flow in only one direction through

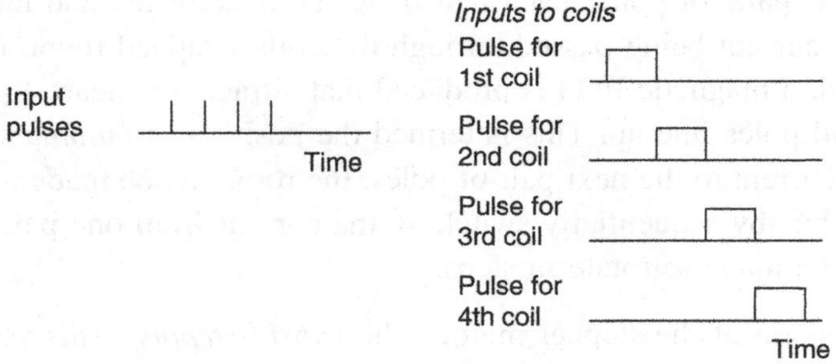

Figure 2.38: Input and outputs of the drive system.

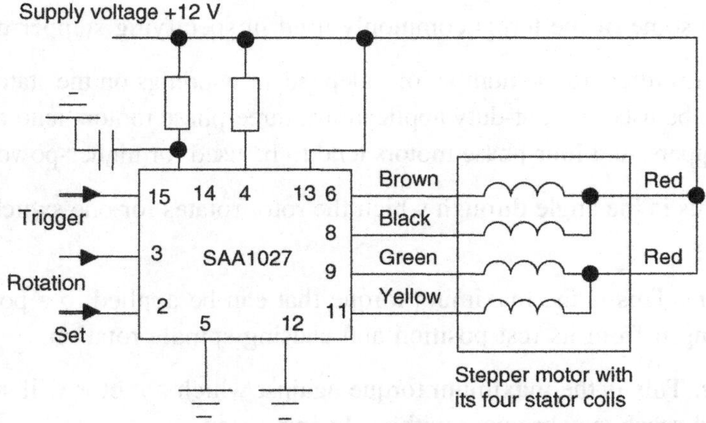

Figure 2.39: Driver circuit connections with the integrated circuit SAA1027.

any particular motor terminal; they're called *bipolar* if the current can flow in either direction through any particular motor terminal. The stepper motor will rotate through one step each time the trigger input goes from low to high. The motor runs clockwise when the rotation input is low and anticlockwise when high. When the set pin is made low, the output resets. In a control system, these input pulses might be supplied by a microprocessor.

2.3 Examples of Applications

The following are some examples of control systems designed to illustrate the use of a range of input and output devices.

2.3.1 A Conveyor Belt

Consider a conveyor belt that is to be used to transport goods from a loading machine to a packaging area (Figure 2.40). When an item is loaded onto the conveyor belt, a contact switch might be used to indicate that the item is on the belt and to start the conveyor motor. The motor then has to keep running until the item reaches the far end of the conveyor and falls off into the packaging area. When it does this, a switch might be activated that has the effect of switching off the conveyor motor. The motor is then to remain off until the next item is loaded onto the belt. Thus the inputs to a PLC controlling the conveyor are from two switches and the output is to a motor.

2.3.2 A Lift

Consider a simple goods lift to move items from one level to another. For example, it might lift bricks from the ground level to the height where some bricklayers are working. The lift is to move upward when a push button is pressed at the ground level to send the lift upward or a push button is pressed at the upper level to request the lift to move upward, but in both cases there is a condition that has to be met that a limit switch indicates that the access gate to the lift platform is closed. The lift is to move downward when a push button is pressed at the upper level to send the lift downward or a push button is pressed at the lower level to request the lift to move downward, but in both cases there is a condition that has to be met that a limit switch indicates that the access gate to the lift platform is closed. Thus the inputs to the control system are electrical on/off signals from push button switches and limit switches. The output from the control system is the signal to control the motor.

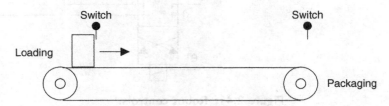

Figure 2.40: Conveyor.

2.3.3 A Robot Control System

Figure 2.41 shows how directional control valves can be used for a control system of a robot. When there is an input to solenoid A of valve 1, the piston moves to the right and causes the gripper to close. If solenoid B is energized with A deenergized, the piston

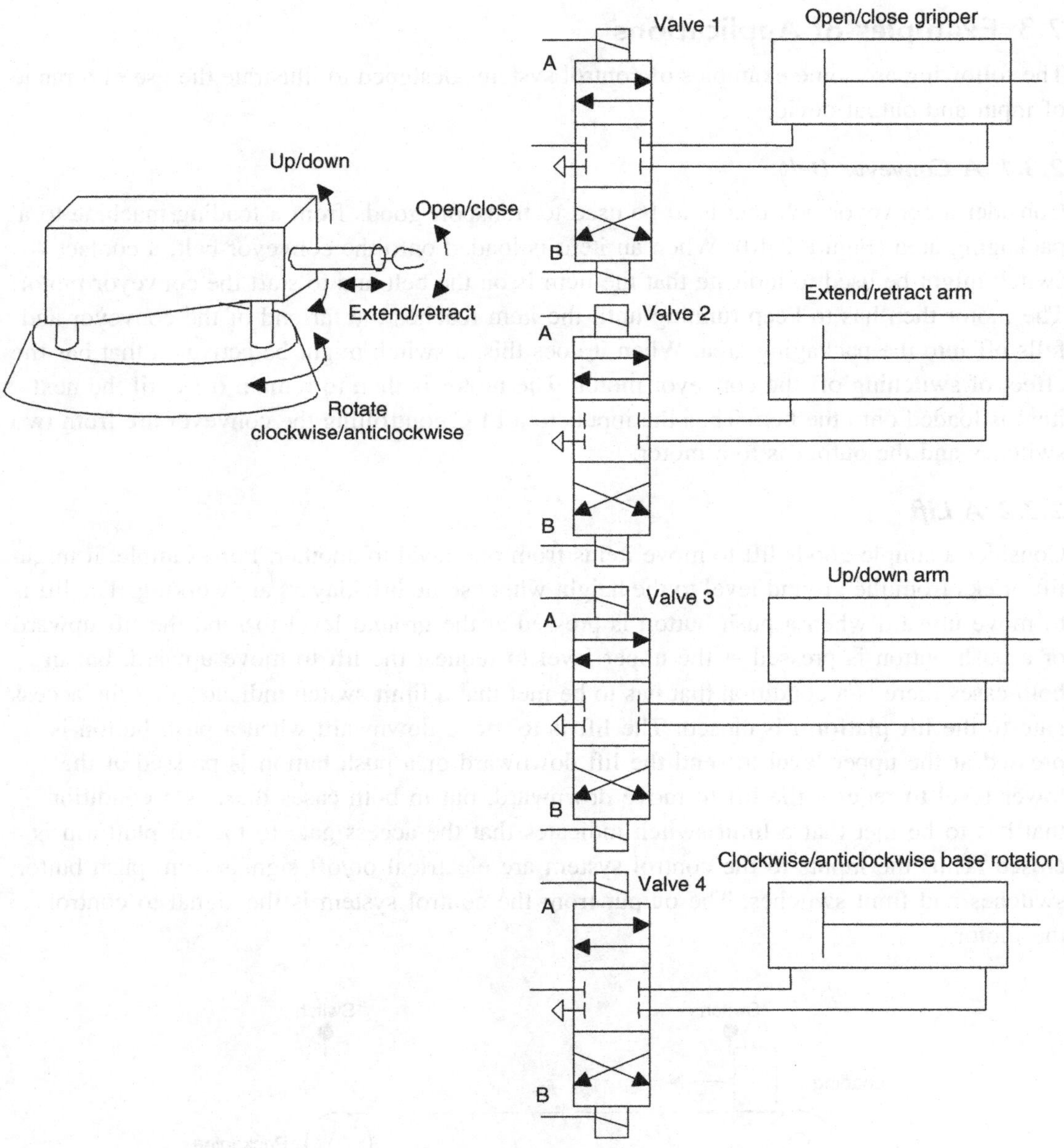

Figure 2.41: Robot controls.

moves to the left and the gripper opens. When both solenoids are deenergized, no air passes to either side of the piston in the cylinder and the piston keeps its position without change. Likewise, inputs to the solenoids of valve 2 are used to extend or retract the arm. Inputs to the solenoids of valve 3 are used to move the arm up or down. Inputs to the solenoids of valve 4 are used to rotate the base in either a clockwise or anticlockwise direction.

2.3.4 Liquid-Level Monitoring

Figure 2.42 shows a method that could be used to give an on/off signal when the liquid in a container reaches a critical level. A magnetic float, a ring circling the sensor probe, falls as the liquid level falls and opens a reed switch when the critical level is reached. The reed switch is in series with a 39 Ω resistor so that this is switched in parallel with a 1 kΩ resistor by the action of the reed switch. Opening the reed switch thus increases the resistance from about 37 Ω to 1 kΩ. Such a resistance change can be transformed by signal conditioning to give suitable on/off signals.

2.3.5 Packages on Conveyor Belt Systems

In some situations, the requirement is to check whether there is a nontransparent item on the belt at a particular position. This can be done using a light emitter on one side of the belt and

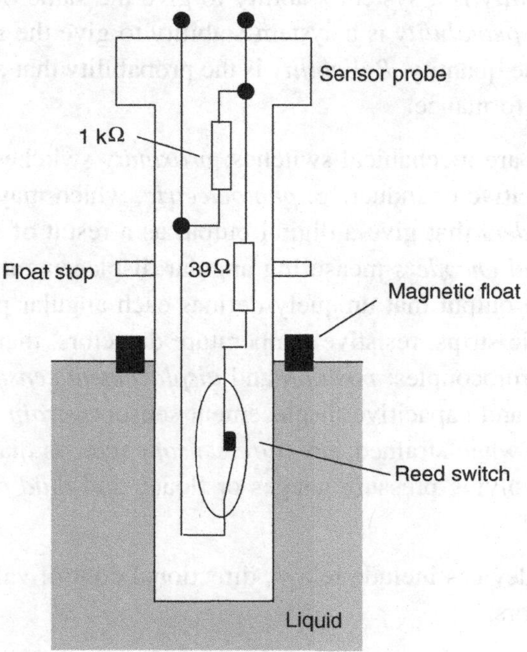

Figure 2.42: Liquid-level monitoring.

a photoelectric sensor on the other, there then being an interruption of the light beam when the item is at the required position. If the item had been transparent, such as a bottle, the photoelectric sensor might have been positioned to pick up reflected light to determine when the item is in the required position.

Summary

The term *sensor* refers to an input device that provides a usable output in response to a specified input. The term *transducer* is generally used for a device that converts a signal from one form to a different physical form.

Common terms used to specify the performance of sensors are as follows: *Accuracy* is the extent to which the value indicated by a measurement system or element might be wrong. *Error* is the difference between the result of a measurement and the true value. *Nonlinearity error* is the error that occurs as a result of assuming a linear relationship between input and output. *Hysteresis error* is the difference in output given for the same measured quantity according to whether that value was reached by a continuously increasing change or a continuously decreasing change. *Range* consists of the limits between which an input can vary. *Response time* is the time that elapses after the input is abruptly increased from zero to a constant value up to the time it reaches some specified percentage of the steady-state value. *Sensitivity* indicates how much the output changes when the quantity being measured changes by a given amount. *Stability* is a system's ability to give the same output for a given input over a period of time. *Repeatability* is a system's ability to give the same value for repeated measurements of the same quantity. *Reliability* is the probability that a system will operate up to an agreed level of performance.

Commonly used sensors are mechanical switches; *proximity* switches, which may be eddy current, reed, capacitive or inductive; *photoelectric*, which may be transmissive or reflective types; *encoders* that give a digital output as a result of angular or linear displacement, *incremental encoders* measuring angular displacement and *absolute encoders* giving a binary output that uniquely defines each angular position; *temperature sensors* such as bimetallic strips, resistive temperature detectors, thermistors, thermodiodes, thermotransistors, or thermocouples; *position* and *displacement sensors* such as potentiometers, LVDTs, and capacitive displacement sensors; *strain gauges*, which give a resistance change when strained; *pressure sensors* such as diaphragm gauges; *liquid-level detectors* involving pressure gauges or floats; and *fluid flow meters* such as the orifice flow meter.

Commonly used output devices include relays, directional control valves with cylinders, DC motors, and stepper motors.

Problems

Problems 1 through 14 have four answer options: A, B, C, or D. Choose the correct answer from the answer options.

1. *Decide whether each of these statements is true (T) or false (F).* A limit switch:
   (i) Can be used to detect the presence of a moving part.
   (ii) Is activated by contacts making or breaking an electrical circuit.
   A. (i) T (ii) T
   B. (i) T (ii) F
   C. (i) F (ii) T
   D. (i) F (ii) F

2. *Decide whether each of these statements is true (T) or false (F).* A thermistor is a temperature sensor that gives resistance changes that are:
   (i) A nonlinear function of temperature.
   (ii) Large for comparatively small temperature changes.
   A. (i) T (ii) T
   B. (i) T (ii) F
   C. (i) F (ii) T
   D. (i) F (ii) F

3. A diaphragm pressure sensor is required to give a measure of the gauge pressure present in a system. Such a sensor will need to have a diaphragm with:
   A. A vacuum on one side.
   B. One side open to the atmosphere.
   C. The pressure applied to both sides.
   D. A controlled adjustable pressure applied to one side.

4. The change in resistance of an electrical resistance strain gauge with a gauge factor of 2.0 and resistance 100 Ω when subject to a strain of 0.001 is:
   A. 0.0002 Ω
   B. 0.002 Ω
   C. 0.02 Ω
   D. 0.2 Ω

5. An incremental shaft encoder gives an output that is a direct measure of:
   A. The diameter of the shaft.
   B. The change in diameter of the shaft.
   C. The change in angular position of the shaft.
   D. The absolute angular position of the shaft.

6. *Decide whether each of these statements is true (T) or false (F).* Input devices that give an analog input for displacement include a:
   (i) Linear potentiometer.
   (ii) Linear variable differential transformer.
   A. (i) T (ii) T
   B. (i) T (ii) F
   C. (i) F (ii) T
   D. (i) F (ii) F

Problems 7 and 8 refer to Figure 2.43, which shows the symbol for a directional valve.

7. *Decide whether each of these statements is true (T) or false (F).* The valve has:
   (i) 4 ports
   (ii) 2 positions
   A. (i) T (ii) T
   B. (i) T (ii) F
   C. (i) F (ii) T
   D. (i) F (ii) F

8. *Decide whether each of these statements is true (T) or false (F).* In the control positions:
   (i) A is connected to T and P to B.
   (ii) P is connected to A and B to T.
   A. (i) T (ii) T
   B. (i) T (ii) F
   C. (i) F (ii) T
   D. (i) F (ii) F

9. *For the arrangement shown in Figure 2.44, decide whether each of these statements is true (T) or false (F).*
   (i) When a current passes through the solenoid, the cylinder extends.
   (ii) When the current ceases, the cylinder remains extended.
   A. (i) T (ii) T
   B. (i) T (ii) F
   C. (i) F (ii) T
   D. (i) F (ii) F

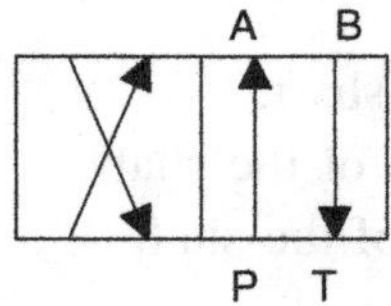

Figure 2.43: Diagram for Problems 7 and 8.

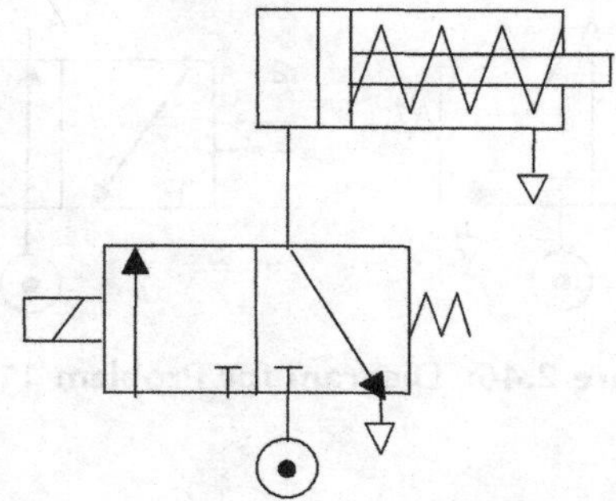

Figure 2.44: Diagram for Problem 9.

10. *For the arrangement shown in Figure 2.45, decide whether each of these statements is true (T) or false (F).*
    (i) When solenoid A is energized, the cylinder extends.
    (ii) When solenoid B is energized, the cylinder extends.
    A. (i) T (ii) T
    B. (i) T (ii) F
    C. (i) F (ii) T
    D. (i) F (ii) F

11. *For the two 3/2 valves shown in Figure 2.46, decide whether each of these statements is true (T) or false (F).*
    (i) When the solenoid in valve 1 is energized, A is vented.
    (ii) When the solenoid in valve 2 is energized, A is vented.
    A. (i) T (ii) T
    B. (i) T (ii) F
    C. (i) F (ii) T
    D. (i) F (ii) F

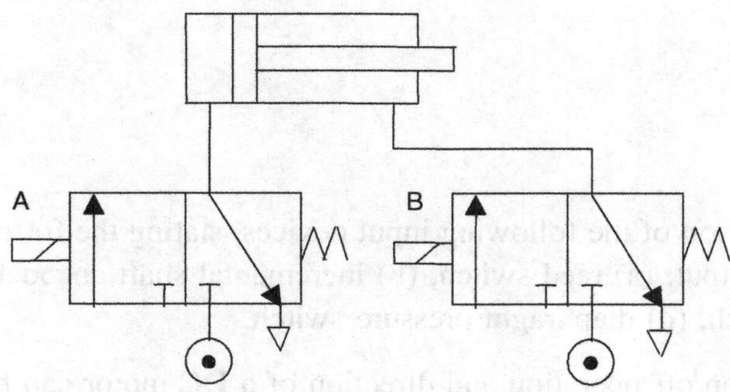

Figure 2.45: Diagram for Problem 10.

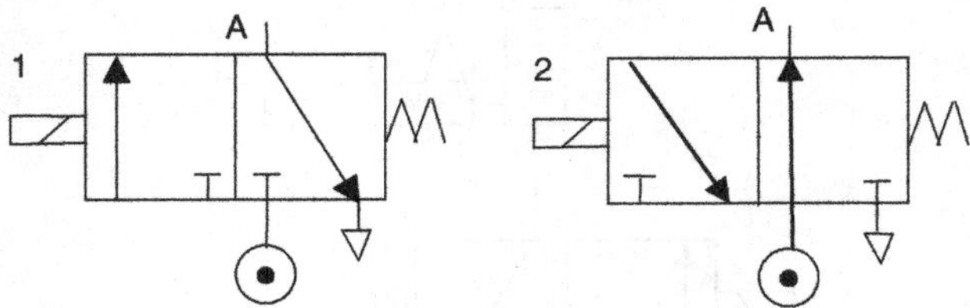

Figure 2.46: Diagram for Problem 11.

12. *Decide whether each of these statements is true (T) or false (F).* A stepper motor has a step angle of 1.8°. This means that:
    (i) Each pulse input to the motor rotates the motor shaft by 1.8°.
    (ii) The motor shaft takes 1 s to rotate through 1.8°.
    A. (i) T (ii) T
    B. (i) T (ii) F
    C. (i) F (ii) T
    D. (i) F (ii) F

13. A stepper motor has a step angle of 7.5°. The digital input rate required to produce a rotation of 10 rev/s is:
    A. 48 pulses per second
    B. 75 pulses per second
    C. 480 pulses per second
    D. 750 pulses per second

14. *Decide whether each of these statements is true (T) or false (F).* A proximity switch is required for detecting the presence of a nonmetallic object. Types of switches that might be suitable are:
    (i) Eddy current type.
    (ii) Capacitive type.
    A. (i) T (ii) T
    B. (i) T (ii) F
    C. (i) F (ii) T
    D. (i) F (ii) F

15. Explain the operation of the following input devices, stating the form of the signal being sensed and the output: (a) reed switch, (b) incremental shaft encoder, (c) photoelectric transmissive switch, (d) diaphragm pressure switch.

16. Explain how the on/off operation and direction of a DC motor can be controlled by switches.

17. Explain the principle of the stepper motor and state the different types available.

18. Select sensors that might be suitable for the following applications: (a) counting boxes moving along a conveyor belt, (b) verifying the level of milk in a plastic bottle moving along a conveyor belt, (c) determining when the piston in a cylinder has reached a particular point in its extension; (d) determining when a metal plate has reached the right position under a tool.

19. The following is part of the specification of a stepper motor. Explain the significance of the terms: phases 4, step angle 7.5°, current per phase 130 mA, resistance per phase 94 Ω, inductance per phase 43 mH, suitable driver SAA1027.

20. Suggest a way by which a spindle could be controlled to position a mechanism at 5° intervals.

21. A range of opaque bottles of various sizes moves along a conveyor belt. Suggest a method that could be used to (a) detect the different sizes and (b) push bottles off the belt.

Lookup Tasks

22. Look up the specifications of thermistors and select one that might be suitable for monitoring temperatures of about 40°C.

23. Look up the specification of the MPX100AP pressure sensor and write an outline of its possible use and capabilities.

24. Look up the specification of the LM3911N integrated temperature sensor and write an outline of its possible use and capabilities.

CHAPTER 3

Digital Systems

Digital systems work with inputs, which are essentially just off/on signals, with the two signal levels represented by 0 and 1. These are termed *binary digits*. The number system used for everyday calculations is the *denary* or *decimal system*. This is based on the use of 10 digits: 0, 1, 2, 3, 4, 5, 6, 7, 8, 9. With a number represented by this system, the digit position in the number indicates the weight attached to each digit, the weight increasing by a factor of 10 as we proceed from right to left. Hence we have:

	10^3	10^2	10^1	10^0
	Thousands	Hundreds	Tens	Units
Denary	1000	100	10	1

Thus if we have the denary number 1234, we have 1 with a place value of 10^3, 2 with a place value of 10^2, 3 with a place value of 10^1, and 4 with a place value of 10^0. Counting can, however, be done to any base. The denary system is convenient mainly because we have 10 fingers. If we had only two fingers, our system for everyday counting would probably have been different. Computers, and hence PLC systems, are based on counting in twos because it is convenient for their systems, their two digits being effectively just the off and on signals. When working with PLCs, other base number systems are also used; for example, input and output addresses are often specified using the octal system, that is, base 8. However, the PLC itself works with binary numbers. In this chapter we take a look at the various number systems.

We also take an introductory look at *logic systems*. A *Combinational logic systems* take binary inputs and combine them to give a binary output. The relationship between the inputs and the output can be described by *truth tables*. With such systems, the output of a particular combination of inputs is determined only by their state at the instant of time concerned. However, with *sequential logic systems* the output is influenced by the history of the past inputs as well as by the present inputs. Both combinational logic and sequential logic systems are introduced in this chapter.

W. Bolton: Programmable Logic Controllers, Sixth Edition. http://dx.doi.org/10.1016/B978-0-12-802929-9.00003-0

3.1 The Binary System

The *binary system* is based on just two digits: 0 and 1. These are termed *bi*nary digi*ts*, or *bits*. When a number is represented by this system, the digit position in the number indicates the weight attached to each digit, the weight increasing by a factor of 2 as we proceed from right to left.

	2^3	2^2	2^1	2^0
	bit 3	bit 2	bit 1	bit 0
Binary	1000	100	10	1

Bit 0 is termed the *least significant bit* (LSB) and the highest bit in a binary number is termed the *most significant bit* (MSB). For example, with the binary number 1010, the LSB is the bit at the right end of the number (0 in this example). The MSB is the bit at the left end of the number (1 in this example).

The conversion of a binary number to a denary number involves the addition of the powers of 2 indicated by the place position of a number in the overall number. Thus for the binary number 1010, we have 1 with a place value of 2^3, 0 with a place value of 2^2, 1 with a place value of 2^1, and 0 with a place value of 2^0, and so the conversion to a denary number is as follows:

	2^3	2^2	2^1	2^0
Binary	1	0	1	0
Denary	$2^3 = 8$	0	$2^1 = 2$	0

Thus the denary equivalent is 10.

The conversion of a denary number to a binary number involves looking for the appropriate powers of 2. We can do this by successive divisions by 2, noting the remainders at each division. Thus with the denary number 31:

$31 \div 2 = 15$ remainder 1; this gives the LSB

$15 \div 2 = 7$ remainder 1

$7 \div 2 = 3$ remainder 1

$3 \div 2 = 1$ remainder 1; this gives the MSB

The binary number is 11111. The first division gives the LSB because we have just divided 31 by 2, that is, 2^1, and found 1 left over for the 2^0 digit. The last division gives the MSB because the 31 has then been divided by 2 four times, that is, 2^4, and the remainder is 1.

3.2 Octal and Hexadecimal

Binary numbers are used in computers because the two states represented by 0 and 1 are easy to deal with in switching circuits, where they can represent off and on. A problem with binary numbers is that a comparatively small binary number requires a large number of digits. For example, the denary number 9, which involves just a single digit, requires four digits when written as the binary number 1001. The denary number 181, involving three digits, is 10110101 in binary form and requires eight digits. For this reason, octal or hexadecimal numbers are sometimes used to make numbers easier to handle and act as a "halfway house" between denary numbers and the binary numbers with which computers work. Thus, for example, Allen-Bradley uses octal numbering in its PLCs for input and output addresses.

3.2.1 Octal System

The *octal system* is based on eight digits: 0, 1, 2, 3, 4, 5, 6, 7. When a number is represented by this system, the digit position in the number indicates the weight attached to each digit, the weighting increasing by a factor of 8 as we proceed from right to left. Thus we have:

	8^3	8^2	8^1	8^0
Octal	1000	100	10	1

To convert denary numbers to octal, we successively divide by 8 and note the remainders. Thus the denary number 15 divided by 8 gives 1 with remainder 7; thus the denary number 15 is 17 in the octal system. To convert from octal to denary, we multiply the digits by the power of 8 appropriate to its position in the number. For example, the octal number 365 is $3 \times 8^2 + 6 \times 8^1 + 5 \times 8^0 = 245$. To convert from binary into octal, the binary number is written in groups of three bits starting with the least significant bit. For example, the binary number 11010110 would be written as:

11 010 110

Each group is then replaced by the corresponding digit from 0 to 7. For example, the 110 binary number is 6, the 010 is 2, and the 11 is 3. Thus the octal number is 326. As another example, the binary number 100111010 is:

100 111 010 *Binary*

4 7 2 *Octal*

Octal-to-binary conversion involves converting each octal digit into its 3-bit equivalent. Thus, for the octal number 21, we have 1 as 001 and 2 as 010:

2 1 *Octal number*

010 001 *Binary number*

and so the binary number is 010001.

3.2.2 Hexadecimal System

The *hexadecimal system (hex)* is based on 16 digits/symbols: 0, 1, 2, 3, 4, 5, 6, 7, 8, 9, A, B, C, D, E, F. When a number is represented by this system, the digit position in the number indicates that the weight attached to each digit increases by a factor of 16 as we proceed from right to left. Thus we have:

	16^3	16^2	16^1	16^0
Hex	1000	100	10	1

For example, the decimal number 15 is F in the hexadecimal system. To convert from denary numbers into hex we successively divide by 16 and note the remainders. Thus the denary number 156, when divided by 16, gives 9 with remainder 12, and so in hex is 9C. To convert from hex to denary, we multiply the digits by the power of 16 appropriate to its position in the number. Thus hex 12 is $1 \times 16^1 + 2 \times 16^0 = 18$. To convert binary numbers into hexadecimal numbers, we group the binary numbers into fours starting from the least significant number. Thus, for the binary number 1110100110 we have:

11 1010 0110 *Binary number*

3 A 6 *Hex number*

For conversion from hex to binary, each hex number is converted to its 4-bit equivalent. Thus, for the hex number 1D we have 0001 for the 1 and 1101 for the D:

1 D *Hex number*

0001 1101 *Binary number*

Thus the binary number is 0001 1101.

3.3 Binary Coded Decimals

Because the external world tends to deal mainly with numbers in the denary system and computers with numbers in the binary system, there is always the problem of conversion. There is, however, no simple link between the position of digits in a denary number and the position of digits in a binary number. An alternative method that is often used is the *binary coded decimal system* (BCD). With this system, each denary digit is coded separately in binary. For example, the denary number 15 has the 5 converted into the binary number 0101 and the 1 into 0001:

1 5 *Denary number*

0001 0101 *Binary number*

to give the number 0001 0101 in BCD. With the BCD system, the largest decimal number that can be displayed is a 9, and so the four binary digits are 1001.

To convert a BCD number to a denary number, each group of four binary numbers is separately converted to a denary number. For example, the BCD number 0011 1001 has a denary number of 3 for 0011 and 9 for 1001, and so the denary number is 39.

0011 1001 *BCD number*

3 9 *Denary number*

Numeric data is often entered into PLCs by rotary or thumb-wheel switches with a 0 to 9 range. Thus there may be a bank of such switches, one giving, say, the hundreds, one the tens, and one the ones. The output from each switch is then converted, independently, into binary to give the overall result of a binary coded decimal number. Some PLCs have a function that can be called up to convert such BCD numbers to binary numbers; in other PLCs it has to be done by programming.

3.4 Numbers in the Binary, Octal, Hex, and BCD Systems

Table 3.1 gives examples of numbers in the denary, binary, octal, hex, and BCD systems.

Table 3.1: Examples of Numbers in Various Systems

Denary	Binary	Octal	Hex	BCD
0	00000	0	0	0000 0000
1	00001	1	1	0000 0001
2	00010	2	2	0000 0010
3	00011	3	3	0000 0011
4	00100	4	4	0000 0100
5	00101	5	5	0000 0101
6	00110	6	6	0000 0110
7	00111	7	7	0000 0111
8	01000	10	8	0000 1000
9	01001	11	9	0000 1001
10	01010	12	A	0001 0000
11	01011	13	B	0001 0001
12	01100	14	C	0001 0010
13	01101	15	D	0001 0011
14	01110	16	E	0001 0100
15	01111	17	F	0001 0101
16	10000	20	10	0001 0110
17	10001	21	11	0001 0111

3.5 Binary Arithmetic

Addition of binary numbers uses the following rules:

$$0 + 0 = 0$$

$$0 + 1 = 1 + 0 = 1$$

$$1 + 1 = 10$$

$$1 + 1 + 1 = 11$$

Consider the addition of the binary numbers 01110 and 10011.

```
        01110
        10011
    ----------
Sum    100001
```

For bit 0 in the sum, $0 + 1 = 1$. For bit 1 in the sum, $1 + 1 = 10$, and so we have 0 with 1 carried to the next column. For bit 2 in the sum, 1 + 0 + the carried 1 = 10. For bit 3 in the sum, 1 + 0 + the carried 1 = 10. We continue this process through the various bits and end up with 100001.

Subtraction of binary numbers follows these rules:

$$0 - 0 = 0$$

$$1 - 0 = 1$$

$$1 - 1 = 0$$

When evaluating $0 - 1$, a 1 is borrowed from the next column on the left that contains a 1. The following example illustrates this method with the subtraction of 01110 from 11011:

```
              11011
              01110
          ---------
Difference    01101
```

For bit 0 we have $1 - 0 = 1$. For bit 1 we have $1 - 1 = 0$. For bit 2 we have $0 - 1$. We thus borrow 1 from the next column and so have $10 - 1 = 1$. For bit 3 we have $0 - 1$ (remember, we borrowed the 1). Again borrowing 1 from the next column, we then have $10 - 1 = 1$. For bit 4 we have $0 - 0 = 0$ (remember, we borrowed the 1).

3.5.1 Signed Numbers

The binary numbers considered so far contain no indication as to whether they are negative or positive and are thus said to be *unsigned*. Since there is generally a need to handle both positive and negative numbers, there needs to be some way of distinguishing between them. This can be done by adding a *sign bit*. When a number is said to be *signed*, its MSB is used to indicate the sign of the number; a 0 is used if the number is positive and a 1 is used if it is negative. Thus for an 8-bit number we have:

xxxx xxxx
↑
Sign bit

When we have a positive number, we write it in the normal way, with a 0 preceding it. Thus a positive binary number of 10110 is written as 010110. A negative number of 10110 is written as 110110. However, this is not the most useful way of writing negative numbers for ease of manipulation by computers.

3.5.2 One's and Two's Complements

A more useful way of writing signed negative numbers is to use the two's complement method. A binary number has two complements, known as the *one's complement* and the *two's complement*. The one's complement of a binary number is obtained by changing all the 1s in the unsigned number into 0s and the 0s into 1s. Thus if we have the binary number 101101, the one's complement of it is 010010. The two's complement is obtained from the one's complement by adding 1 to the LSB of the one's complement. Thus the two's complement of 101101 becomes 010011.

When we have a negative number, to obtain the signed two's complement, we obtain the two's complement and then sign it with a 1. Consider the representation of the decimal number −6 as a signed two's complement number when the total number of bits is eight. We first write the binary number for +6, that is, 0000110, then obtain the one's complement of 1111001, add 1 to give 1111010, and finally sign it with a 1 to indicate it is negative. The result is thus 11111010.

Unsigned binary number when sign ignored	000 0110
One's complement	111 1001
Add 1	1
Unsigned two's complement	111 1010
Signed two's complement	1111 1010

Table 3.2 lists some signed two's complements for denary numbers, given to 4 bits.

Table 3.2: Signed Two's Complements

Denary Number	Signed Two's Complement
−5	1011
−4	1100
−3	1101
−2	1110
−1	1111

When we have a positive number, we sign the normal binary number with a 0, that is, we write only negative numbers in the two's complement form. A consequence of adopting this method of writing negative and positive numbers is that when we add the signed binary equivalent of +4 and –4, that is, 0000 0100 and 111 1100, we obtain (1)0000 0000 and so zero within the constraints of the number of bits used, the (1) being neglected.

Subtraction of a positive number from a positive number can be considered to be the addition of a negative number to a positive number. Thus we obtain the signed two's complement of the negative number and then add it to the signed positive number. Hence, for the subtraction of the denary number 6 from the denary number 4, we can consider the problem as being (+4) + (−6). Hence we add the signed positive number to the signed two's complement for the negative number.

Binary form of +4	0000 0100
(−6) as signed two's complement	1111 1010
Sum	1111 1110

The MSB, that is, the sign, of the outcome is 1 and so the result is negative. This is the 8-bit signed two's complement for −2.

If we wanted to add two negative numbers, we would obtain the signed two's complement for each number and then add them. Whenever a number is negative, we use the signed two's complement; when it's positive, we use just the signed number.

3.5.3 Floating Point Numbers

Before we discuss floating point numbers, let's consider *fixed point numbers*. Fixed point numbers are numbers for which there is a fixed location of the point separating integers from fractional numbers. Thus, 15.3 is an example of a denary fixed point number, 1010.1100 an example of a fixed point binary number, and DE.2A an example of a fixed point hexadecimal number. We have, with the 8-bit binary number, four digits before the binary point and four digits after it. When two such binary numbers are added by a computing system, the procedure is to recognize that the fixed point is fixed the same in both numbers, so we can ignore it for the addition, carry out the addition of the numbers, and then insert in the result the binary point in its fixed position. For example, suppose we want to add 0011.1010 and 0110.1000; we drop the binary point to give:

$$0011\ \ 1010 + 0110\ \ 1000 = 1010\ \ 0010$$

Inserting the binary point then gives 1010.0010.

Using fixed points does present problems. If we are concerned with very large or very small numbers, we could end up with a large number of zeros between the integers and the point, that is, 0.000 000 000 000 023. For this reason, *scientific notation* is used for such numbers. Thus, the above number might be written as 0.23×10^{-13} or 2.3×10^{-14} or 23×10^{-15}. Likewise, the binary number 0.0000 0111 0010 might be represented as 110010×2^{-12} (the 12 would also be in binary format) or 11001.0×2^{-11} (the 11 being in binary format). Such notation is said to have a *floating point*.

A floating point number is in the form $a \times r^e$, where a is termed the *mantissa*, r the *radix* or *base*, and e the *exponent* or *power*. With binary numbers the base is understood to be 2, that is, we have $a \times 2^e$, and when we know we are dealing with binary numbers we need not store the base with the number. Thus a computing system needs, in addition to storing the sign, that is, whether positive or negative, to store the mantissa and the exponent.

Because with floating point numbers it is possible to store a number in several different ways—for example, 0.1×10^2 and 0.01×10^3—with computing systems such numbers are normalized. This means that they are all put in the form $0.1 \times r^e$. Thus, with binary numbers we have 0.1×2^e; if we had 0.00001001 it would become 0.1001×2^{-4}. To take account of the sign of a binary number, we then add a sign bit of 0 for a positive number and 1 for a negative number. Thus the number 0.1001×2^{-4} becomes 1.1001×2^{-4} if negative and 0.1001×2^{-4} if positive.

Unlike fixed point numbers, floating point numbers cannot be directly added unless the exponents are the same. Thus to carry out addition we need to make the exponents the same.

3.6 PLC Data

Most PLCs operate with a 16-bit word, with the term *word* meaning the group of bits constituting some information. This allows a positive number in the range 0 to +65535, that is, 1111 1111 1111 1111, to be represented, or a signed number in the range −32768 to +32767 in two's complement, the MSB then representing the sign. Such signed numbers are referred to as *integers*, with the symbol INT being used with inputs and outputs in programs of such 16-bit words. The term SINT is used for *short integer numbers*, for which only 8 bits are used, such numbers giving the range −128 to +127. The term DINT is used for *double-integer numbers*, for which 32 bits are used, such numbers giving the range -2^{31} to $+2^{31} - 1$. LINT is used for *long integer numbers*, for which 64 bits are used, such numbers giving the range -2^{63} to $+2^{63} - 1$. Where numbers are not signed, the symbols UINT, USINT, UDINT, and ULINT are used with integers, short integers, double integers, and long integers.

Decimal fractions are referred to as *real* or *floating point numbers* and are represented by the symbol REAL for inputs and outputs in programs. These consist of two 16-bit words; so we might have 1.234567E+03 for the number $1.234\,567 \times 10^{+3}$, the E indicating that the number that follows is the exponent. The term LREAL is used for *long real numbers*, in which 64 bits are used.

The term BOOL is used for *Boolean type data*, such data being on/off values, that is, 0 or 1, and thus represented by single bits.

Time duration, such as for the duration of a process, is represented by the IEC standard using the symbols *d* for days, *h* for hours, *m* for minutes, *s* for seconds, and *ms* for milliseconds, as, for example, T#12d2h5s3ms or TIME#12d2h5s for 12 days, 2 hours, 5 seconds, and 3 milliseconds. Note that # is the symbol used to indicate that what follows is a numerical quantity.

3.7 Combinational Logic Systems

Consider a system that might be used as an "interlock" to safeguard the operation of a machine. The machine is to start only if two safety conditions are realized: the workpiece is in position and the safety guard is in position. The workpiece in position can be regarded as input A to a system and the safety guard in position as input B (Figure 3.1).

For the input conditions to be expressed in binary form, we require there to be just two possibilities for each input. In this case, if we phrase the question to be posed of each input as having a YES or NO answer, we have just two conditions, which we can write as 1 for YES and 0 for NO. Thus input A can be phrased as follows: "Is the workpiece in position?" and the answer is YES or NO. Input B can be phrased as: "Is the safeguard in position?" and the answer is YES or NO. For this system we require there to be an output when input A is 1 and input B is 1. This relationship between the inputs and output can be tabulated as a *truth table* showing all the possible combinations of inputs, the combination of which leads to a 1, that is, YES, output, or a 0, that is, NO, output. Table 3.3 is the truth table for this system.

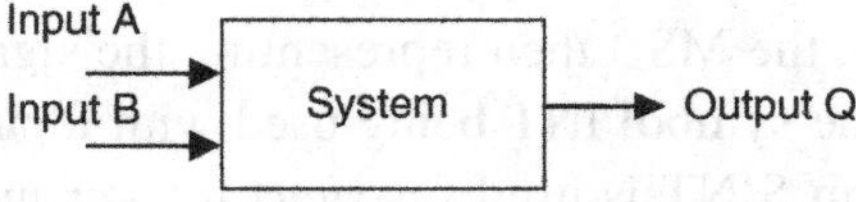

Figure 3.1: Machine interlock system.

Table 3.3: A Truth Table

Input A	Input B	Output Q
0	0	0
1	0	0
0	1	0
1	1	1

Each input can take only two values, represented by 0 or 1, and are described as *two-state variables* or *logical variables*. The complete system constructed with such variable is termed a *logic system* or *logic gates*. If the output of such a system depends only on the present states of the inputs, as with the machine "interlock," it is termed a *combinational logic system*.

Useful combinational logic systems, which we will meet in Chapter 5, are the AND gate, the OR gate, the NOT gate, the NAND gate, the NOR gate, and the XOR gate. The machine "interlock" system is an example of an AND gate in that input A *and* input B have to be 1 for the output to be 1.

3.8 Sequential Logic Systems

With a *sequential logic system*, the present output is influenced by the history of its past inputs as well as by its present input. This is unlike a combinational logic system, where the output *only* depends on the current state of its inputs. A binary counter can be regarded as a sequential logic system in that the binary output depends on the present input and the sum of the previous inputs. It thus has a "memory."

Most sequential systems are based on a small number of sequential logic systems called *bistables*, so-called because they have two stable conditions and can be switched from one to the other by appropriate inputs. Once the circuit has switched, it remains in the other stable state until another input pulse has been received to force it to return to the original state. Basically bistables are a memory device; they can "remember" the effect of an input after the input has been removed.

A latch and a flip-flop, so called because it can, on command, *flip* into one stable state or *flop* back again to the other, are bistables. A latch is triggered by the voltage level applied to its input, provided that it has been enabled by its clock input being, generally, high. Flip-flops are devices that change state at either the leading edge or the trailing edge of an enable/clock pulse.

3.8.1 Latches

The *clocked D latch* has a data input D, outputs Q and $\bar{Q}$, and an enable/clock input CLK (Figure 3.2). Q is always the complement of $\bar{Q}$. The logical state of the Q output will follow any changes in the logical state of the D input as long as the clock input remains high. When

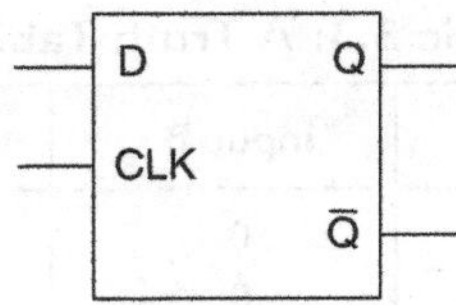

Figure 3.2: The clocked D latch.

the clock input goes low, the logical state of the D input at that moment will be retained as the Q output, no matter what changes occur at the D input. When the clock input goes high again, the output Q will again follow any changes in the logical state of the D input. The latch is said to be *transparent* when the clock is high.

Truth tables can be drawn for latches, but they must take into account the effect of applying a pulse on the clock input. For this reason they are often referred to as *function tables*. Table 3.4 is the function table for the clocked D latch. Q^+ is the state of the Q output after a clock-triggering input.

The *clocked SR latch* has two input terminals, S for set and R for reset; outputs Q and $\bar{Q}$; and an enable/clock input CLK (Figure 3.3). Q is always the complement of $\bar{Q}$. When both S and R are held low, the logical state of the outputs will not change. When S is 1 and R is 0, the logical state of the output Q will become 1, no matter what its value was before. This is termed the *set* operation. If S is 0 and R is 1, the Q output will be 0, whatever its value was before. This is the *reset* operation. If both S and R are 1, the operation of the latch is

Table 3.4: Clocked D-Latch

CLK	D	Q	Q^+	
0	0	0	0	No change in output, held at previous value.
0	0	1	1	
0	1	0	0	
0	1	1	1	
1	0	0	0	Output changes to new value.
1	0	1	0	
1	1	0	1	
1	1	1	1	

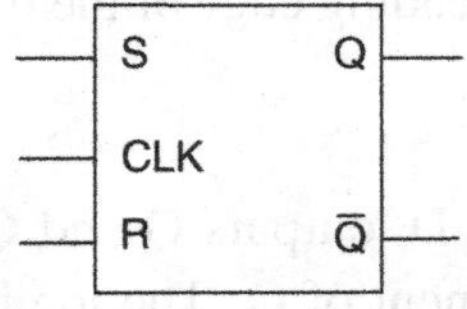

Figure 3.3: The clocked SR latch.

Table 3.5: Clocked SR Latch

CLK	S	R	Q	Q^+	
0	0	0	0	0	No change, held.
0	0	0	1	1	
0	1	0	0	0	
0	1	0	1	1	
0	0	1	0	0	
0	0	1	1	1	
0	1	1	0	0	
0	1	1	1	1	
1	0	0	0	0	
1	0	0	1	1	
1	1	0	0	1	Set
1	1	0	1	1	
1	0	1	0	0	Reset
1	0	1	1	0	
1	1	1	0	x	Indeterminate
1	1	1	1	x	

unpredictable, and so this combination of inputs should not be allowed to occur. Table 3.5 shows the function table.

3.8.2 Flip-Flops

A JK flip-flop has two data input terminals, J and K; a clock input; and two output terminals Q and $\bar{Q}$ (Figure 3.4). A JK flip-flop will change its output state at a clock transition, either at a leading edge or the trailing edge of the clock pulse. A JK flip-flop always changes state when J = K = 1 and might be said to *toggle*. Table 3.6 is the function table.

Summary

The *denary* number system is based on the use of 10 digits 0, 1, 2, 3, 4, 5, 6, 7, 8, 9. The *binary* system is based on just two: 0 and 1. The *octal* system is based on eight digits: 0, 1, 2, 3, 4, 5, 6, 7. The *hexadecimal* system is based on the use of 16 digits: 0, 1, 2, 3, 4, 5,

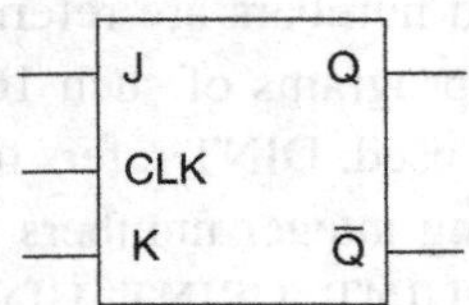

Figure 3.4: The clocked JK flip-flop.

Table 3.6: Clocked JK Flip-Flop

J	K	Q	Q^+	
0	0	0	0	No change, held.
0	0	1	1	
1	0	0	0	
1	0	1	1	
0	0	1	1	
1	0	0	1	Set
1	0	1	1	
0	1	0	0	Reset
0	1	1	0	
1	1	0	1	Toggle
1	1	1	0	

6, 7, 8, 9, A, B, C, D, E, F. The *binary coded decimal* (BCD) system has each denary digit coded separately into binary.

For the addition of two binary numbers, we have $0 + 0 = 0$, $0 + 1 = 1$, and $1 + 1=10$. For subtraction, we have $0 - 0 = 0$, $1 - 0 = 1$, and $1 - 1 = 0$. Binary numbers that give no indication of whether they are negative or positive are termed *unsigned*. When a number is signed, the most significant bit indicates the sign of the number—0 if positive and 1 if negative. A binary number has two complements. The *one's complement* is obtained by changing all the 1s in the unsigned number into 0s and the 0s into 1s. The *two's complement* is obtained from the one's complement by adding 1 to the least significant bit. When we have a negative number, the signed two's complement number is obtained signing the two's complement with a 1.

A floating point number is in the form $a \times r^e$, where a is termed the *mantissa*, r the *radix* or *base*, and e the *exponent* or *power*. With binary numbers the base is understood to be 2, that is, we have $a \times 2^e$; when we know we are dealing with binary numbers we need not store the base with the number. Thus a computing system needs to store the mantissa and the exponent in addition to storing the sign, whether positive or negative.

Most PLCs operate with a 16-bit word, the term *word* being used for the group of bits constituting some information. Signed numbers are referred to as *integers*, with the symbol INT used with inputs and outputs in programs of such 16-bit words. SINT is used for short integer numbers where only 8 bits are used. DINT refers to double-integer numbers for which 32 bits are used. LINT is used for long integer numbers where 64 bits are used. Where numbers are not signed, the symbols UINT, USINT, UDINT, and ULINT are used with integers, short integers, double integers, and long integers.

A system constructed with inputs and outputs represented by 0 or 1 is termed a *logic system*. If the output of such a system depends only on the present states of the inputs, it is termed a *combinational logic system*. The relationship between the inputs and output can be tabulated as a *truth table* showing all the possible combinations of inputs that lead to a 1 output and from which a 0 output.

With a *sequential logic system*, the present output is influenced by the history of its past inputs as well as by its present input, most being based on systems called *bistables* because they have two stable conditions and can be switched from one to the other by appropriate inputs. Once the circuit has switched, it remains in the other stable state until another input pulse has been received to force it to return to the original state. A latch and a flip-flop are bistables. A latch is triggered by the voltage level applied to its input, provided it has been enabled by its clock input being, generally, high. Flip-flops are devices that change state at either the leading edge or the trailing edge of an enable/clock pulse. Truth tables can be drawn for latches, but they must take account of the effect of applying a pulse on the clock input. For this reason they are often referred to as *function tables*.

Problems

1. Convert the following binary numbers to denary numbers: (a) 000011, (b) 111111, (c) 001101.
2. Convert the following denary numbers to binary numbers: (a) 100, (b) 146, (c) 255.
3. Convert the following hexadecimal numbers to denary numbers: (a) 9F, (b) D53, (c) 67C.
4. Convert the following denary numbers to hexadecimal numbers: (a) 14, (b) 81, (c) 2562.
5. Convert the following hexadecimal numbers to binary numbers: (a) E, (b) 1D, (c) A65.
6. Convert the following octal numbers to denary numbers: (a) 372, (b) 14, (c) 2540.
7. Convert the following denary numbers to octal numbers: (a) 20, (b) 265, (c) 400.
8. Convert the following octal numbers to binary numbers: (a) 270, (b) 102, (c) 673.
9. Convert the following decimal numbers to BCD equivalents: (a) 20, (b) 35, (c) 92.
10. Convert the following denary numbers to signed two's complement binary 8-bit format: (a) –1, (b) –35, (c) –125.
11. Convert the following signed two's complement binary 8-bit numbers to their denary equivalents: (a) 1111 0000, (b) 1100 1001, (c) 1101 1000.
12. Convert the following binary numbers to normalized floating point numbers: (a) 0011 0010, (b) 0000 1100, (c) 1000.0100.

13. Explain what is meant by the terms combinational logic systems and sequential logic systems.

14. For the following truth tables, which combination of inputs will lead to a 1 output?

(a)

Input A	Input B	Output Q
0	0	0
1	0	0
0	1	0
1	1	1

(b)

Input A	Input B	Output Q
0	0	0
1	1	1
0	1	1
1	1	1

(c)

Input A	Input B	Output Q
0	0	1
1	0	1
0	1	1
1	1	0

15. For a clocked D latch, what inputs will be needed to change the Q output from 0 to 1?

16. Give the function table for a clocked SR latch when the input to S is also applied to R but via a NOT gate, such a gate converting 0 to 1 and 1 to 0.

Lookup Tasks

17. Look up the specifications for (a) an AND gate and (b) a transparent D-type latch.

CHAPTER 4

I/O Processing

This chapter continues the discussion of inputs and outputs from Chapter 2 and is a brief consideration of the processing of the signals from input and output devices. The input/output (I/O) unit provides the interface between the PLC controller and the outside world and must therefore provide the necessary signal conditioning to get the signal to the required level and also to isolate it from possible electrical hazards such as high voltages. This chapter includes the forms of typical input/output modules and, in an installation where sensors are some distance from the PLC processing unit, their communication links to the PLC.

4.1 Input/Output Units

Input signals from sensors and outputs required for actuating devices can be:

- *Analog*. A signal for which the size is related to the size of the quantity being sensed.
- *Discrete*. Essentially just an on/off signal.
- *Digital*. A sequence of pulses.

The CPU, however, must have an input of digital signals of a particular size, normally 0 to 5 V. The output from the CPU is digital, normally 0 to 5 V. Thus there is generally a need to manipulate input and output signals so that they are in the required form.

The input/output (I/O) units of PLCs are designed so that a range of input signals can be changed into 5 V digital signals and so that a range of outputs are available to drive external devices. It is this built-in facility to enable a range of inputs and outputs to be handled that makes PLCs so easy to use. The following is a brief indication of the basic circuits used for input and output units. In the case of rack instruments, they are mounted on cards that can be plugged into the racks, and so the input/output characteristics of the PLC can thus be changed by changing the cards. A single box form of PLC has input/output units incorporated by the manufacturer.

4.1.1 Input Units

The terms *sourcing* and *sinking* refer to the manner in which DC devices are interfaced with the PLC (see Section 1.3.5). For a PLC input unit with sourcing, it is the source of the

W. Bolton: Programmable Logic Controllers, Sixth Edition. http://dx.doi.org/10.1016/B978-0-12-802929-9.00004-2

current supply for the input device connected to it (Figure 4.1a). With sinking, the input device provides the current to the input unit (Figure 4.1b).

Figures 4.2 and 4.3 show the basic input unit circuits for DC and AC inputs. Optoisolators (see Section 1.3.4) are used to provide protection. With the AC input unit, a rectifier bridge network is used to rectify the AC so that the resulting DC signal can provide the signal for use by the optoisolator to give the input signals to the CPU of the PLC. Individual status lights are provided for each input to indicate when the input device is providing a signal.

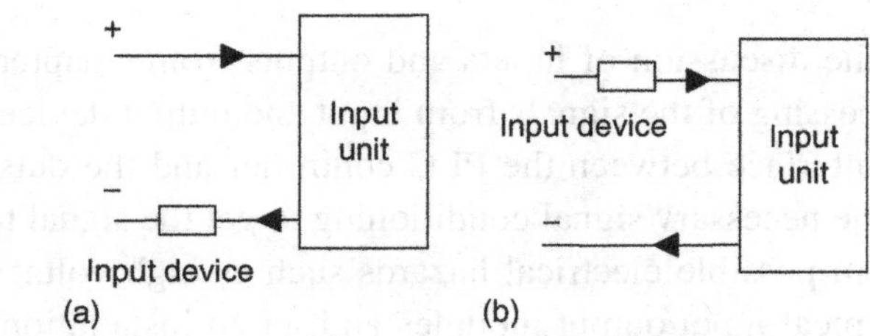

Figure 4.1: Input unit: (a) sourcing and (b) sinking.

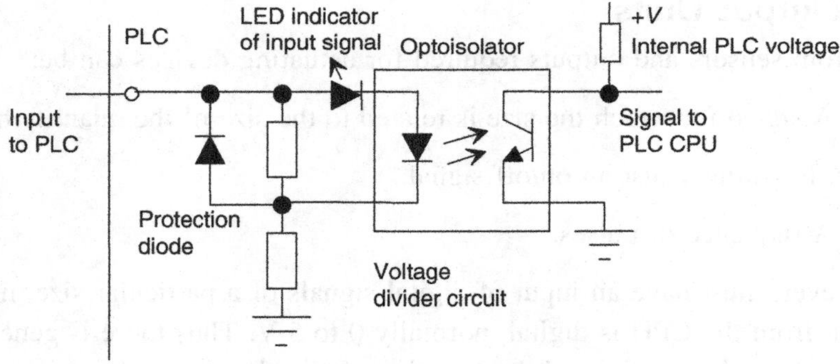

Figure 4.2: DC input unit.

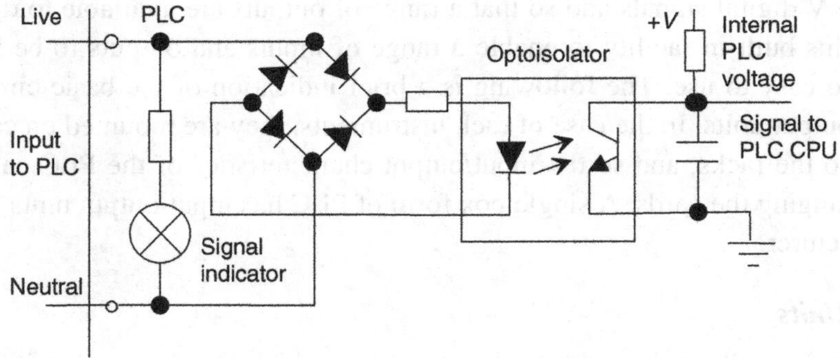

Figure 4.3: AC input unit.

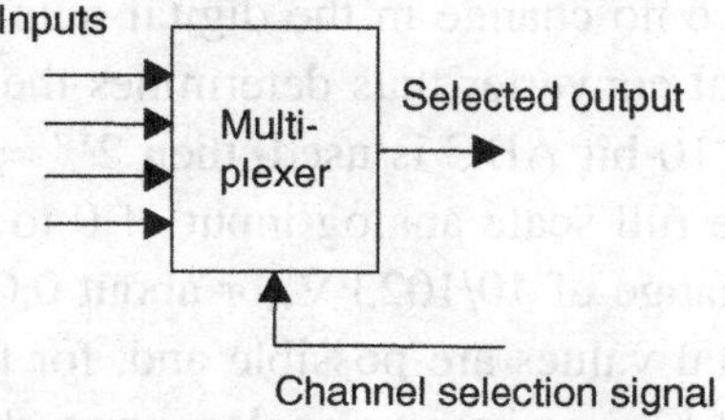

Figure 4.4: Multiplexer.

When analog signals are inputted to a PLC, the input channel needs to convert the signal to a digital signal using an analog-to-digital converter. With a rack-mounted system this may be achieved by mounting a suitable analog input card in the rack. So that one analog card is not required for each analog input, multiplexing is generally used (Figure 4.4). This involves more than one analog input being connected to the card and then electronic switches used to select each input in turn. Cards are typically available containing 4, 8, or 16 analog inputs.

Figure 4.5a illustrates the function of an analog-to-digital converter (ADC). A single analog input signal gives rise to on/off output signals along perhaps eight separate wires. The eight signals then constitute the so-termed digital *word* corresponding to the analog input signal level. With such an 8-bit converter there are $2^8 = 256$ different digital values possible; these are 0000 0000 to 1111 1111, that is, 0 to 255. The digital output goes up in steps (Figure 4.5b) and the analog voltages required to produce each digital output are termed *quantization levels.*

If the binary output is to change, the analog voltage has to change by the difference in analog voltage between successive levels. The term *resolution* is used for the smallest change in analog voltage that will give rise to a change in 1 bit in the digital output. With an 8-bit ADC, if, say, the full-scale analog input signal varies between 0 and 10 V, a step of one digital bit involves an analog input change of 10/255 V or about 0.04 V. This means that a change of

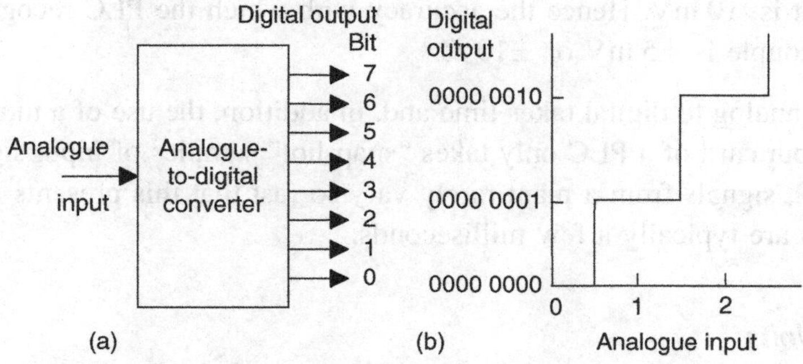

Figure 4.5: (a) Function of an analog-to-digital converter, and (b) an illustration of the relationship between the analog input and the digital output.

0.03 V in the input will produce no change in the digital output. The number of bits in the output from an analog-to-digital converter thus determines the *resolution*, and hence *accuracy*, that is possible. If a 10-bit ADC is used, then $2^{10} = 1024$ different digital values are possible and, for the full-scale analog input of 0 to 10 V, a step of one digital bit involves an analog input change of 10/1023 V, or about 0.01 V. If a 12-bit ADC is used, then $2^{12} = 4096$ different digital values are possible and, for the full-scale analog input of 0 to 10 V, a step of one digital bit involves an analog input change of 10/4095 V, or about 2.4 mV. In general, the resolution of an n-bit ADC is $1/(2^n - 1)$; this is sometimes approximated to 2^{-n}.

The following illustrates the analog-to-digital conversion for an 8-bit converter when the analog input is in the range 0 to 10 V:

Analog Input (V)	Digital Output (V)
0.00	0000 0000
0.04	0000 0001
0.08	0000 0010
0.12	0000 0011
0.16	0000 0100
0.20	0000 0101
0.24	0000 0110
0.28	0000 0111
0.32	0000 1000
etc.	

To illustrate this idea, consider a thermocouple used as a sensor with a PLC and giving an output of 0.5 mV per °C. What will be the accuracy with which the PLC will activate the output device if the thermocouple is connected to an analog input with a range of 0 to 10 V DC and using a 10-bit analog-to-digital converter? With a 10-bit converter, there are $2^{10} =$ 1024 bits covering the 0 to 10 V range. Thus a change of 1 bit corresponds to 10/1023 V or about 0.01 V, that is, 10 mV. Hence the accuracy with which the PLC recognizes the input from the thermocouple is ±5 mV or ±10°C.

Conversion from analog to digital takes time and, in addition, the use of a multiplexer means that an analog input card of a PLC only takes "snapshot" samples of input signals. For most industrial systems, signals from a plant rarely vary so fast that this presents a problem. Conversion times are typically a few milliseconds.

4.1.2 Output Units

With a PLC output unit, when it provides the current for the output device (Figure 4.6a) it is said to be *sourcing*, and when the output device provides the current to the output unit, it is said to be *sinking* (Figure 4.6b). Quite often, sinking input units are used for interfacing with electronic equipment and sourcing output units for interfacing with solenoids.

Output units can be relay, transistor, or triac. Figure 4.7 shows the basic form of a relay output unit, Figure 4.8 that of a transistor output unit, and Figure 4.9 that of a triac output unit.

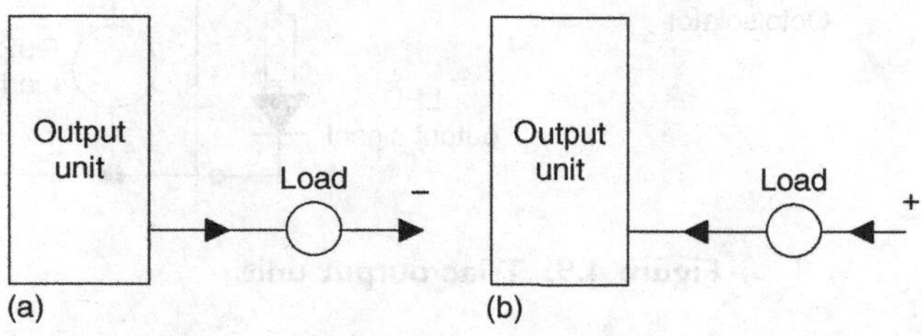

Figure 4.6: Output unit: (a) sourcing, and (b) sinking.

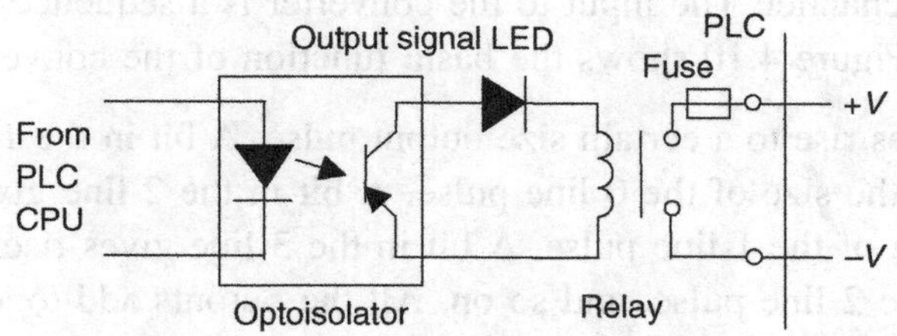

Figure 4.7: Relay output unit.

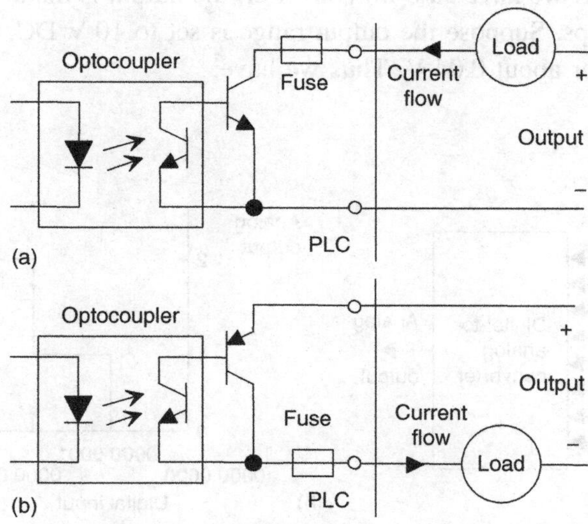

Figure 4.8: Basic forms of transistor output: (a) current sinking, and (b) current sourcing.

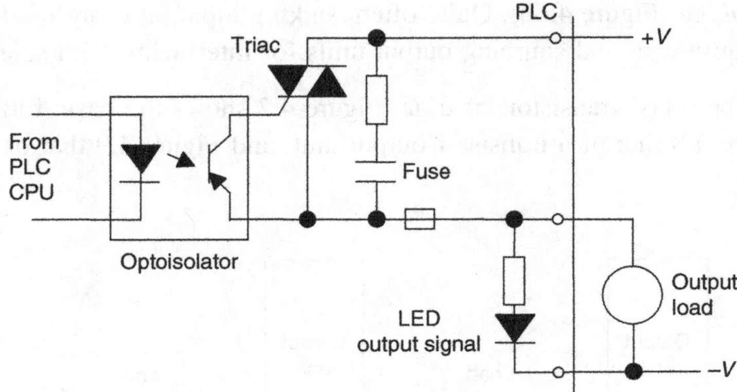

Figure 4.9: Triac output unit.

Analog outputs are frequently required and can be provided by digital-to-analog converters (DACs) at the output channel. The input to the converter is a sequence of bits with each bit along a parallel line. Figure 4.10 shows the basic function of the converter.

A bit in the 0 line gives rise to a certain size output pulse. A bit in the 1 line gives rise to an output pulse of twice the size of the 0 line pulse. A bit in the 2 line gives rise to an output pulse of twice the size of the 1 line pulse. A bit in the 3 line gives rise to an output pulse of twice the size of the 2 line pulse, and so on. All the outputs add together to give the analog version of the digital input. When the digital input changes, the analog output changes in a stepped manner, the voltage changing by the voltage changes associated with each bit. For example, if we have an 8-bit converter, the output is made up of voltage values of $2^8 = 256$ analog steps. Suppose the output range is set to 10 V DC. One bit then gives a change of 10/255 V or about 0.04 V. Thus we have:

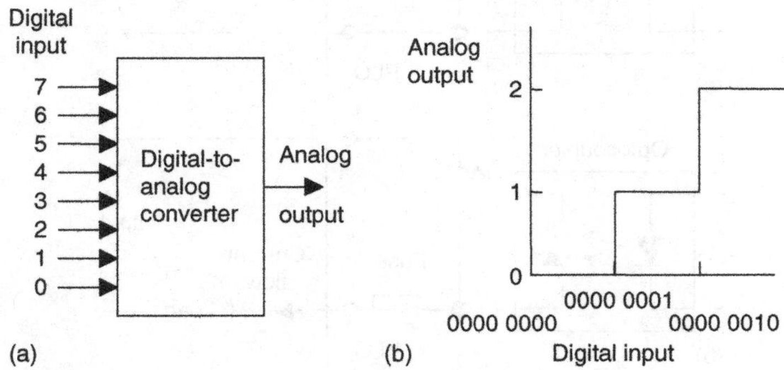

Figure 4.10: (a) DAC function, and (b) digital-to-analog conversion.

Digital Input (V)	Analog Output (V)
00000000	0.00
00000001	0.04
00000010	0.08 + 0.00 = 0.08
00000011	0.08 + 0.04 = 0.12
00000100	0.16
00000101	0.016 + 0.00 + 0.04 = 0.20
00000110	0.016 + 0.08 = 0.24
00000111	0.016 + 0.08 + 0.04 = 0.28
00001000	0.32
etc.	

Analog output modules are usually provided in a number of outputs, such as 4 to 20 mA, 0 to +5 V DC, and 0 to +10 V DC, and the appropriate output is selected by switches on the module. Modules generally have outputs in two forms, one for which all the outputs from that module have a common voltage supply and one that drives outputs with their own individual voltage supplies. Figure 4.11 shows the basic principles of these two forms of output.

4.2 Signal Conditioning

When connecting sensors that generate digital or discrete signals to an input unit, care has to be taken to ensure that voltage levels match. However, many sensors generate analog signals. To avoid having a multiplicity of analog input channels to cope with the wide diversity of analog signals that can be generated by sensors, external signal conditioning is often used to bring analog signals to a common range and so allow a standard form of analog input channel to be used.

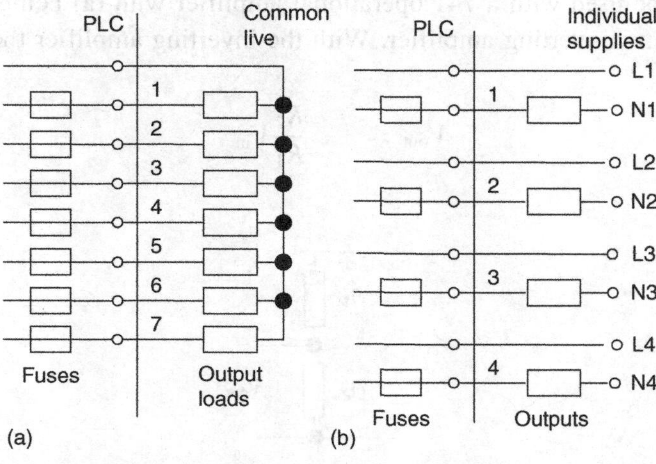

Figure 4.11: Forms of output: (a) common supply, and (b) individual supplies.

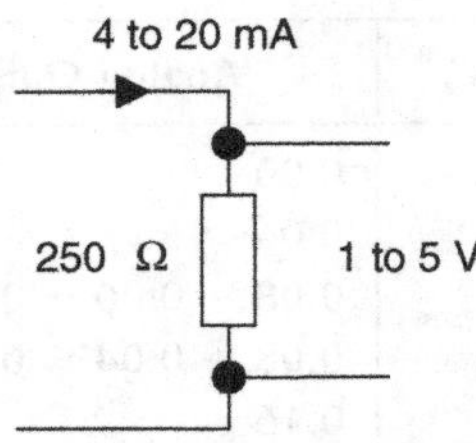

Figure 4.12: Standard analog signal.

A common standard that is used (Figure 4.12) is to convert analog signals to a current in the range 4 to 20 mA by passing it through a 250 Ω resistance to give a 1 to 5 V input signal. Thus, for example, a sensor used to monitor liquid level in the height range 0 to 1 m would have the 0 level represented by 4 mA and the 1 m represented by 20 mA. The use of 4 mA to represent the low end of the analog range serves the purpose of distinguishing between when the sensor is indicating zero and when the sensor is not working and giving zero response for that reason. When this happens the current would be 0 mA. The 4 mA also is often a suitable current to operate a sensor and so eliminate the need for a separate power supply.

4.2.1 Changing Voltage Levels

A potential divider (Figure 4.13) can be used to reduce a voltage from a sensor to the required level; the output voltage level V_{out} is:

$$V_{out} = \frac{R_2}{R_1 + R_2} V_{in}$$

Amplifiers can be used to increase the voltage level; Figure 4.14 shows the basic form of the circuits that might be used with a 741 operational amplifier with (a) being an inverting amplifier and (b) a noninverting amplifier. With the inverting amplifier the output V_{out} is:

$$V_{out} = -\frac{R_2}{R_1} V_{in}$$

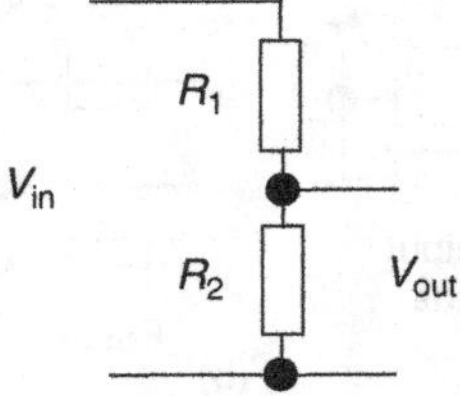

Figure 4.13: Potential divider.

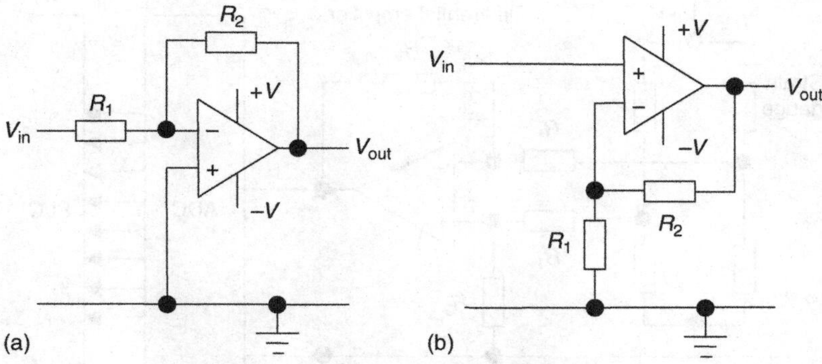

Figure 4.14: Operational amplifier circuits.

and with the noninverting amplifier:

$$V_{out} = \frac{R_1 + R_2}{R_1} V_{in}$$

Often a differential amplifier is needed to amplify the difference between two input voltages. Such is the case when a sensor—for example, a strain gauge—is connected in a Wheatstone bridge and the output is the difference between two voltages or with a thermocouple where the voltage difference between the hot and cold junctions is required. Figure 4.15 shows the basic form of an operational amplifier circuit for this purpose.

The output voltage V_{out} is:

$$V_{out} = \frac{R_2}{R_1}(V_2 - V_1)$$

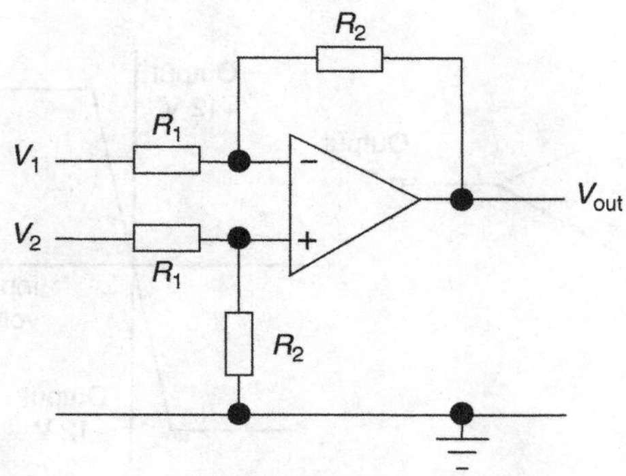

Figure 4.15: Differential amplifier.

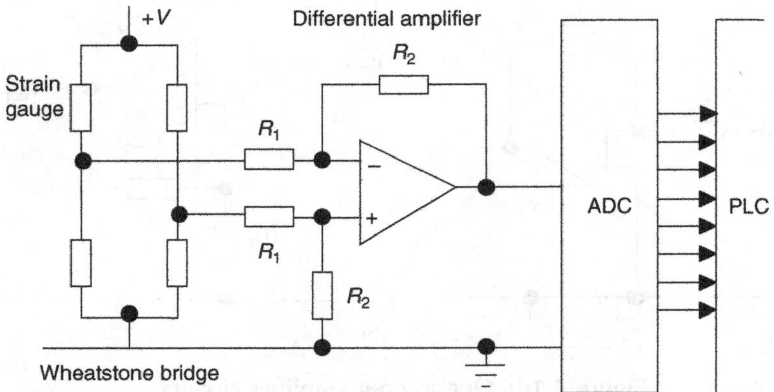

Figure 4.16: Signal conditioning with a strain gauge sensor.

As an illustration of the use of signal conditioning, Figure 4.16 shows the arrangement that might be used for a strain gauge sensor. The sensor is connected in a Wheatstone bridge and the out-of-balance potential difference amplified by a differential amplifier before being fed via an analog-to-digital converter unit, which is part of the analog input port of the PLC.

4.2.2 Op-Amp Comparator

The output of an operational amplifier saturates at about ±12 V, such outputs typically being reached with inputs of about ±10 μV. Figure 4.17 shows the basic form of the characteristic. Thus, operational amplifiers are widely used to give on/off signals based on the relative value of two input signals. One signal is connected to the noninverting terminal and the other to the inverting terminal. The operational amplifier determines whether the signal to the inverting terminal is above or below that to the noninverting terminal, that is, it is a comparator. By reversing the reference and input voltage connections, the output polarity is changed. Such comparators can be used as the basis for on/off control systems. Thus we

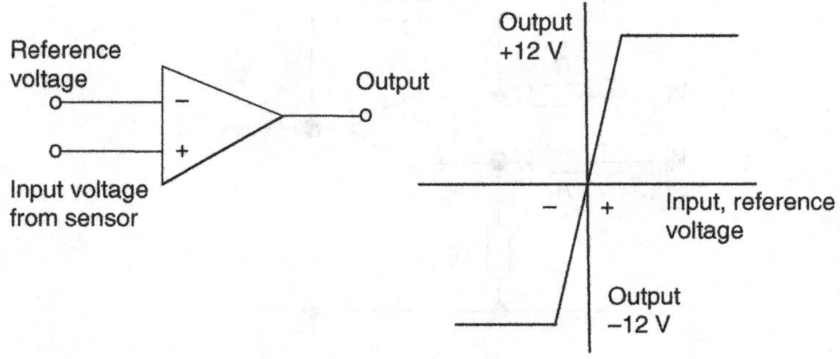

Figure 4.17: Operational amplifier characteristic.

might have a reference voltage and compare the voltage from a sensor with it and so obtain an on/off output depending on whether the voltage from the sensor is above or below the reference voltage.

4.2.3 Output Protection

Outputs involving coils of wire, such as solenoids and motors, are inductors. When there is a current, such inductors store energy in the magnetic field, and when the current is switched off, the magnetic field collapses and produces a back EMF that can be quite large. The simplest method of protection against this back EMF is to place a diode in parallel with the coil. The diode is so connected that current cannot flow through it when the current is energizing the coil but will short-circuit the coil and so suppress the current arising from the back EMF. Such a diode is often referred to as a *flyback diode*.

Some output devices may need series current-limiting resistors. For example, an LED display has generally a maximum current rating of about 10 to 30 mA. With 20 mA the voltage drop across it might be 2.1 V. Thus if we have an input of 5 V to it, 2.9 V has to be dropped across a series resistor. This means a series resistance of $2.9/0.020 = 145\ \Omega$ and so a standard resistor of $150\ \Omega$ might be used. Some LEDs are supplied with built-in resistors.

4.3 Remote Connections

When there are many inputs or outputs located considerable distances away from the PLC, though it would be possible to run cables from each such device to the PLC, a more economic solution is to use input/output modules in the vicinity of the inputs and outputs and use just a single core cable to connect each over the long distances to the PLC, instead of using the multicore cable that would be needed without such distant I/O modules (Figure 4.18).

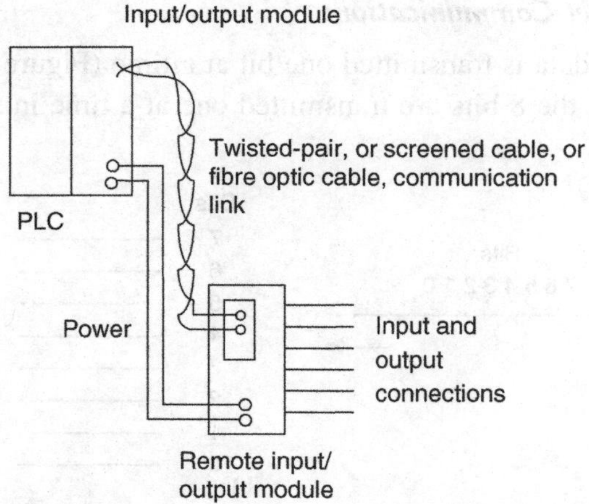

Figure 4.18: Use of remote input/output module.

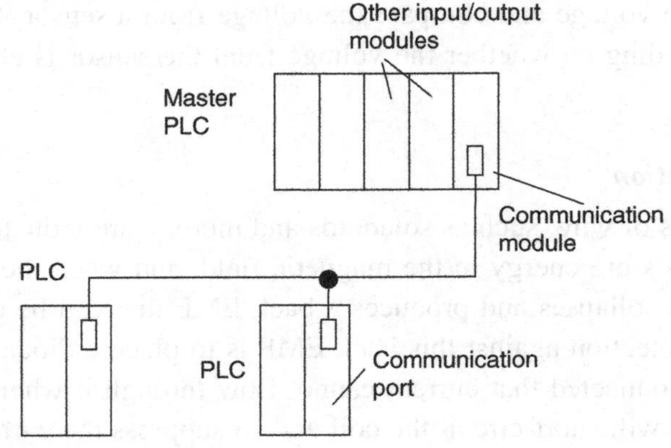

Figure 4.19: Use of remote input/output PLC systems.

In some situations a number of PLCs may be linked together with a master PLC unit sending and receiving input/output data from the other units (Figure 4.19). The distant PLCs do not contain the control program since all the control processing is carried out by the master PLC.

The cables used for communicating data between remote input/output modules and a central PLC, or remote PLCs and the master PLC are typically *twisted-pair cabling*, often routed through grounded steel conduit to reduce electrical "noise." *Coaxial cable* enables higher data rates to be transmitted and does not require the shielding of steel conduit. *Fiber-optic cabling* has the advantage of resistance to noise, small size, and flexibility and is now becoming more widely used.

4.3.1 Serial and Parallel Communications

In *serial communication*, data is transmitted one bit at a time (Figure 4.20a). Thus if an 8-bit word is to be transmitted, the 8 bits are transmitted one at a time in sequence along a cable.

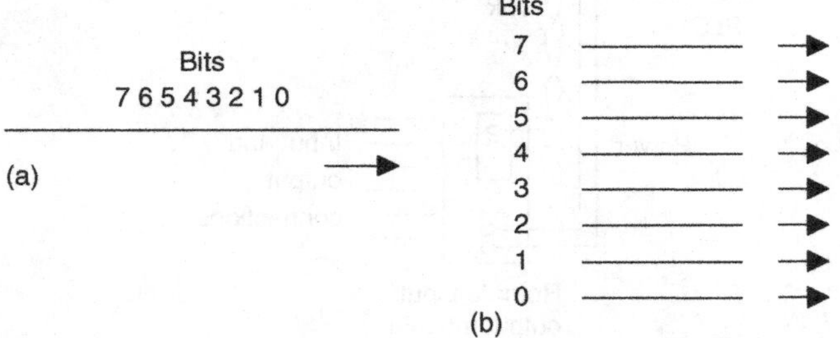

Figure 4.20: (a) Serial communication and (b) parallel communication.

This means that a data word has to be separated into its constituent bits for transmission and then reassembled into the word when received. In *parallel communication*, all the constituent bits of a word are simultaneously transmitted along parallel cables (Figure 4.20b). This allows data to be transmitted over short distances at high speeds.

Serial communication is used for transmitting data over long distances. It is much cheaper to run the single core cable needed for serial communication over a long distance than the multicore cables that would be needed for parallel communication. With a PLC system, serial communication might be used for the connection between a computer, when used as a programming terminal, and a PLC. Parallel communication might be used when connecting laboratory instruments to the system. Internally however, PLCs work with parallel communications for speed. Thus, circuits called *universal asynchronous receivers/ transmitters* (UARTs) have to be used at input/output ports to convert serial communications signals to parallel.

4.3.2 Serial Standards

For successful serial communications to occur, it is necessary to specify:

- The voltage levels to be used for signals, that is, what signal represents a 0 and what represents a 1.
- What the bit patterns being transmitted mean and how the message is built up. Bear in mind that a sequence of words is being sent along the same cable and it is necessary to be able to determine when one word starts and finishes and the next word starts.
- The speed at which the bit pattern is to be sent, that is, the number of bits per second.
- Synchronization of the clocks at each end. This is necessary if, for example, a particular duration transmitted pulse is to be recognized by the receiver as just a single bit rather than two bits.
- Protocols, or flow controls, to enable such information as "able to receive data" or "not ready to receive data" to be received. This is commonly done by using two extra signal wires (termed *handshake wires*), one to tell the receiver the transmitter is ready to send data and the other to tell the transmitter the receiver is ready to receive it.
- Error-checking to enable a bit pattern to be checked to determine whether corruption of the data has occurred during transmission.

The most common standard serial communications interface is the *RS232*. Connections are made via 25-pin D-type connectors (Figure 4.21), usually, though not always, with a male plug on cables and a female socket on the equipment. Not all the pins are used in every application.

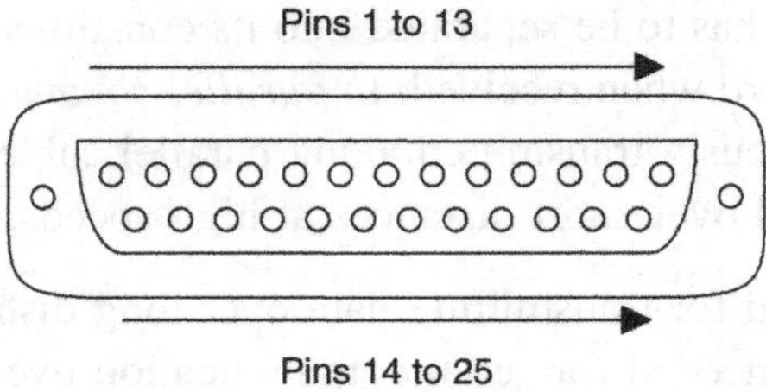

Figure 4.21: A D-type connector.

The minimum requirements are:

Pin 1: Ground connection to the frame of the chassis

Pin 2: Serial transmitted data (output data pin)

Pin 3: Serial received data (input data pin)

Pin 7: Signal ground, which acts as a common signal return path

Configurations that are widely used with interfaces involving computers are as follows:

Pin 1: Ground connection to the frame of the chassis

Pin 2: Serial transmitted data (output data pin)

Pin 3: Serial received data (input data pin)

Pin 4: Request to send

Pin 5: Clear to send

Pin 6: Data set ready

Pin 7: Signal ground, which acts as a common signal return path

Pin 20: Data terminal ready

The signals sent through pins 4, 5, 6, and 20 are used to check that the receiving end is ready to receive a signal, the transmitting end is ready to send and the data is ready to be sent. With RS232, a 1 bit is represented by a voltage between −5 and −25 V, normally −12 V, and a 0 by a voltage between +5 and +25 V, normally +12 V.

The term *baud rate* is used to describe the transmission rate, which is approximately the number of bits transmitted or received per second. However, not all the transmitted bits can be used for data; some have to be used to indicate the start and stop of a serial piece of data, often termed *flags*, and as a check as to whether the data has been corrupted during transmission. Figure 4.22 shows the type of signal that might be sent with RS232. The parity

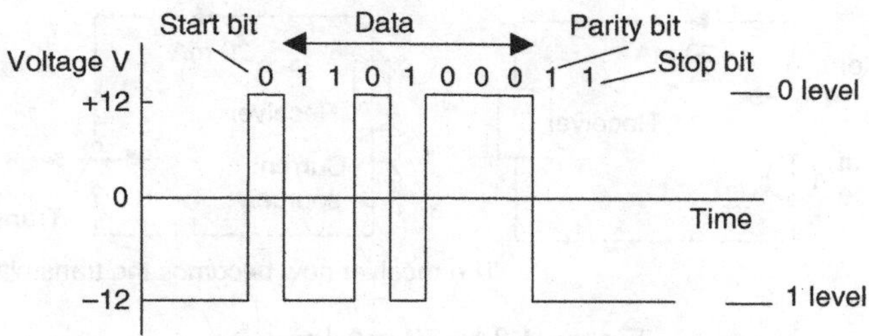

Figure 4.22: RS232 signal levels.

bit is added to check whether corruption has occurred, with, in even parity, a 1 being added to make the number of 1s an even number. To send seven bits of data, 11 bits may be required.

Other standards such as *RS422* and *RS423* are similar to RS232. The distance over which RS232 can be used is limited with noise limiting the transmission of high numbers of bits per second when the length of cable is more than about 15 m. RS422 can be used for longer distances. This method uses a balanced method of transmission. Such circuits require two lines for the transmission, the transmitted signal being the voltage difference between the two lines. Noise affecting both lines equally will have no effect on the transmitted signal. Figure 4.23 shows how, for RS232 and RS422, the data rates that can be transmitted without noise becoming too significant depend on the distance. RS422 lines can be used for much greater distances than RS232.

RS422 uses a balanced interface, employing a voltage on one line with respect to another line to define a 1 signal and the opposite polarity between the two lines to define a 0 signal. RS423 uses an unbalanced interface, involving a voltage on the line relative to a fixed voltage, usually the signal ground, to determine 1 and 0 signals. It has a line for all transmitted signals and a line for all received signals. RS423 is not as fast for transmission as RS422 but has similar distances of transmission. Some devices may be configured to provide either RS422 or RS423 interfaces.

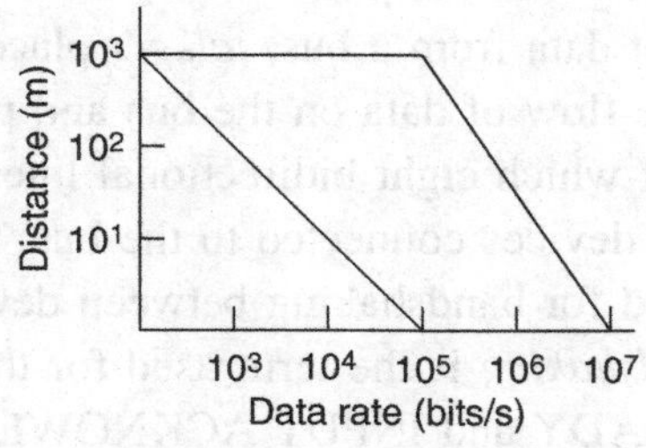

Figure 4.23: Transmission with RS232 and RS422.

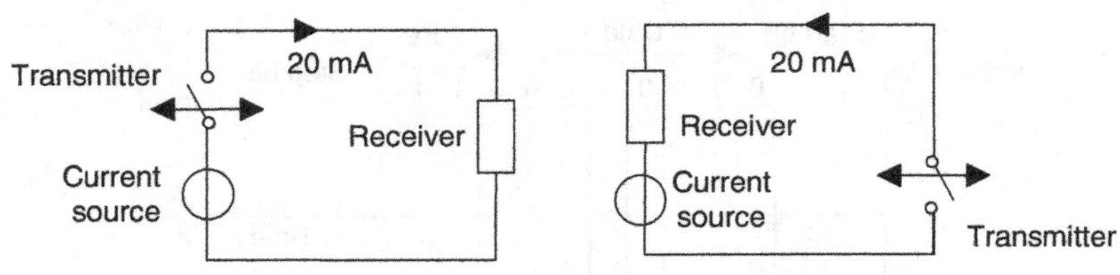

Figure 4.24: 20 mA loop.

An alternative to RS422 or RS423 is the *20 mA loop*, which was an earlier standard and is still widely used for long distance serial communication, particularly in industrial systems where the communication path is likely to suffer from electrical noise (Figure 4.24). Distances up to a few kilometers are possible. This system consists of a circuit, a loop of wire, containing a current source. The serial data is transmitted by the current being switched on and off, a 0 being transmitted as zero current and a 1 as 20 mA. For two-way communications, a pair of separate wires is used for the transmission and receiver loops. The serial data is encoded with a start bit, a right data bit, and two stop bits.

Other buses that are used often in particular situations are the *Inter-IC Communication*, or I^2C, bus, designed by Philips for use in communication between integrated circuits or modules; the Controller Area Network, or *CAN*, bus, developed by Bosch for the engine management systems of cars; the *Universal Serial Bus*, or *USB*, bus, designed to enable monitors, printers, and other input devices to be easily connected to computer systems; and *Firewire*, developed by Apple Computers to give plug-and-play capabilities with computer systems.

4.3.3 Parallel Standards

The standard interface most commonly used for parallel communications is *IEEE-488*. This was originally developed by Hewlett-Packard to link its computers and instruments and was known as the *Hewlett-Packard Instrumentation Bus*. It is now often termed the *General-Purpose Instrument Bus*. This bus provides a means of making interconnections so that parallel data communications can take place among listeners, talkers, and controllers. *Listeners* are devices that accept data from a bus; *talkers* place data, on request, on the bus; and *controllers* manage the flow of data on the bus and provide processing facilities. A bus has a total of 24 lines, of which eight bidirectional lines are used to carry data and commands between the various devices connected to the bus, five lines are used for control and status signals, three are used for handshaking between devices, and eight are ground return lines (Figure 4.25). *Handshaking* is the term used for the transfer of control information, such as DATA READY and INPUT ACKNOWLEDGED signals, between two devices.

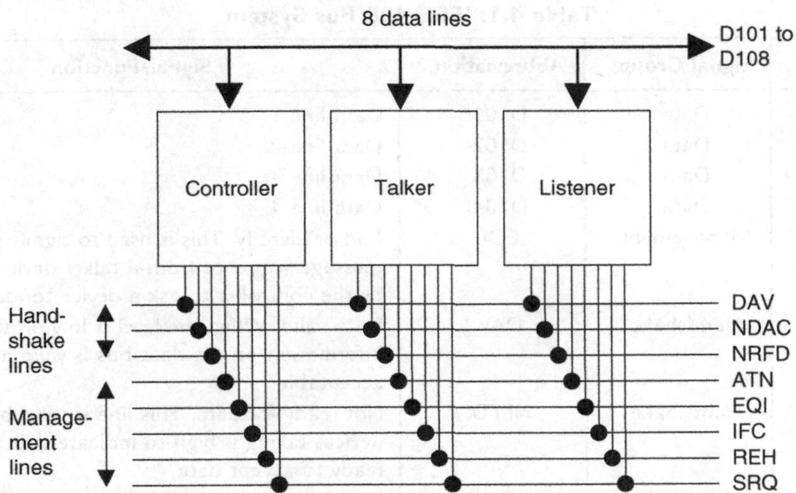

Figure 4.25: The IEEE-488 bus structure.

Commands from the controller are signaled by taking the ATTENTION LINE (ATN) low; otherwise it is high, thus indicating that the data lines contain data. The commands can be directed to individual devices by placing addresses on the data lines. Each device on the bus has its own address. Device addresses are sent via the data lines as a parallel 7-bit word, the lowest 5 bits providing the device address and the other 2 bits providing control information. If both these bits are 0, the commands are sent to all addresses; if bit 6 is 1 and bit 7 a 0, the addressed device is switched to be a listener; if bit 6 is 0 and bit 7 is 1, the device is switched to be a talker.

As illustrated by the function of the ATN line, the management lines each have an individual task in the control of information. The handshake lines are used for controlling the transfer of data. The three lines ensure that the talker will only talk when it is being listened to by listeners. Table 4.1 lists the functions of all the lines and their pin numbers in a 25-way D-type connector.

Figure 4.26 shows the handshaking sequence that occurs when data is put on the data lines. Initially DAV is high, indicating that there is no valid data on the data bus, and NRFD and NDAC are low. When a data word is put on the data lines, NRFD is made high to indicate that all listeners are ready to accept data, and DAV is made low to indicate that new data is on the data lines. When a device accepts a data word, it sets NDAC high to indicate that is has accepted the data and NRFD low to indicate that it is now not ready to accept data. When all the listeners have set NDAC high, the talker cancels the data valid signal, DAV going high. This then results in NDAC being set low. The entire process can then be repeated for another word being put on the data bus.

Table 4.1: IEEE-488 Bus System

Pin	Signal Group	Abbreviation	Signal/Function
1	Data	D101	Data line 1.
2	Data	D102	Data line 2.
3	Data	D103	Data line 3.
4	Data	D104	Data line 4.
5	Management	EOI	End or identify. This is used to signify the end of a message sequence from a talker device or is used by the controller to ask a device to identify itself.
6	Handshake	DAV	Data valid. When the level is low on this line, the information on the data bus is valid and acceptable.
7	Handshake	NRFD	Not ready for data. This line is used by listener devices taking it high to indicate that they are ready to accept data.
8	Handshake	NDAC	Not data accepted. This line is used by listeners taking it high to indicate that data is being accepted.
9	Management	IFC	Interface clear. This is used by the controller to reset all the devices of the system to the start state.
10	Management	SRQ	Service request. This is used by devices to signal to the controller that they need attention.
11	Management	ATN	Attention. This is used by the controller to signal that it is placing a command on the data lines.
12		SHIELD	Shield.
13	Data	D105	Data line 5.
14	Data	D106	Data line 6.
15	Data	D107	Data line 7.
16	Data	D108	Data line 8.
17	Management	REN	Remote enable. This enables a device on the bus to indicate that it is to be selected for remote control rather than by its own control panel.
18		GND	Ground/common.
19		GND	Ground/common.
20		GND	Ground/common.
21		GND	Ground/common.
22		GND	Ground/common.
23		GND	Ground/common.
24		GND	Ground/common.

4.3.4 Protocols

It is necessary to exercise control of the flow of data between two devices so that what constitutes the message, and how the communication is to be initiated and terminated, is defined. This is termed the *protocol*.

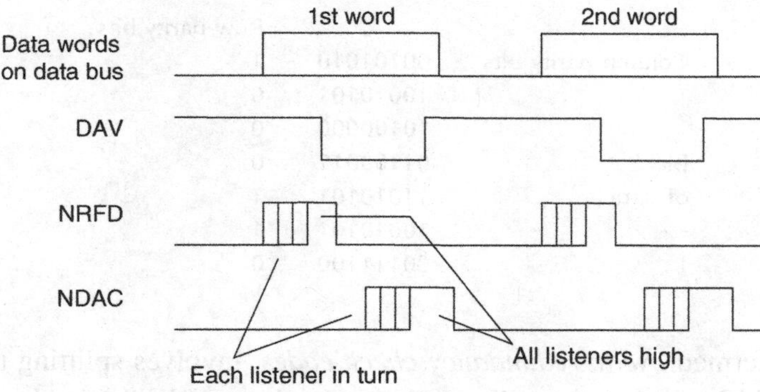

Figure 4.26: Handshaking sequence.

Thus one device needs to indicate to the other to start or stop sending data. This can be done by using handshaking wires connecting transmitting and receiving devices so that a signal along one such wire can tell the receiver that the transmitter is ready to send (RTS) and along another wire that the transmitter is ready to receive, a clear-to-send signal (CTS). RTS and CTS lines are provided for in RS232 serial communication links.

An alternative is to use additional characters on the transmitting wires. With the ENQ/ACK protocol, data packets are sent to a receiver with a query character ENQ. When this character is received, the end of the data packet has been reached. Once the receiver has processed that data, it can indicate it is ready for another block of data by sending back an acknowledge (ACK) signal. Another form, the XON/XOFF, has the receiving device sending a XOFF signal to the sending device when it wants the data flow to cease. The transmitter then waits for an XON signal before resuming transmission.

One form of checking for errors in the message that might occur as a result of transmission is the *parity check*. This is an extra bit added to a message to ensure that the number of bits in a piece of data is always odd or always even. For example, 0100100 is even because there is an even number of 1s; 0110100 is odd because there is an odd number of 1s. To make both these odd parity, the extra bit added at the end in the first case is 1 and in the second case 0, that is, we have 01001001 and 01101000. Thus when a message is sent out with odd bit parity, if the receiver finds that the bits give an even sum, the message has been corrupted during transmission and the receiver can request that the message be repeated.

The parity bit method can detect whether there is an error resulting from a single 0 changing to a 1 or a 1 changing to a 0 but cannot detect two such errors occurring, since there is then no change in parity. To check on such occurrences, more elaborate checking methods have to be used. One method involves storing data words in an array of rows and columns. Parity can then be checked for each row and each column. The following illustrates this concept for seven words using even parity.

		Row parity bits
Column parity bits	00101010	1
↑	10010101	0
	10100000	0
Block	01100011	0
of data	11010101	1
	10010101	1
↓	00111100	0

Another method, termed *cyclic redundancy check codes*, involves splitting the message into blocks. Each block is then treated as a binary number and is divided by a predetermined number. The remainder from this division is sent as the error-checking number on the conclusion of the message and enables a check to be made on the accuracy of the message.

4.3.5 ASCII Codes

The most widely used code for the transmission of characters is the American Standard Code for Information Interchange (ASCII). This is a 7-bit code giving 128 different combinations of bits covering lower- and uppercase alphanumeric characters, punctuation, and 32 control codes. As an illustration, Table 4.2 shows the codes used for capital letters and the digits 0 through 9. Examples of control codes are SOH, for start of heading, that is, the first character of a heading of an information message, as 000 0001; STX, for start of text, as 000 0010; ETX, for end of text, as 000 0011; and EOT, for end of transmission, as 000 0011.

Table 4.2: Examples of ASCII Codes

	ASCII		ASCII		ASCII
A	100 0001	N	100 1110	0	011 0000
B	100 0010	O	100 1111	1	011 0001
C	100 0011	P	101 0000	2	011 0010
D	100 0100	Q	101 0001	3	011 0011
E	100 0101	R	101 0010	4	011 0100
F	100 0110	S	101 0011	5	011 0101
G	100 0111	T	101 0100	6	011 0110
H	100 1000	U	101 0101	7	011 0111
I	100 1001	V	101 0110	8	011 1000
J	100 1010	W	101 0111	9	011 1001
K	100 1011	X	101 1000		
L	100 1100	Y	101 1001		
M	100 1101	Z	101 1010		

4.4 Networks

The increasing use of automation in industry has led to the need for communications and control on a plant-wide basis, with programmable controllers, computers, robots, and CNC machines interconnected. The term *local area network* (LAN) is used to describe a communications network designed to link computers and their peripherals within the same building or site.

Networks can take three basic forms. With the *star* form (Figure 4.27a), the terminals are each directly linked to a central computer, termed the *host* or *master*, with the terminals being called *slaves*. The host contains the memory, processing, and switching equipment to enable the terminals to communicate. Access to the terminals is by the host asking each terminal in turn whether it wants to talk or listen. With the *bus* or *single highway* type of network (Figure 4.27b), each of the terminals is linked into a single cable and so each transmitter/receiver has a direct path to each other transmitter/receiver in the network. Methods, that is, protocols, have to be adopted to ensure that no more than one terminal talks at once; otherwise confusion can occur. A terminal has to be able to detect whether another terminal is talking before it starts to talk. With the *ring* network (Figure 4.27c), a continuous cable, in the form of a ring, links all the terminals. Again, methods have to be employed to enable communications from different terminals without messages becoming mixed up. The single highway and the ring methods are often called *peer to peer* in that each terminal has equal status. Such a system allows many stations to use the same network.

With ring-based networks, two commonly used methods that are employed to avoid two stations talking at once and so giving rise to confusion are token passing and slot passing. With *token passing*, a special bit pattern called a *token* is circulated round the network. When a station wants to transmit into the network, it waits until it receives the token, then transmits the data with the token attached. Another station that wants to transmit cannot do so until the token has been freed by removal from the data by a receiver. With *slot passing*, empty slots are circulated into which stations can deposit data for transmission.

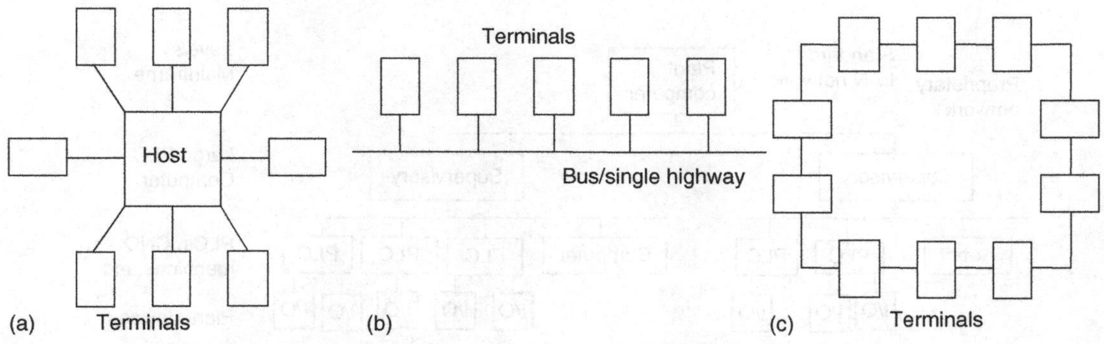

Figure 4.27: Networks: (a) star, (b) bus/single highway, and (c) ring.

Bus systems generally employ the method in which a system that wants to transmit listens to see whether any messages are being transmitted. If no message is being transmitted, a station can take control of the network and transmit its message. This method is known as *carrier sense multiple access* (CSMA). However, we could end up with two stations simultaneously perceiving the network to be clear for transmission and both simultaneously taking control and sending messages. The result would be a "collision" of their transmitted data, resulting in corruption. If such a situation is detected, both stations cease transmitting and wait a random time before attempting to again transmit. This is known as *carrier sense multiple access with collision detection* (CSMA/CD).

PLC manufacturers adopt different forms of network systems and methods of communication for use with their PLCs. For example, Mitsubishi uses a network termed MelsecNET, Allen-Bradley has Data Highway Plus, General Electric uses GENET, Texas Instruments uses TIWAY, and Siemens has PROFIBUS DP. Most, like Allen-Bradley, employ peer-to-peer forms. With Siemens, PROFIBUS DP is a star, that is, a master/slave form.

4.4.1 Distributed Systems

Often PLCs figure in an hierarchy of communications (Figure 4.28). Thus at the lowest level we have input and output devices such as sensors and motors interfaced through I/O interfaces with the next level. The next level involves controllers such as small PLCs or small computers, linked through a network, with the next level of larger PLCs and computers exercising local area control. These in turn may be part of a network, with a large mainframe company computer controlling all.

There is increasing use made of systems that can both control and monitor industrial processes. This involves control and the gathering of data. The term SCADA, which stands for *supervisory control and data acquisition system*, is widely used for such a system.

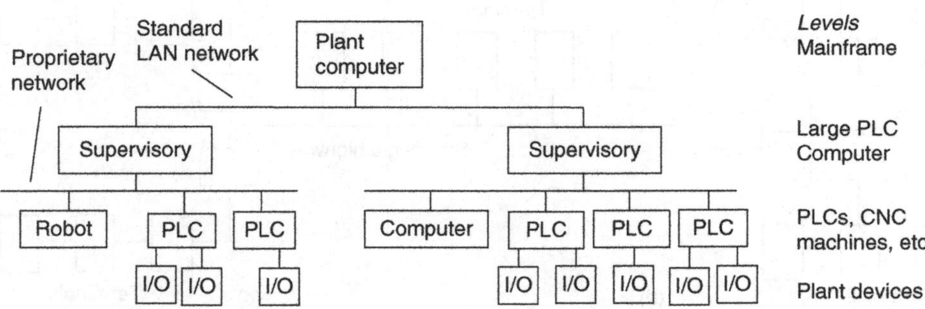

Figure 4.28: Control hierarchy.

4.4.2 Network Standards

Interconnecting several devices can present problems of compatibility; for example, they may operate at different baud rates or use different protocols. To facilitate communications between devices, the International Standards Organization (ISO) in 1979 devised a model to be used for standardization for open systems interconnection (OSI); the model is termed the *ISO OSI model*. A communication link between items of digital equipment is defined in terms of physical, electrical, protocol, and user standards, the ISO OSI model breaking this down into seven layers (Figure 4.29).

The function of each layer in the model is:

Layer 1: Physical medium. This layer is concerned with the coding and physical transmission of information. Its functions include synchronizing data transfer and transferring bits of data between systems.

Layer 2: Data link. This layer defines the protocols for sending and receiving information between systems that are directly connected to each other. Its functions include assembling bits from the physical layer into blocks and transferring them, controlling the sequence of data blocks, and detecting and correcting errors.

Layer 3: Network. This layer defines the switching that routes data between systems in the network.

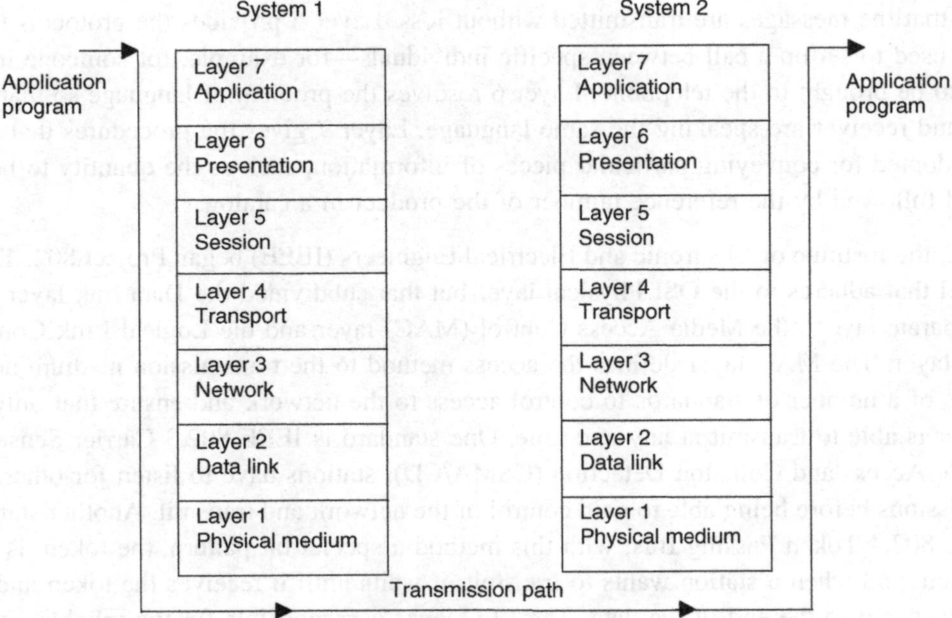

Figure 4.29: ISO/OSI model.

Layer 4: Transport. This layer defines the protocols responsible for sending messages from one end of the network to the other. It controls message flow.

Layer 5: Session. This layer provides the function to set up communications between users at separate locations.

Layer 6: Presentation. This layer ensures that information is delivered in an understandable form.

Layer 7: Application. This layer has the function of linking the user program into the communication process and is concerned with the meaning of the transmitted information.

Each layer is self-contained and only deals with the interfaces of the layer immediately above and below it; it performs its tasks and transfers its results to the layer above or the layer below. It thus enables manufacturers of products to design products operable in a particular layer that will interface with the hardware of other manufacturers.

To illustrate the function of each layer, consider the analogy of making a telephone call. The physical medium is the telephone line, and layer 1 has to ensure that the voice signal is converted into an electrical signal for transmission and then, at the other end of the line, back into an electrical signal. Layer 1 thus defines the types of connectors and the signal levels required. Layer 2 ensures that words that are not clearly received are transferred back to the sender for retransmission. Layer 3 provides the mechanism for dialing the number of the person to be called to make the connection between sender and receiver. Layer 4 is used to ensure that the messages are transmitted without loss. Layer 5 provides the protocols that can be used to set up a call between specific individuals—for example, for someone in an office to be brought to the telephone. Layer 6 resolves the problem of language so that both caller and receiver are speaking the same language. Layer 7 gives the procedures that are to be adopted for conveying particular pieces of information, such as the quantity to be ordered followed by the reference number of the product in a catalog.

In 1980, the Institute of Electronic and Electrical Engineers (IEEE) began Project 802. This is a model that adheres to the OSI Physical layer but that subdivided the Data link layer into two separate layers: the Media Access Control (MAC) layer and the Logical Link Control (LLC) layer. The MAC layer defines the access method to the transmission medium and consists of a number of standards to control access to the network and ensure that only one user is able to transmit at any one time. One standard is IEEE 802.3 Carrier Sense Multiple Access and Collision Detection (CSMA/CD); stations have to listen for other transmissions before being able to gain control of the network and transmit. Another standard is IEEE 802.4 Token Passing Bus; with this method a special bit pattern, the token, is circulated, and when a station wants to transmit, it waits until it receives the token and then attaches it to the end of the data. The LLC layer is responsible for the reliable transmission of data packets across the Physical layer.

The IEC 61131 standard for programmable logic controllers (see Chapter 1.5.1) includes section 61131-5, which is concerned with communications. This defines standards for the communication facilities with PLCs whether they are used as servers, i.e. providing information and responding to service requests, or as a client, requesting information and initiating service requests. To allow IEC compliant PLCs to exchange information and control signals, a number of standard communication blocks are defined:

- CONNECT so that a channel can be established between a calling PLC and a remote PLC by calling it up and supplying it with the full network address of the PLC.
- STATUS and USTATUS to provide the facility to read the status of remote PLCs to ensure that they are fully operational. The STATUS block provides the request for remote status information from a remote PLC, the USTATUS block enables a PLC to receive status information from a remote PLC.
- READ, USEND and URCV concern the reading of the values of variables from remote PLCs. The READ block polls a remote PLC for values of one of more variables. The USEND block is used by a PLC to transmit data to a particular instance of a URCV block existing in a remote PLC.
- WRITE, SEND and RCV for the control of interactions between PLCs. The WRITE block writes one or more values to one or more variables within a remote PLC. The SEND block is for a request to a remote PLC to send data, the RCV block being when it receives it.
- NOTIFY and ALARM for a PLC to signal remote PLCs alarm condition with NOTIFY being a report of an alarm message and ALARM when an alarm message is reported with an acknowledgment that the alarm has been received.
- REMOTE_VAR can be used to obtain a specific address of a named variable.

4.5 Examples of Commercial Systems

The following are examples of systems that may be met with installations involving PLCs.

4.5.1 MAP

By 1990 General Motors in the United States had a problem automating its manufacturing activities; the company needed all its systems to be able to talk to each other. It thus developed a standard communications system for factory automation applications, called the *manufacturing automation protocol* (MAP). The system applied to all systems on the shop floor, such as robot systems, PLCs, and welding systems. Table 4.3 shows the MAP model and its relationship to the ISO model. In order for non-OSI equipment to operate on the MAP system, *gateways* may be used. These are self-contained units or interface boards

Table 4.3: MAP

ISO Layer		MAP
7	Application	ISO file transfer, MMS, FTAM, CASE
6	Presentation	
5	Session	ISO session kernel
4	Transport	ISO transport class 4
3	Network	ISO Internet
2	Data link	IEEE 802.2 class 1; IEEE 802.4 token bus
1	Physical	IEEE 802.4 broadband
	Transmission	10 mbps coaxial cable with RF modulators

Note: MMS = manufacturing message service, FTAM = file transfer, CASE = common applications service; each of these provides a set of commands that will be understood by devices and the software used.

that fit in the device so that messages from a non-OSI network/device may be transmitted through the MAP broadband token bus to other systems.

The Application layer supports the *Manufacturing Message Service* (MMS), which defines the interactions between PLCs and numerically controlled machines and robots.

For the data link, methods are needed to ensure that only the user of the network is able to transmit at any one time, and for MAP the method used is token passing. The term *broadband* is used for a network in which information is modulated onto a radio frequency carrier that is then transmitted through the coaxial cable.

MAP is not widely used; a more commonly used system is the *Ethernet*. This is a single bus system with CSMA/CD used to control access. It uses coaxial cable with a maximum length of 500 m; up to 1024 stations can be accommodated, and repeaters that restore signal amplitude, waveform, and timing can be used to extend this capability (Figure 4.30). Each station is connected to the bus via a transceiver, which clamps onto the bus cable. The term *vampire tap* is used for the clamp on to the cable since stations can be connected or removed without disrupting system operation.

The term *baseband* is used when the signal is transmitted as just a series of voltage levels directly representing the bits being transmitted.

4.5.2 Ethernet

Ethernet does not have a master station, each connected station being of equal status, and so we have peer-to-peer communication. A station that wants to send a message on the bus will determine whether the bus is clear and, when it is, put its message frame on the bus. There is the slight probability that more than one station will sense an idle bus and attempt

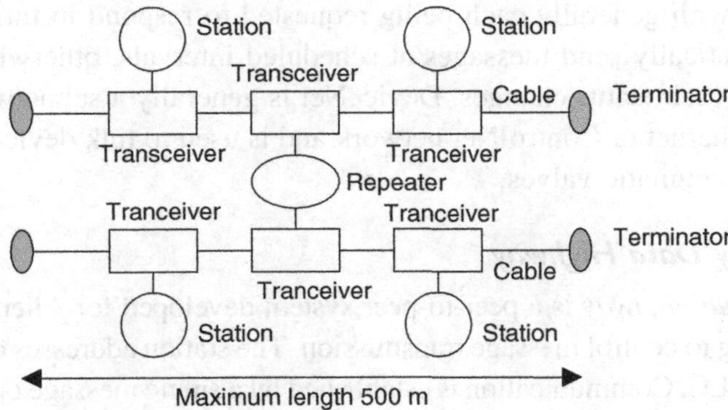

Figure 4.30: Baseband Ethernet with repeaters.

to transmit. Thus each sender monitors the bus during transmission and detects when the signal on the bus does not match its own output. When such a "collision" is detected, the transmission continues for a short while in order to give time to other stations to detect the collision and then the station attempts to retransmit at a later time. Each message includes a bit sequence to indicate the destination address, source address, the data to be transmitted, and a message check sequence. The message check sequence contains the cycle redundancy check (see Section 4.3.4). At each receiving station the frame's destination address is checked to determine whether the frame is destined for it. If so, it accepts the message. Ethernet is widely used where systems involve PLCs having to communicate with computers. The modular Allen-Bradley PLC-5 can be configured for use with a range of communication networks by the addition of suitable modules (refer back to Figure 1.15), including a module enabling use with Ethernet. Ethernet is faster than MAP because the token-passing method of MAP is slower than the method used with Ethernet. PLC manufacturers often have their own networks, in addition to generally offering the option of Ethernet.

4.5.3 ControlNet

This is a network used by Allen-Bradley. Data is placed on the network with no indication as to who it is for. All the stations using this data can thus simultaneously accept it at the same time. This reduces the number of messages needed to be placed on the network and so increases the network speed. This allows PLC racks and their data to be shared equally among several processors and not be just dedicated to one. Network access is controlled by a timing algorithm called Concurrent Time Domain Multiple Access (CTDMA), which determines a node's ability to transmit in the network.

4.5.4 DeviceNet

This is based on the Controller Area Network (CAN), a system that has been widely used with cars (see Section 4.3.2). Each device in the network is requested to send or receive an update of its status, with generally each being requested to respond in turn. Devices are configured to automatically send messages at scheduled intervals, otherwise sending messages only when their status changes. DeviceNet is generally a subnetwork of a PLC that is connected to an Ethernet or ControlNet network and is used to link devices such as sensors, motor starters, and pneumatic valves.

4.5.5 Allen-Bradley Data Highway

The *Allen-Bradley data highway* is a peer-to-peer system developed for Allen-Bradley PLCs and uses token passing to control message transmission. The station addresses of each PLC are set by switches on each PLC. Communication is established by a single message on the data highway, specifying the sending and receiving addresses and the length of block to be transferred.

4.5.6 PROFIBUS

Process Field Bus (PROFIBUS) is a system that was developed in Germany and is used by Siemens with its PLCs. PROFIBUS DP (Decentralized Periphery) is a device-level bus that usually operates with a single DP master and several slaves. Several such DP systems can be installed on one PROFIBUS network. The transmissions are via RS485 (similar to RS422; see Section 4.3.2) or glass fiber optics. Such a system is comparable to DeviceNet. PROFIBUS PA (Process Automation) is an extension of PROFIBUS DP for data transmission from devices such as sensors and actuators. PROFIBUS DP can be connected to PROFIBUS PA using a DP/PA coupler if the work can be operated at 45.45 kbits/s; otherwise a DP/PA link has to be used to convert the data transfer rate of PROFIBUS DP to that of PROFIBUS PA.

4.5.7 Factory-Floor Network

A factory-floor network can use a number of network systems. Thus there may be Ethernet to provide the information layer for data collections and program maintenance, with the next layer down being ControlNet, to deal with real-time input/output processing, and, at the lowest layer, DeviceNet, to deal with sensors and drives. PLCs would take instructions from the Ethernet layer and exercise control through the ControlNet layer.

4.6 Processing Inputs

A PLC is continuously running through its program and updating it as a result of the input signals. Each such loop is termed a *cycle*. PLCs could be operated by each input being examined as it occurred in the program, its effect on the program determined, and the output correspondingly changed. This mode of operation is termed *continuous updating*.

Because there is time spent interrogating each input in turn with continuous updating, the time taken to examine several hundred input/output points can become comparatively long. To allow more rapid execution of a program, a specific area of RAM is used as a buffer store between the control logic and the input/output unit. Each input/output has an address in this

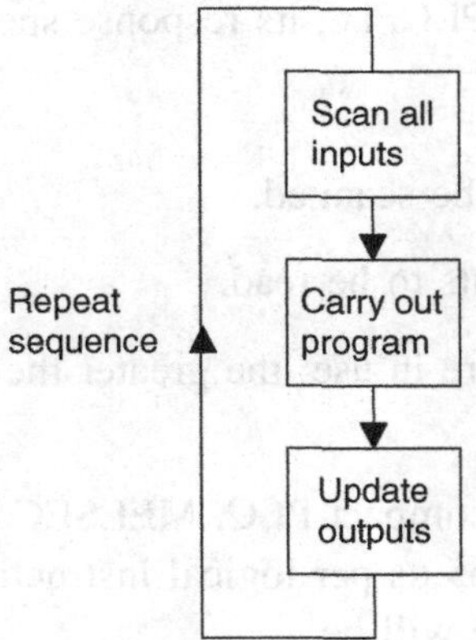

Figure 4.31: PLC operation.

memory. At the start of each program cycle the CPU scans all the inputs and copies their status into the input/output addresses in RAM. As the program is executed, the stored input data is read, as required, from RAM and the logic operations are carried out. The resulting output signals are stored in the reserved input/output section of RAM. At the end of each program cycle all the outputs are transferred from RAM to the appropriate output channels. The outputs then retain their status until the next updating. This method of operation is termed *mass I/O copying*. The sequence can be summarized as follows (Figure 4.31):

1. Scan all the inputs and copy into RAM.
2. Fetch, decode, and execute all program instructions in sequence, copying output instructions to RAM.
3. Update all outputs.
4. Repeat the sequence.

The time taken to complete a cycle of scanning inputs and updating outputs according to the program instructions, that is, the *cycle time*, though relatively quick, is not instantaneous and means that the inputs are not watched all the time, but instead that samples of their states are taken periodically. A typical cycle time is on the order of 10 to 50 ms. This means that the inputs and outputs are updated every 10 to 50 ms and thus there can be a delay of this order in the system reaction. It also means that if a very brief input cycle appears at the wrong moment in the cycle, it could be missed. In general, any input must be present for longer than the cycle time. Special modules are available for use in such circumstances.

Consider a PLC with a cycle time of 40 ms. What is the maximum frequency of digital impulses that can be detected? The maximum frequency will be if one pulse occurs every 40 ms, that is, a frequency of 1/0.04 = 25 Hz.

The cycle or scanning time for a PLC, i.e. its response speed, is determined by:

1. The CPU used.
2. The size of the program to be scanned.
3. The number of inputs/outputs to be read.
4. The system functions that are in use; the greater the number, the slower the scanning time.

As an illustration, the Mitsubishi compact PLC, MELSEC FX3U (see Section 1.4), has a quoted program cycle time of 0.065 μs per logical instruction. Thus the more complex the program, the longer the cycle time will be.

4.7 I/O Addresses

The PLC has to be able to identify each particular input and output. It does this by allocating addresses to each input and output. With a small PLC this is likely to be just a number, prefixed by a letter to indicate whether it is an input or an output. Thus for the Mitsubishi PLC we might have inputs with addresses X400, X401, X402, and so on and outputs with addresses Y430, Y431, Y432, and so on, the X indicating an input and the Y an output. Toshiba uses a similar system.

With larger PLCs that have several racks of input and output channels, the racks are numbered. With the Allen-Bradley PLC-5, the rack containing the processor is given the number 0 and the addresses of the other racks are numbered 1, 2, 3, and so on, according to how setup switches are set. Each rack can have a number of modules, and each one deals with a number of inputs and/or outputs. Thus addresses can be of the form shown in Figure 4.32. For example, we might have an input with address I:012/03. This would indicate an input, rack 01, module 2, and terminal 03.

With the Siemens SIMATIC S7, the inputs and outputs are arranged in groups of eight. Each such group is termed a *byte*, and each input or output within a group of eight is termed a *bit*. The inputs and outputs thus have their addresses in terms of the byte and bit numbers, effectively giving a module number followed by a terminal number, a full stop (.) separating the two numbers. Figure 4.33 shows the system. Thus I0.1 is an input at bit 1 in byte 0, and Q2.0 is an output at bit 0 in byte 2.

The GEM-80 PLC assigns inputs and output addresses in terms of the module number and terminal number within that module. The letter A is used to designate inputs, and B outputs. Thus A3.02 is an input at terminal 02 in module 3, and B5.12 is an output at terminal 12 in module 5.

In addition to using addresses to identify inputs and outputs, PLCs also use their addressing systems to identify internal, software-created devices, such as relays, timers, and counters.

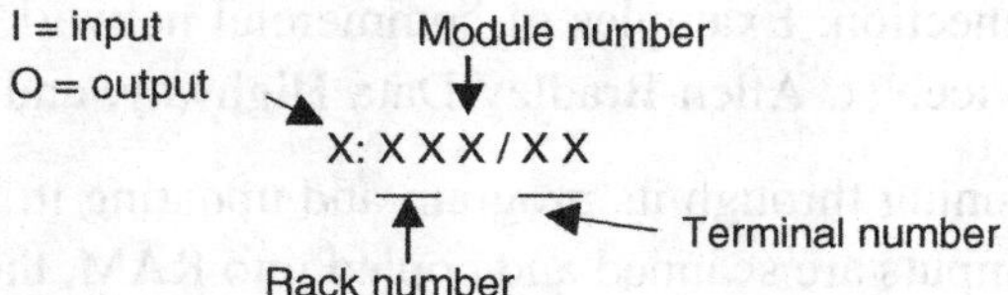

Figure 4.32: Allen-Bradley PLC-5 addressing.

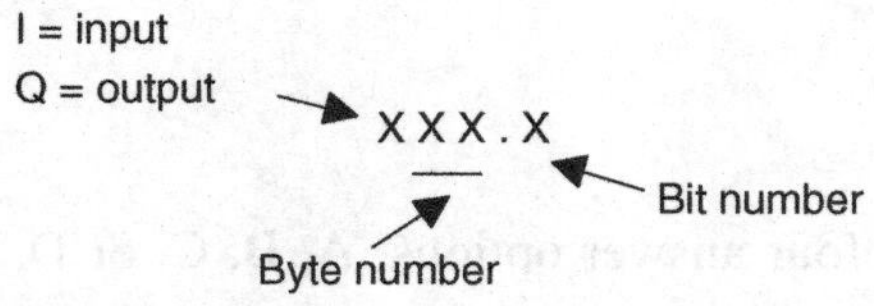

Figure 4.33: Siemens SIMATIC S7 addressing.

Summary

The input/output units of PLCs are designed so that a range of input signals can be changed into 5 V digital signals and a range of output signals are available. For a PLC input unit with sourcing, it is the source of the current supply for the input device connected to it; with sinking, the input device provides the current to the input unit. For a PLC output unit with sourcing, it provides the current to the output device, and for sinking, the output device produces the current for the PLC output. Output units can be relay, transistor, or triac.

For inputs, signal conditioning is generally used to convert analog signals to a current in the range 4 to 20 mA and, thus, by passing through a 250 Ω resistor, to a 1 to 5 V input signal. This might be achieved by a potential divider or perhaps an operational amplifier. An operational amplifier can be used to compare two signals and give an on/off signal based on their relative values.

Serial communication is when data is transmitted one bit at a time. Parallel communication occurs when a data word is separated into its constituent bits and each bit is simultaneously transmitted along parallel cables. The most common serial standard is RS232; other standards are RS422 and RS423. The 20 mA loop can be used for serial communication. The most common parallel standard interface is IEEE-488. Protocols are necessary to exercise control of the flow of data between devices. The most commonly used code for the transmission of characters is ASCII.

The term *local area network* (LAN) describes a communications network designed to link computers and their peripherals within the same building or site. Networks can take three forms: star, bus, or ring. Often PLCs figure in a hierarchy of communications, with input and output devices at the lowest level, at the next level small PLCs or computers, and at the next level, larger PLCs and computers. The ISO OSI model has been devised for standardization for open systems interconnection. Examples of commercial network systems are MAP, Ethernet, ControlNet, DeviceNet, Allen-Bradley Data Highway, and PROFIBUS.

A PLC is continuously running through its program and updating it. It does this by mass I/O copying, in which all the inputs are scanned and copied into RAM, then fetched and decoded, and all program instructions are executed in sequence and output instructions copied to RAM. Then all the outputs are updated before repeating the sequence. The PLC has to be able to identify each particular input and output, and it does this by allocating addresses to each input and output.

Problems

Problems 1 through 15 have four answer options: A, B, C, or D. Choose the correct answer from the answer options.

1. An ADC is used to sample the output voltage from a pressure sensor. If the output from the sensor is 0 V when the pressure is 0 kPa and 10 V when it is 10 kPa, the minimum number of ADC bits needed to resolve the sensor output if the sensor error is not to exceed 0.01 kPa is:
   A. 4
   B. 8
   C. 10
   D. 12
2. A 12-bit ADC can be used to represent analog voltages over its input range with:
   A. 12 different binary numbers
   B. 24 different binary numbers
   C. 144 different binary numbers
   D. 4096 different binary numbers
3. For an analog input range of 0 to 10 V, the minimum size ADC needed to register a change of 0.1 V is:
   A. 4-bit
   B. 6-bit
   C. 8-bit
   D. 12-bit
4. An inverting operational amplifier circuit has an input resistance of 10 kΩ and feedback resistance of 100 kΩ. The closed-loop gain of the amplifier is:
   A. −100
   B. −10
   C. +10
   D. +100

Problems 5 and 6 refer to an operational amplifier with a closed loop gain of 100 and an input resistance of 47 kΩ.

5. The feedback resistor for an inverting op-amp amplifier will be:
   A. 4.65 kΩ
   B. 4.7 kΩ
   C. 465 kΩ
   D. 470 kΩ

6. The feedback resistor for a noninverting op-amp amplifier will be:
   A. 4.65 kΩ
   B. 4.7 kΩ
   C. 465 kΩ
   D. 470 kΩ

7. *Decide whether each of these statements is true (T) or false (F).* A serial communication interface:
   (i) Involves data being transmitted and received one bit at a time.
   (ii) Is a faster form of transmission than parallel communication.
   A. (i) T (ii) T
   B. (i) T (ii) F
   C. (i) F (ii) T
   D. (i) F (ii) F

8. *Decide whether each of these statements is true (T) or false (F).* The RS232 communications interface:
   (i) Is a serial interface.
   (ii) Is typically used for distances up to about 15 m.
   A. (i) T (ii) T
   B. (i) T (ii) F
   C. (i) F (ii) T
   D. (i) F (ii) F

Problems 9 and 10 refer to the following, which shows the bits on an RS232 data line being used to transmit the data 1100001:

0110000111
X YZ

9. *Decide whether each of these statements is true (T) or false (F).* The extra bits X and Z at the beginning and the end are:
   (i) To check whether the message is corrupted during transmission.
   (ii) To indicate where the data starts and stops.

A. (i) T (ii) T
B. (i) T (ii) F
C. (i) F (ii) T
D. (i) F (ii) F

10. *Decide whether each of these statements is true (T) or false (F).* Bit Y is:
(i) The parity bit showing odd parity.
(ii) The parity bit showing even parity.
A. (i) T (ii) T
B. (i) T (ii) F
C. (i) F (ii) T
D. (i) F (ii) F

11. *Decide whether each of these statements is true (T) or false (F).* The parallel data communication interface:
(i) Enables data to be transmitted over short distances at high speeds.
(ii) Has a common standard known as IEEE-488.
A. (i) T (ii) T
B. (i) T (ii) F
C. (i) F (ii) T
D. (i) F (ii) F

12. *Decide whether each of these statements is true (T) or false (F).* For communications over distances of the order of 100 to 300 m with a high transmission rate:
(i) The RS232 interface can be used.
(ii) The 20 mA current loop can be used.
A. (i) T (ii) T
B. (i) T (ii) F
C. (i) F (ii) T
D. (i) F (ii) F

13. *Decide whether each of these statements is true (T) or false (F).* With input/output processing, mass input/output copying:
(i) Scans all the inputs and copies their states into RAM.
(ii) Is a faster process than continuous updating.
A. (i) T (ii) T
B. (i) T (ii) F
C. (i) F (ii) T
D. (i) F (ii) F

14. The cycle time of a PLC is the time it takes to:
    A. Read an input signal.
    B. Read all the input signals.
    C. Check all the input signals against the program.
    D. Read all the inputs, run the program, and update all outputs.

15. *Decide whether each of these statements is true (T) or false (F).* A PLC with a long cycle time is suitable for:
    (i) Short duration inputs.
    (ii) High-frequency inputs.
    A. (i) T (ii) T
    B. (i) T (ii) F
    C. (i) F (ii) T
    D. (i) F (ii) F

16. Specify (a) the odd parity bit and (b) the even parity bit to be used when the data 1010100 is transmitted.

17. Explain the purpose of using a parity bit.

18. Explain the continuous updating and the mass input/output copying methods of processing inputs/outputs.

19. What input resistance and feedback resistance can be used with an inverting operational amplifier circuit to give a gain of –100?

20. Compare the star, bus and ring forms of network and the methods used to avoid problems with messages.

21. What are the functions of (a) PROFIBUS DP and PROFIBUS PA, (b) ControlNet, and DeviceNet?

22. A network is said to involve token passing. What does this mean?

Lookup Tasks

23. Look up the network systems that the PLCs of a particular manufacturer are designed to operate with.

CHAPTER 5

Ladder and Functional Block Programming

Programs for microprocessor-based systems have to be loaded in *machine code*, a sequence of binary code numbers to represent the program instructions. However, *assembly language* based on the use of mnemonics can be used; for example, LD is used to indicate the operation required to load the data that follows the LD, and a computer program called an *assembler* is used to translate the mnemonics into machine code. Programming can be made even easier by the use of the so-called *high-level languages*, such as C, BASIC, Pascal, FORTRAN, and COBOL. These languages use prepackaged functions, represented by simple words or symbols descriptive of the function concerned. For example, with C language the symbol & is used for the logic AND operation. However, the use of these methods to write programs requires some skill in programming, and PLCs are intended to be used by engineers without any great knowledge of programming. As a consequence, *ladder programming* (LAD) was developed as a means of writing programs that can then be converted into machine code by software for use with the PLC microprocessor. This method of writing programs became adopted by most PLC manufacturers, but each tended to develop its own version, and so an international standard has been adopted for ladder programming and, indeed, all the methods used for programming PLCs. The standard, published in 1993, is IEC 61131-3 (see Section 1.4.2). *Functional block programming* (FBD) is another method of programming.

This chapter is an introduction to programming a PLC using ladder diagrams and functional block diagrams. Here we are concerned with the basic techniques involved in developing ladder and function block programs to represent basic switching operations involving the logic functions of AND, OR, EXCLUSIVE OR, NAND, and NOR, as well as latching. Later chapters continue with ladder programming involving other elements.

5.1 Ladder Diagrams

As an introduction to ladder diagrams, consider the simple wiring diagram for an electrical circuit in Figure 5.1a. The diagram shows the circuit for switching on or off an electric motor. We can redraw this diagram in a different way, using two vertical lines to represent

W. Bolton: Programmable Logic Controllers, Sixth Edition. http://dx.doi.org/10.1016/B978-0-12-802929-9.00005-4

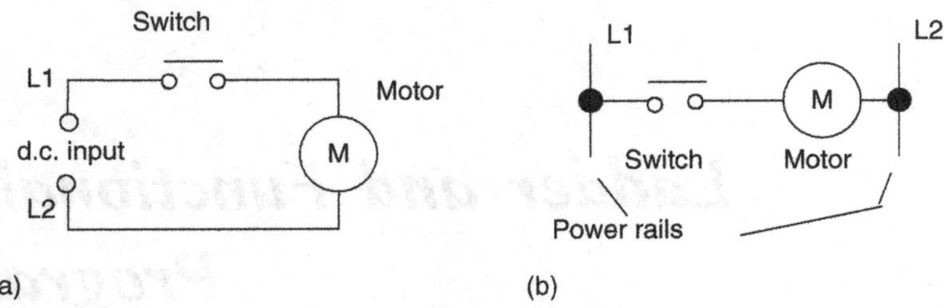

Figure 5.1: Ways of drawing the same electrical circuit.

the input power rails and stringing the rest of the circuit between them. Figure 5.1b shows the result. Both circuits have the switch in series with the motor and supplied with electrical power when the switch is closed. The circuit shown in Figure 5.1b is termed a *ladder diagram*.

With such a diagram, the power supply for the circuits is always shown as two vertical lines, with the rest of the circuit as horizontal lines. The power lines, or *rails*, as they are often called, are like the vertical sides of a ladder, with the horizontal circuit lines similar to the rungs of the ladder. The horizontal rungs show only the control portion of the circuit; in the case of Figure 5.1b it is just the switch in series with the motor. Circuit diagrams often show the relative physical location of the circuit components and how they are actually wired. With ladder diagrams, no attempt is made to show the actual physical locations, and the emphasis is on clearly showing how the control is exercised.

Figure 5.2 shows an example of a ladder diagram for a circuit that is used to start and stop a motor using push buttons. In the normal state, push button 1 is open and push button 2 closed. When button 1 is pressed, the motor circuit is completed and the motor starts. Also, the holding contacts wired in parallel with the motor close and remain closed as long as the motor is running. Thus when the push button 1 is released, the holding contacts maintain the circuit and hence the power to the motor. To stop the motor, push button 2 is pressed. This disconnects the power to the motor, and the holding contacts open. Thus when push button 2 is released, there is still no power to the motor. Thus we have a motor that is started by pressing button 1 and stopped by pressing button 2.

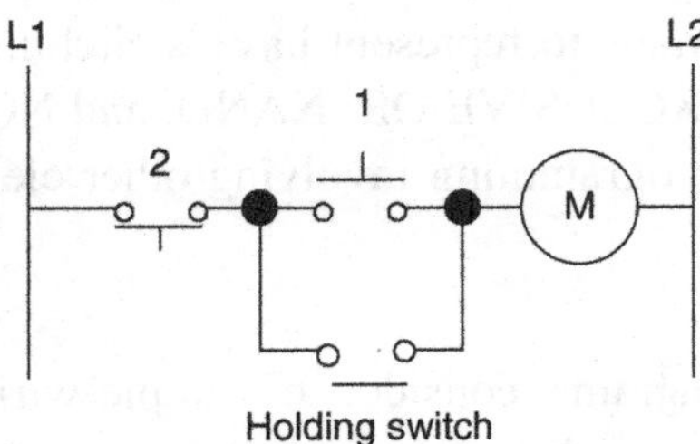

Figure 5.2: Stop/start switch.

5.1.1 PLC Ladder Programming

A very commonly used method of programming PLCs is based on the use of *ladder diagrams*. Writing a program is then equivalent to drawing a switching circuit. The ladder diagram consists of two vertical lines representing the power rails. Circuits are connected as horizontal lines, that is, the rungs of the ladder, between these two verticals.

In drawing a ladder diagram, certain conventions are adopted:

- The vertical lines of the diagram represent the power rails between which circuits are connected. The power flow is taken to be from the left-hand vertical across a rung.
- Each rung on the ladder defines one operation in the control process.
- A ladder diagram is read from left to right and from top to bottom. Figure 5.3 shows the scanning motion employed by the PLC. The top rung is read from left to right. Then the second rung down is read from left to right and so on. When the PLC is in its run mode, it goes through the entire ladder program to the end, the end rung of the program being clearly denoted, and then promptly resumes at the start (see Section 4.4). This procedure of going through all the rungs of the program is termed a *cycle*. The end rung might be indicated by a block with the word END or RET, for return, since the program promptly returns to its beginning. The *scan time* depends on the number of runs in the program, taking about 1 ms per 1000 bytes of program and so typically ranging from about 10 ms up to 50 ms.
- Each rung must start with an input or inputs and must end with at least one output. The term *input* is used for a control action, such as closing the contacts of a switch. The term *output* is used for a device connected to the output of a PLC, such as a motor. As the

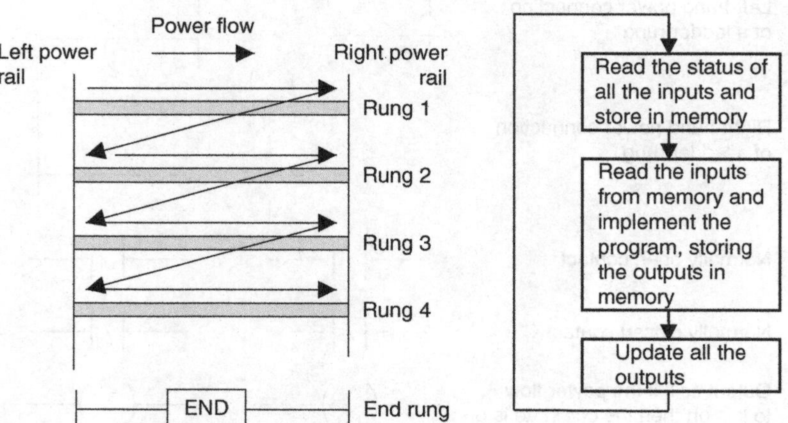

Figure 5.3: Scanning the ladder program.

program is scanned, the outputs are not updated instantly, but the results stored in memory and all the outputs are updated simultaneously at the end of the program scan (see Section 4.6).

- Electrical devices are shown in their normal condition. Thus a switch that is normally open until some object closes it is shown as open on the ladder diagram. A switch that is normally closed is shown closed.
- A particular device can appear in more than one rung of a ladder. For example, we might have a relay that switches on one or more devices. The same letters and/or numbers are used to label the device in each situation.
- The inputs and outputs are all identified by their addresses; the notation used depends on the PLC manufacturer. This is the address of the input or output in the memory of the PLC (see Section 4.6).

Figure 5.4 shows standard IEC 61131-3 symbols that are used for input and output devices. Some slight variations occur between the symbols when used in semigraphic form and

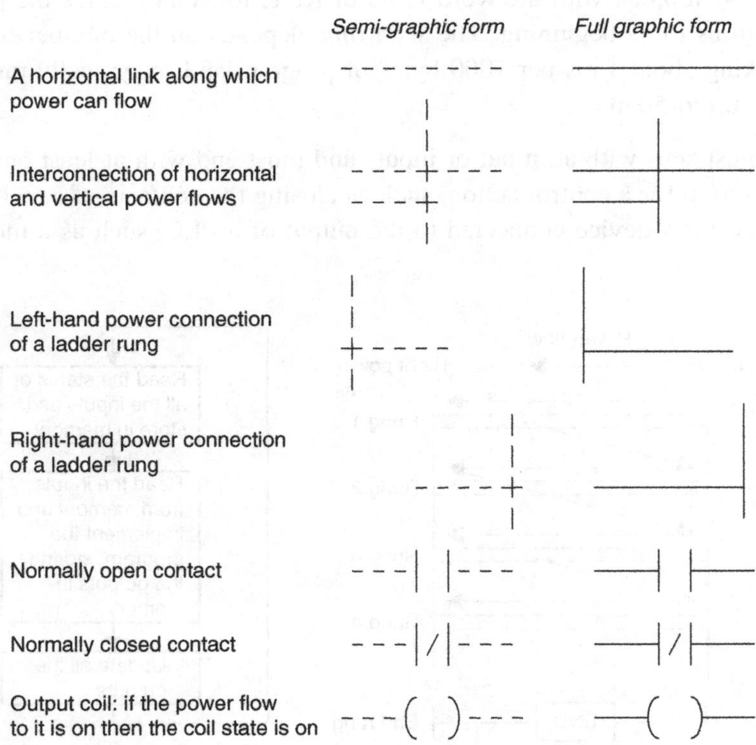

Figure 5.4: Basic symbols.

when in full graphic, the semigraphic form being the one created by simply typing using the normal keyboard, whereas the graphic form is the result of using drawing tools. Note that inputs are represented by various symbols representing normally open or normally closed contacts. The action of the input is equivalent to opening or closing a switch. Output coils are represented by just one form of symbol. (More symbols are introduced in later chapters.) The name of the associated variable with its address is displayed directly above the symbol (for example, for an input start switch, X400, and for an output Motor 1, Y430).

To illustrate the drawing of the rung of a ladder diagram, consider a situation where energizing an output device, such as a motor, depends on a normally open start switch being activated by being closed. The input is thus the switch and the output the motor. Figure 5.5a shows the ladder diagram. Starting with the input, we have the normally open symbol | | for the input contacts. There are no other input devices and the line terminates with the output, denoted by the symbol ( ). When the switch is closed, that is, there is an input, the output of the motor is activated. Only while there is an input to the contacts is there an output. If there had been a normally closed switch |/| with the output (Figure 5.5b), there would have been an output until that switch was opened. Only while there was no input to the contacts would there have been an output.

In drawing ladder diagrams, the names of the associated variable and addresses of each element are appended to its symbol. The more descriptive the name, the better, such as *pump motor control switch* rather than just *input*, and *pump motor* rather than just *output*. Thus Figure 5.6 shows how the ladder diagram of Figure 5.5a would appear using (a) Mitsubishi, (b) Siemens, (c) Allen-Bradley, and (d) Telemecanique notations for the addresses. Thus Figure 5.6a indicates that this rung of the ladder program has an input from address X400 and an output to address Y430. When connecting the inputs and outputs to the PLC, the relevant ones must be connected to the terminals with these addresses.

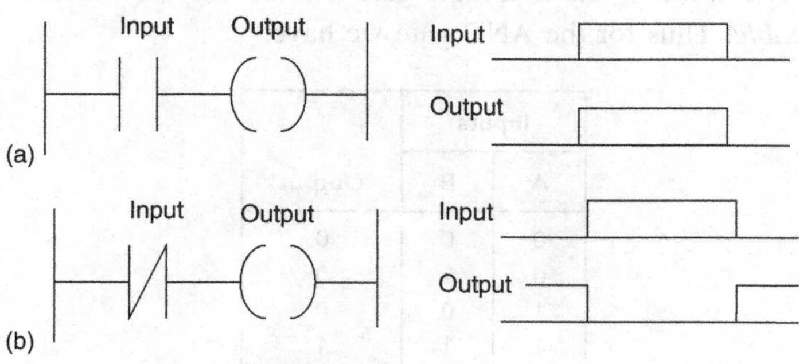

Figure 5.5: A ladder rung.

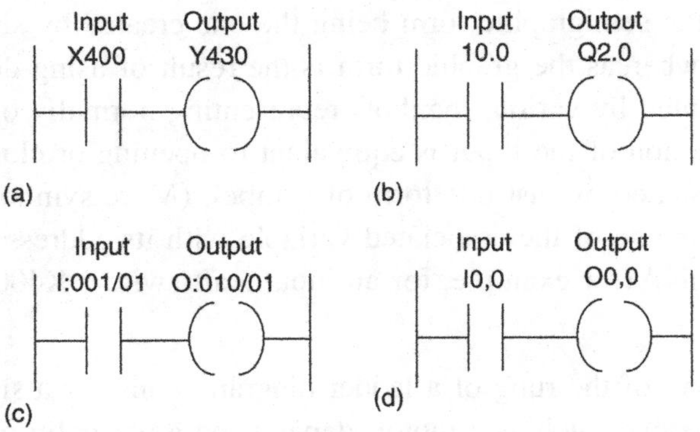

Input X400 Output Y430 (a)

Input 10.0 Output Q2.0 (b)

Input I:001/01 Output O:010/01 (c)

Input I0,0 Output O0,0 (d)

Figure 5.6: Notation: (a) Mitsubishi, (b) Siemens, (c) Allen-Bradley, and (d) Telemecanique.

5.2 Logic Functions

There are many control situations requiring actions to be initiated when a certain combination of conditions is realized. Thus, for an automatic drilling machine (as illustrated in Figure 1.1a), there might be the condition that the drill motor is to be activated upon activation of the limit switches that indicate the presence of the workpiece and the drill position as being at the surface of the workpiece. Such a situation involves the AND logic function, condition A *and* condition B having both to be realized for an output to occur. This section is a consideration of such logic functions.

5.2.1 AND

Figure 5.7a shows a situation in which an output is not energized unless two normally open switches are both closed. Switch A *and* switch B must both be closed, which thus gives an AND logic situation. We can think of this as representing a control system with two inputs, A and B (Figure 5.7b). Only when A *and* B are both on is there an output. Thus if we use 1 to indicate an on signal and 0 to represent an off signal, for there to be a 1 output, we must have A *and* B both 1. Such an operation is said to be controlled by a *logic gate*, and the relationship between the inputs to a logic gate and the outputs is tabulated in a form known as a *truth table*. Thus for the AND gate we have:

Inputs		
A	B	Output
0	0	0
0	1	0
1	0	0
1	1	1

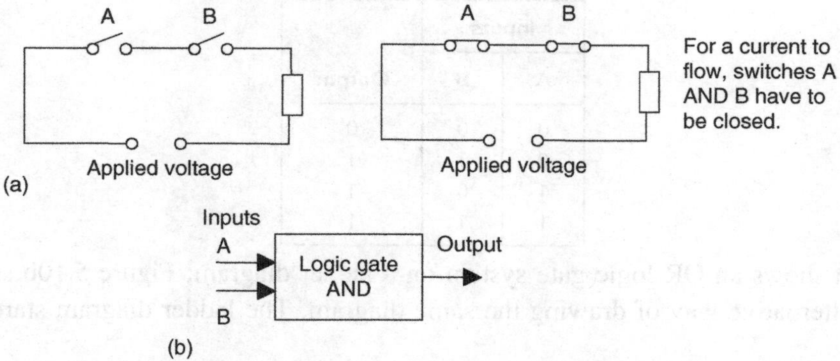

Figure 5.7: (a) An AND circuit, and (b) an AND logic gate.

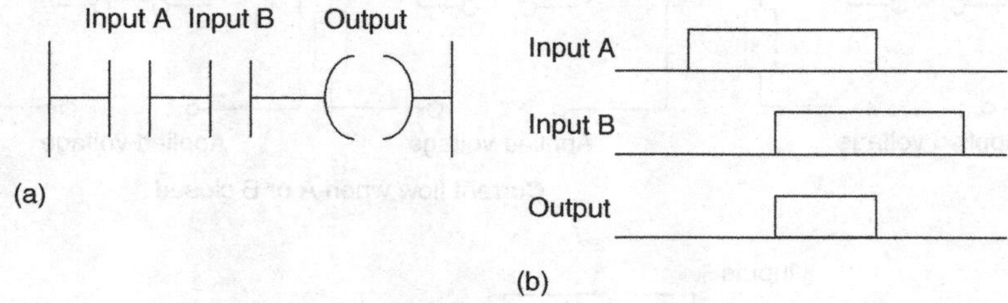

Figure 5.8: An AND gate with a ladder diagram rung.

An example of an AND gate is an interlock control system for a machine tool so that it can only be operated when the safety guard is in position and the power switched on.

Figure 5.8a shows an AND gate system on a ladder diagram. The ladder diagram starts with | |, a normally open set of contacts labeled input A, to represent switch A and in series with it | |, another normally open set of contacts labeled input B, to represent switch B. The line then terminates with ( ) to represent the output. For there to be an output, both input A and input B have to occur, that is, input A and input B contacts have to be closed (Figure 5.8b). In general:

> "*On a ladder diagram, contacts in a horizontal rung, that is, contacts in series, represent the logical AND operations.*"

5.2.2 OR

Figure 5.9a shows an electrical circuit in which an output is energized when switch A *or* B, both normally open, are closed. This describes an OR logic gate (Figure 5.9b) in that input A *or* input B must be on for there to be an output. The truth table is as follows:

Inputs		Output
A	B	
0	0	0
0	1	1
1	0	1
1	1	1

Figure 5.10a shows an OR logic gate system on a ladder diagram; Figure 5.10b shows an equivalent alternative way of drawing the same diagram. The ladder diagram starts with | |,

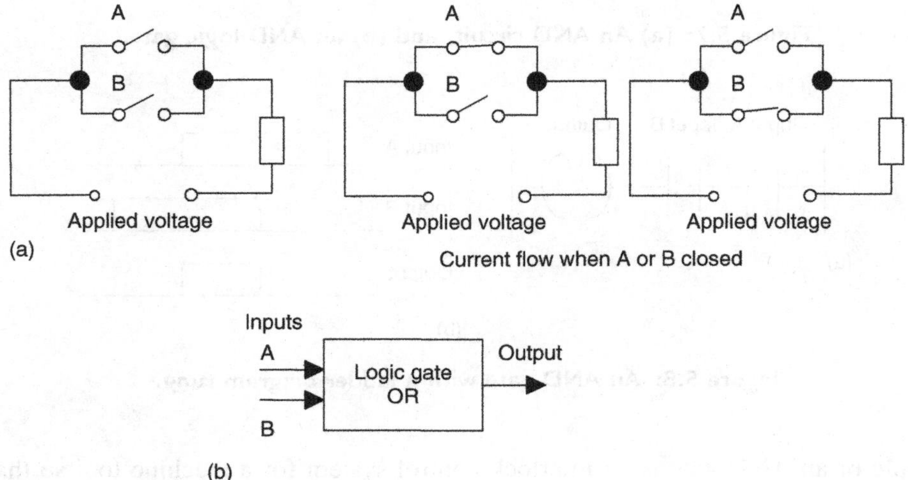

Figure 5.9: (a) An OR electrical circuit, and (b) an OR logic gate.

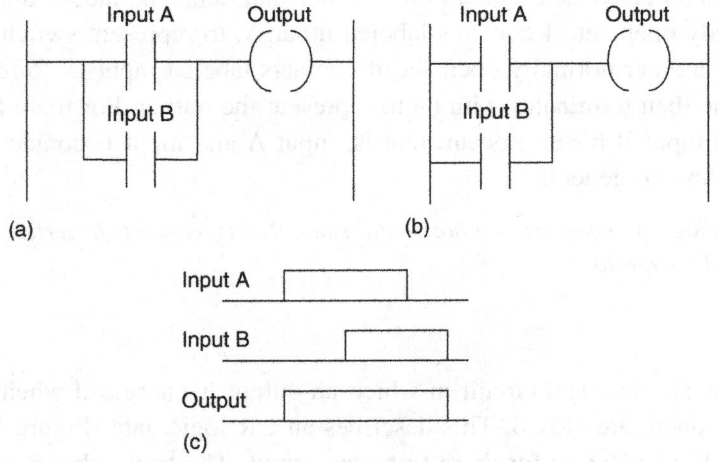

Figure 5.10: An OR gate using a ladder diagram.

normally open contacts labeled input A, to represent switch A and in parallel with it | |, normally open contacts labeled input B, to represent switch B. Either input A *or* input B must be closed for the output to be energized (Figure 5.10c). The line then terminates with ( ) to represent the output. In general:

> "*Alternative paths provided by vertical paths from the main rung of a ladder diagram, that is, paths in parallel, represent logical OR operations.*"

An example of an OR gate control system is a conveyor belt transporting bottled products to packaging where a deflector plate is activated to deflect bottles into a reject bin if either the weight is not within certain tolerances or there is no cap on the bottle.

5.2.3 NOT

Figure 5.11a shows an electrical circuit controlled by a switch that is normally closed. When there is an input to the switch, it opens and there is then no current in the circuit. This example illustrates a NOT gate in that there is an output when there is no input and no output when there is an input (Figure 5.11c). The gate is sometimes referred to as an *inverter*. The truth table is as follows:

Input	
A	Output
0	1
1	0

Figure 5.11b shows a NOT gate system on a ladder diagram. The input A contacts are shown as being normally closed. This input is in series with the output ( ). With no input

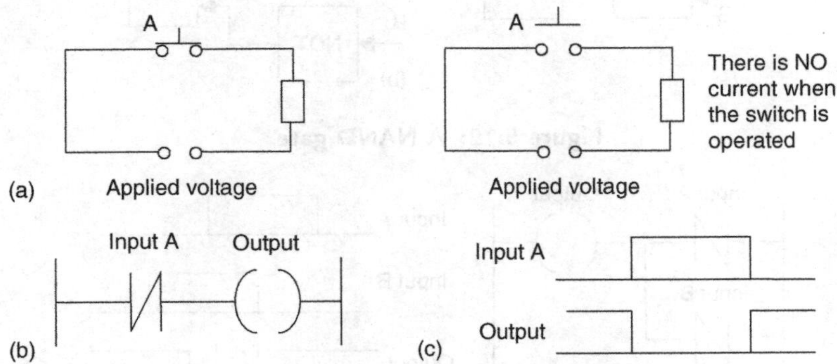

Figure 5.11: (a) A NOT circuit, (b) a NOT logic gate with a ladder rung, and (c) a high output when no input to A.

to input A, the contacts are closed and so there is an output. When there is an input to input A, it opens and there is then no output.

An example of a NOT gate control system is a light that comes on when it becomes dark, that is, when there is no light input to the light sensor there is an output.

5.2.4 NAND

Suppose we follow an AND gate with a NOT gate (Figure 5.12a). The consequence of having the NOT gate is to invert all the outputs from the AND gate. An alternative that gives exactly the same result is to put a NOT gate on each input and then follow that with an OR gate (Figure 5.12b). The same truth table occurs, namely:

Inputs		
A	**B**	**Output**
0	0	1
0	1	1
1	0	1
1	1	0

Either input A or input B (or both) have to be 0 for there to be a 1 output. There is an output when either input A or input B (or both) are not 1. The combination of these gates is termed a *NAND gate*.

Figure 5.13 shows a ladder diagram that gives a NAND gate. When either input A is 0 or input B is 0 (or both are 0), the output is 1. When the inputs to both input A and input B are 1, the output is 0.

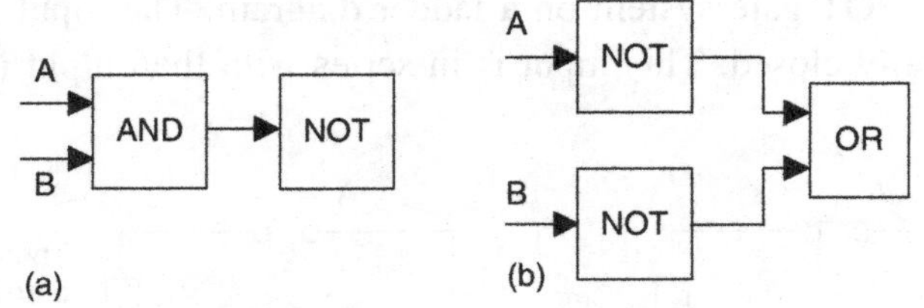

Figure 5.12: A NAND gate.

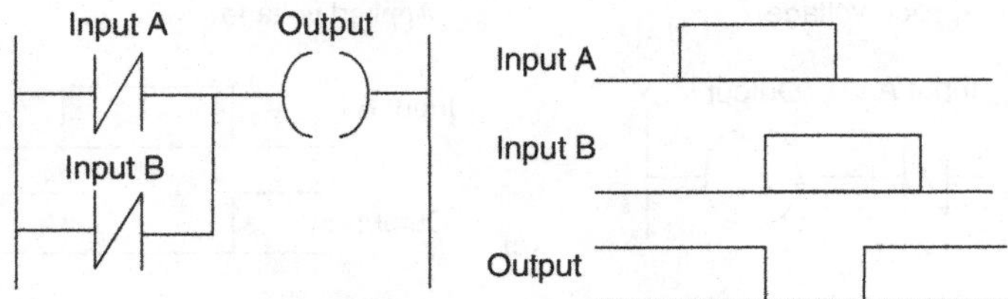

Figure 5.13: A NAND gate using a ladder diagram.

An example of a NAND gate control system is a warning light that comes on if, with a machine tool, the safety guard switch and the limit switch signaling the presence of the workpiece have not been activated.

5.2.5 NOR

Suppose we follow an OR gate by a NOT gate (Figure 5.14a). The consequence of having the NOT gate is to invert the output of the OR gate. An alternative, which gives exactly the same results, is to put a NOT gate on each input and then an AND gate for the resulting inverted inputs (Figure 5.14b). The following is the resulting truth table:

Inputs		
A	**B**	**Output**
0	0	1
0	1	0
1	0	0
1	1	0

The combination of OR and NOT gates is termed a *NOR gate*. There is an output when neither input A nor input B is 1.

Figure 5.15 shows a ladder diagram of a NOR system. When input A and input B are both not activated, there is a 1 output. When either input A or input B are 1, there is a 0 output.

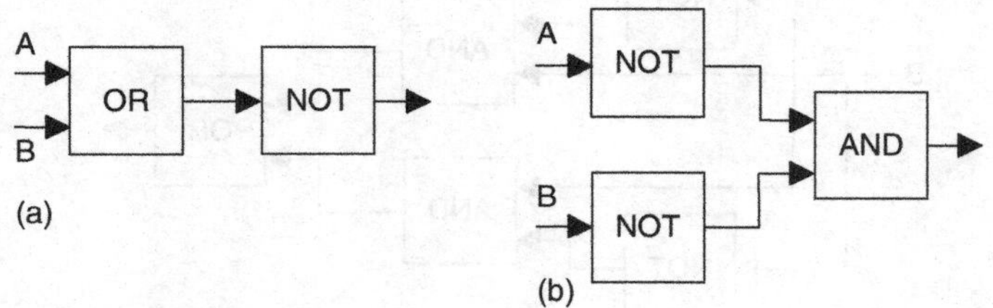

Figure 5.14: A NOR gate.

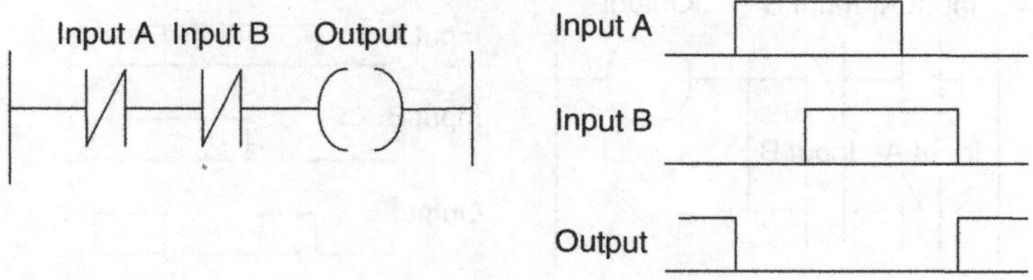

Figure 5.15: A NOR gate using a ladder diagram.

5.2.6 Exclusive OR (XOR)

The OR gate gives an output when either or both of the inputs are 1. However, sometimes there is a need for a gate that gives an output when either of the inputs is 1 but not when both are 1, that is, has the truth table:

Inputs		Output
A	B	
0	0	0
0	1	1
1	0	1
1	1	0

Such a gate is called an *EXCLUSIVE OR*, or *XOR*, gate. One way of obtaining such a gate is by using NOT, AND, and OR gates as shown in Figure 5.16.

Figure 5.17 shows a ladder diagram for an XOR gate system. When input A and input B are not activated, there is 0 output. When just input A is activated, the upper branch results in the output being 1. When just input B is activated, the lower branch results in the output being 1. When both input A and input B are activated, there is no output. In this example of a logic gate, input A and input B have two sets of contacts in the circuits, one set being normally open and the other normally closed. With PLC programming, each input may be considered to have as many sets of contacts as necessary.

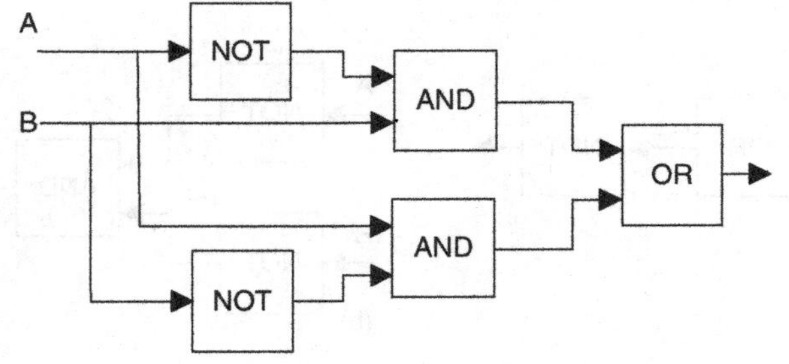

Figure 5.16: An XOR gate.

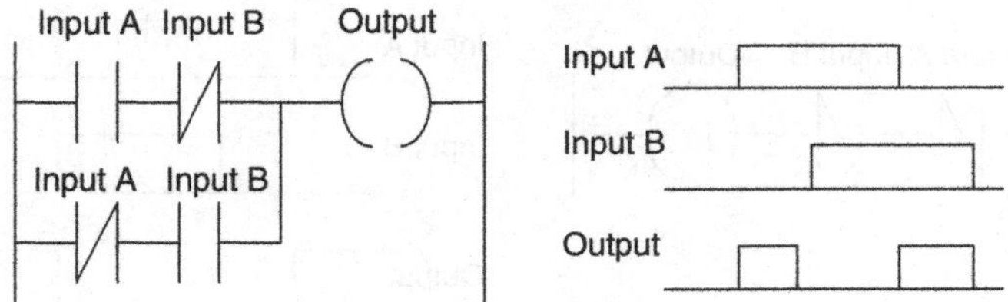

Figure 5.17: An XOR gate using a ladder diagram.

5.3 Latching

There are often situations in which it is necessary to hold an output energized, even when the input ceases. A simple example of such a situation is a motor that is started by pressing a push-button switch. Though the switch contacts do not remain closed, the motor is required to continue running until a stop push-button switch is pressed. The term *latch circuit* is used for the circuit that carries out such an operation. It is a self-maintaining circuit in that, after being energized, it maintains that state until another input is received.

An example of a latch circuit is shown in Figure 5.18. When the input A contacts close, there is an output. However, when there is an output, another set of contacts associated with the output closes. These contacts form an OR logic gate system with the input contacts. Thus, even if input A opens, the circuit will still maintain the output energized. The only way to release the output is by operating the normally closed contact B.

As an illustration of the application of a latching circuit, consider a motor controlled by stop and start push-button switches and for which one signal light must be illuminated when the power is applied to the motor and another when it is not applied. Figure 5.19 shows a ladder diagram with Mitsubishi notation for the addresses. X401 is closed when the program is started. When X400 is momentarily closed, Y430 is energized and its contacts close. This results in latching as well as the switching off of Y431 and the switching on of Y432. To switch the motor off, X401 is pressed and opens. Y430 contacts open in the top rung and third rung but close in the second rung. Thus Y431 comes on and Y432 goes off.

Latching is widely used with startups so that the initial switching on of an application becomes latched on.

5.4 Multiple Outputs

With ladder diagrams, there can be more than one output connected to a contact. Figure 5.20 shows a ladder program with two output coils. When the input contacts close, both the coils give outputs.

For the ladder rung shown in Figure 5.21, output A occurs when input A occurs. Output B occurs only when both input A and input B occur.

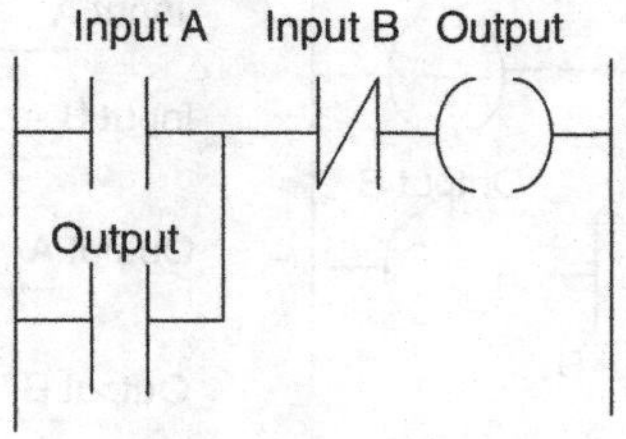

Figure 5.18: Latch circuit.

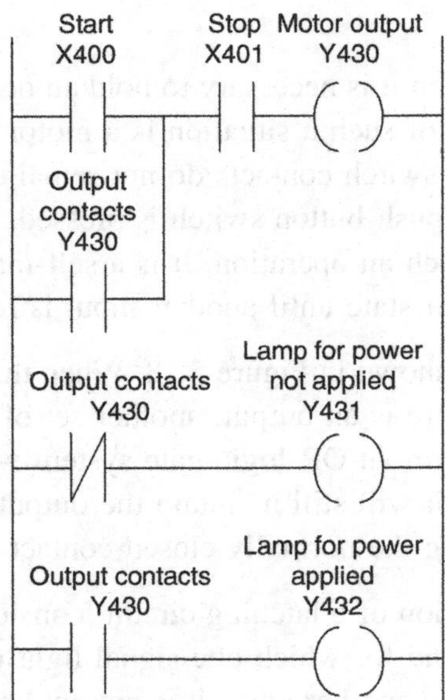

Figure 5.19: A motor on/off, with signal lamps, ladder diagram. Note that the stop contacts X401 are shown as being programmed as open. If the stop switch used is normally closed, X401 receives a startup signal to close. This gives a safer operation than programming X401 as normally closed.

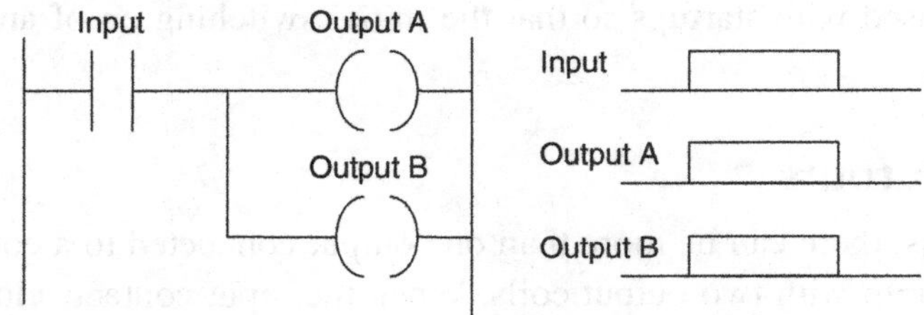

Figure 5.20: Ladder rung with two outputs.

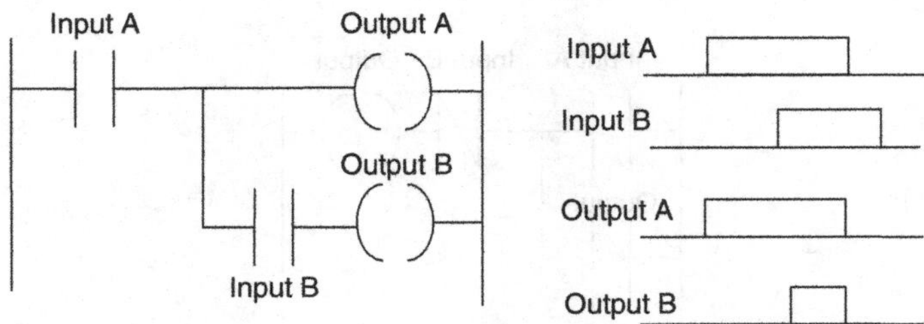

Figure 5.21: Ladder rung with two inputs and two outputs.

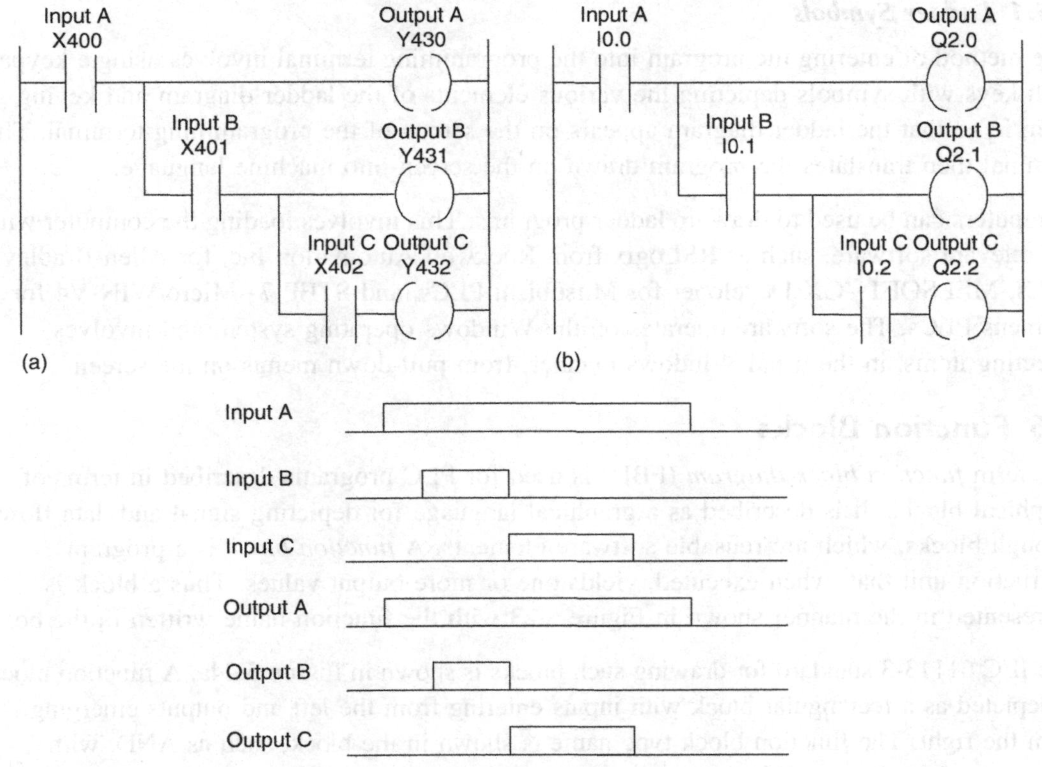

Figure 5.22: Sequenced outputs.

Such an arrangement enables a sequence of outputs to be produced, the sequence being in the sequence in which contacts are closed. Figure 5.22 illustrates this idea with the same ladder program in Mitsubishi and Siemens notations. Outputs A, B, and C are switched on as the contacts in the sequence given by the contacts A, B, and C are being closed. Until input A is closed, none of the other outputs can be switched on. When input A is closed, output A is switched on. Then, when input B is closed, output B is switched on. Finally, when input C is closed, output C is switched on.

5.5 Entering Programs

Each horizontal rung on the ladder represents an instruction in the program to be used by the PLC. The entire ladder gives the complete program. There are several methods that can be used for keying the program into a programming terminal. Whatever method is used to enter the program into a programming terminal or computer, the output to the memory of the PLC has to be in a form that can be handled by the microprocessor. This is termed *machine language* and is just binary code, such as 0010100001110001.

5.5.1 Ladder Symbols

One method of entering the program into the programming terminal involves using a keypad with keys with symbols depicting the various elements of the ladder diagram and keying them in so that the ladder diagram appears on the screen of the programming terminal. The terminal then translates the program drawn on the screen into machine language.

Computers can be used to draw up ladder programs. This involves loading the computer with the relevant software, such as RSLogix from Rockwell Automation Inc. for Allen-Bradley PLCs, MELSOFT–GX Developer for Mitsubishi PLCs, and STEP 7–Micro/WIN V4 for Siemens PLCs. The software operates on the Windows operating system and involves selecting items, in the usual Windows manner, from pull-down menus on the screen.

5.6 Function Blocks

The term *function block diagram* (FBD) is used for PLC programs described in terms of graphical blocks. It is described as a graphical language for depicting signal and data flows through blocks, which are reusable software elements. A *function block* is a program instruction unit that, when executed, yields one or more output values. Thus a block is represented in the manner shown in Figure 5.23 with the function name written in the box.

The IEC 61113-3 standard for drawing such blocks is shown in Figure 5.24a. A function block is depicted as a rectangular block with inputs entering from the left and outputs emerging from the right. The function block type name is shown in the block, such as AND, with the name of the function block in the system shown above it, for example Timer1. Names of function block inputs are shown within the block at the appropriate input and output points. Cross-diagram connectors are used to indicate where graphical lines would be difficult to draw without cluttering up or complicating a diagram and show where an output at one point is used as an input at another. Figure 5.24b shows an example of a function block diagram.

Function blocks can have standard functions, such as those of the logic gates, counters, or timers, or have functions defined by the user, such as a block to obtain an average value of inputs (Figure 5.24c).

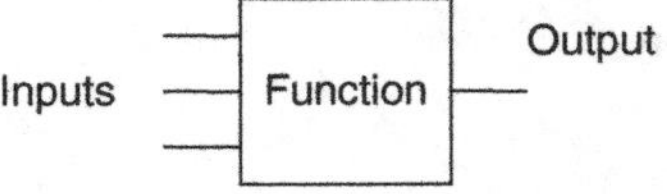

Figure 5.23: Function block.

5.6.1 Logic Gates

Programs are often concerned with logic gates. Two forms of standard circuit symbols are used for logic gates, one originating in the United States and the other an international standard form (IEEE/ANSI) that uses a rectangle with the logic function written inside it.

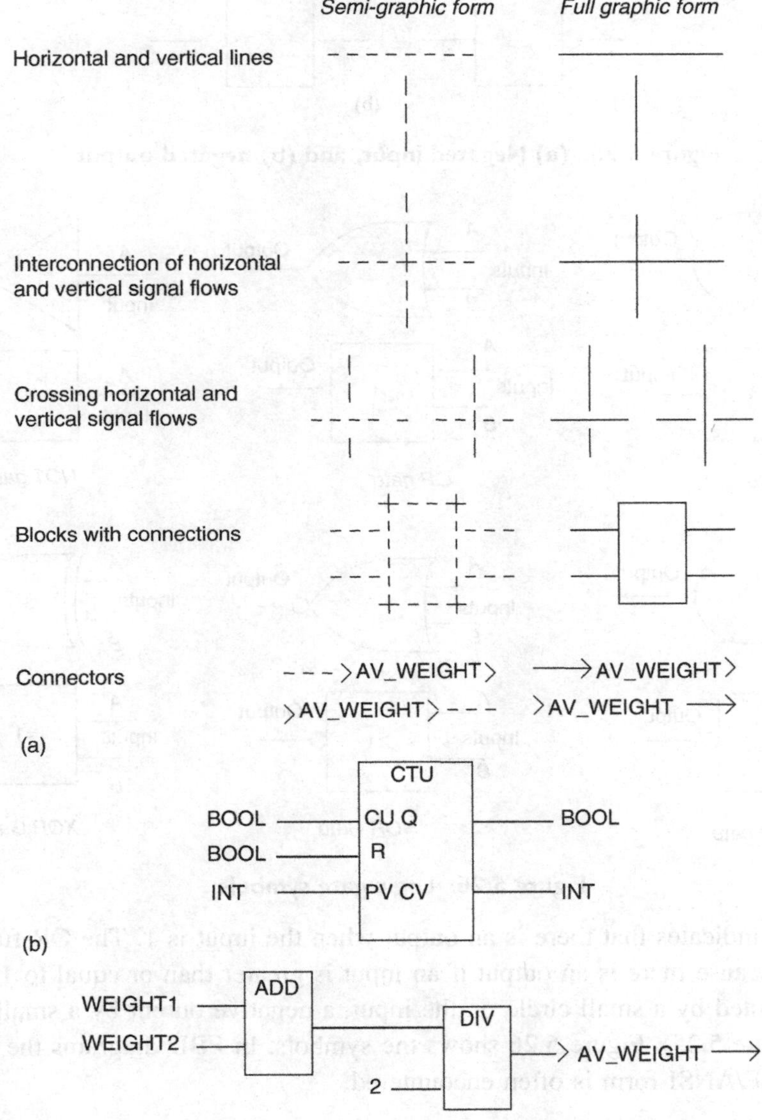

Figure 5.24: Function block diagram representation. (a) Example of a standard function block: an up-counter CTU that gives an output at Q when the input of pulses at the CU count has reached the set value, which is set by PV. Each time there is a pulse input, the output CV is incremented by 1. The input R is to reset the counter. The input and output labels are used to indicate the type of signal involved: BOOL for Boolean and INT for integer. See Chapter 10 for the use of such a block. (b) Example of user-defined blocks to give the average of two weights to be used further in some other function block.

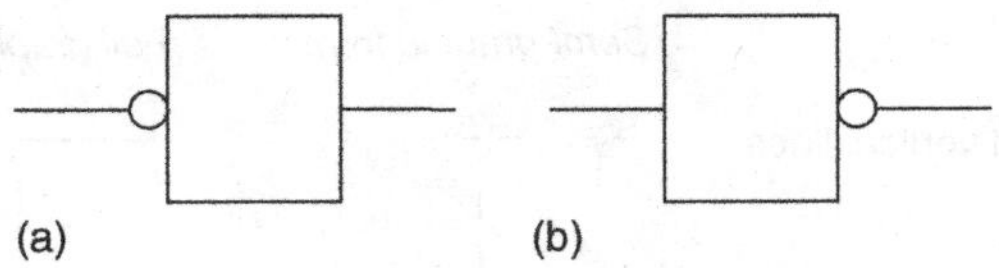

Figure 5.25: (a) Negated input, and (b) negated output.

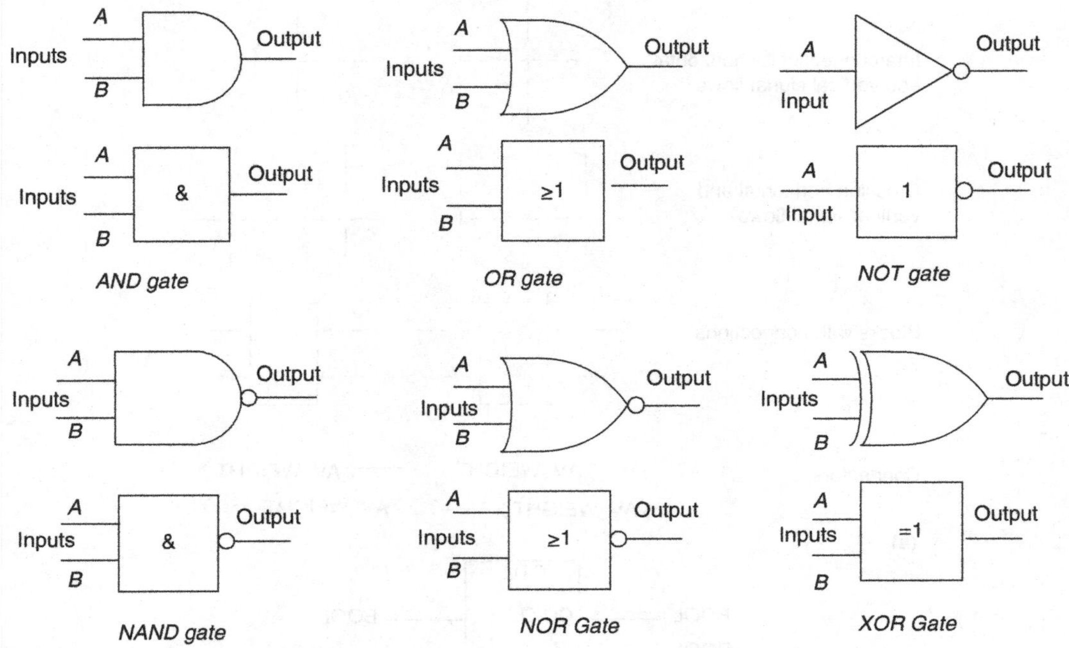

Figure 5.26: Logic gate symbols.

The 1 in a box indicates that there is an output when the input is 1. The OR function is given by ≥1 because there is an output if an input is greater than or equal to 1. A negated input is represented by a small circle on the input, a negative output by a small circle on the output (Figure 5.25). Figure 5.26 shows the symbols. In FBD diagrams the notation used in the IEEE/ANSI form is often encountered.

Figure 5.27 shows the effect of such functional blocks in PLC programs.

To illustrate the form of such a diagram and its relationship to a ladder diagram, Figure 5.28 shows an OR gate. When either the A or B input is 1, there is an output.

Figure 5.29 shows a ladder diagram and its function block equivalent in Siemens notation. The = block is used to indicate an output from the system.

Figure 5.30 shows a ladder diagram involving the output with contacts acting as an input. The function block diagram equivalent can be shown as a feedback loop.

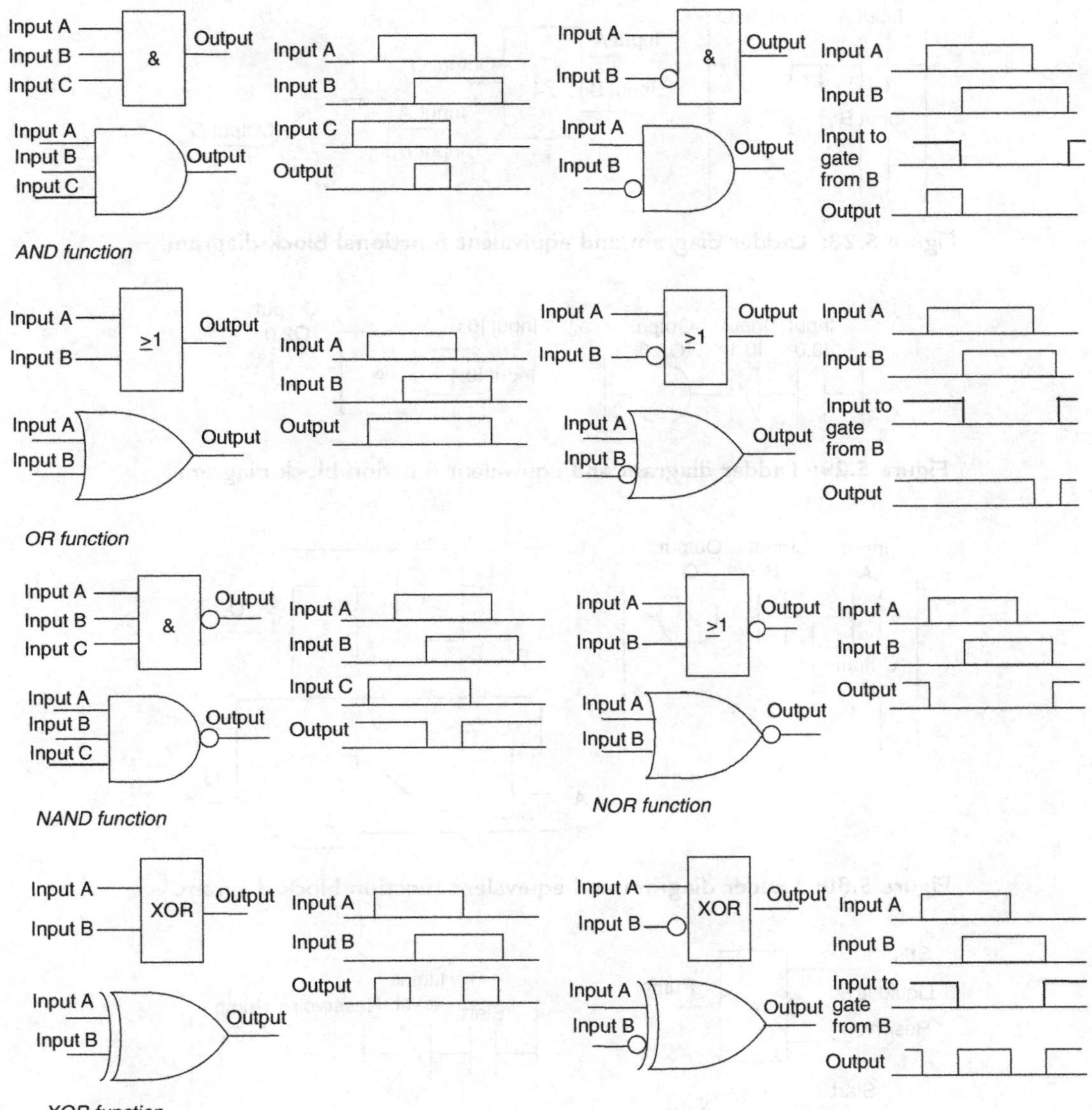

Figure 5.27: Functional logic gate blocks.

Consider the development of a function block diagram and ladder diagram for an application in which a pump is required to be activated and pump liquid into a tank when the start switch is closed, the level of liquid in the tank is below the required level, and there is liquid in the reservoir from which it is to be pumped. What is required is an AND logic situation between the start switch input and a sensor input that is on when the liquid in the tank is below the required level. We might have a switch that is on until the liquid is at the

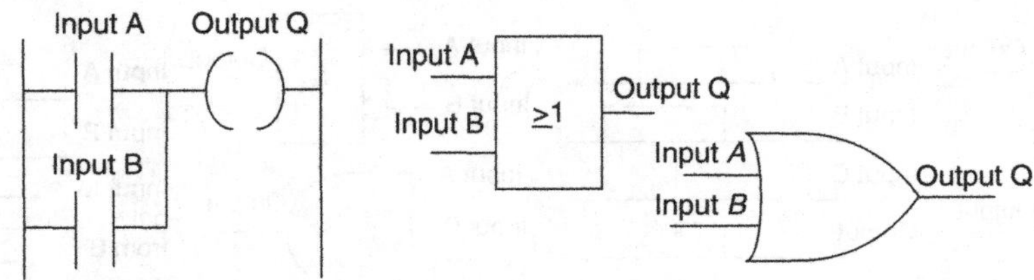

Figure 5.28: Ladder diagram and equivalent functional block diagram.

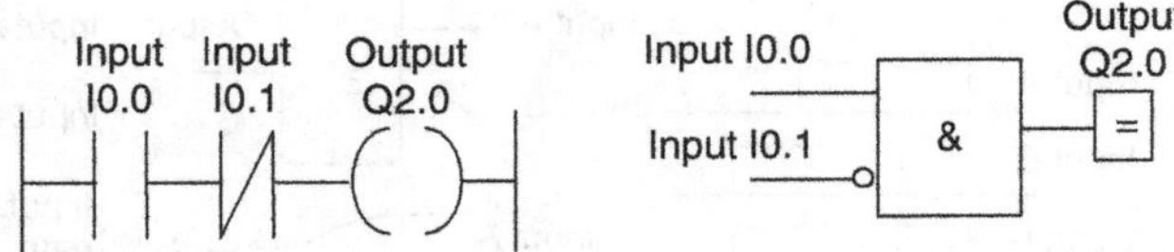

Figure 5.29: Ladder diagram and equivalent function block diagram.

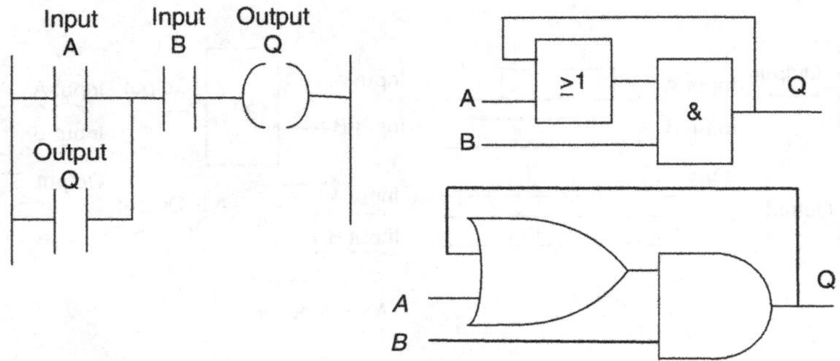

Figure 5.30: Ladder diagram and equivalent function block diagram.

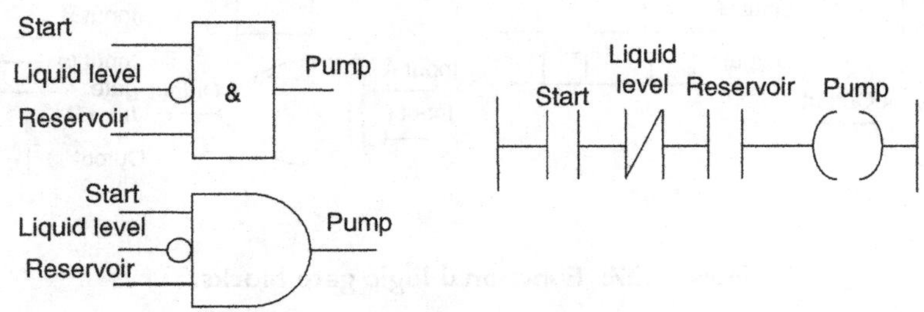

Figure 5.31: Pump application.

required level. These two elements are then in an AND logic situation with a switch indicating that there is liquid in the reservoir. Suppose this switch gives an input when there is liquid. The function block diagram and the equivalent ladder diagram are then of the form shown in Figure 5.31.

5.6.2 Boolean Algebra

Ladder programs can be derived from Boolean expressions since we are concerned with a mathematical system of logic. In Boolean algebra there are just two digits, 0 and 1. When we have an AND operation for inputs A and B, we can write:

$$A \cdot B = Q$$

where Q is the output. Thus Q is equal to 1 only when A = 1 and B = 1. The OR operation for inputs A and B is written as:

$$A + B = Q$$

Thus Q is equal to 1 when A = 1 or B = 1. The NOT operation for an input A is written as:

$$\bar{A} = Q$$

Thus when A is not 1, there is an output.

As an illustration of how we can relate Boolean expressions with ladder diagrams, consider the expression:

$$A + B \cdot C = Q$$

This tells us that we have A or the term B and C giving the output Q. Figure 5.32 shows the ladder and functional block diagrams. Written in terms of Mitsubishi notation, the preceding expression might be:

$$X400 + X401 \cdot X402 = Y430$$

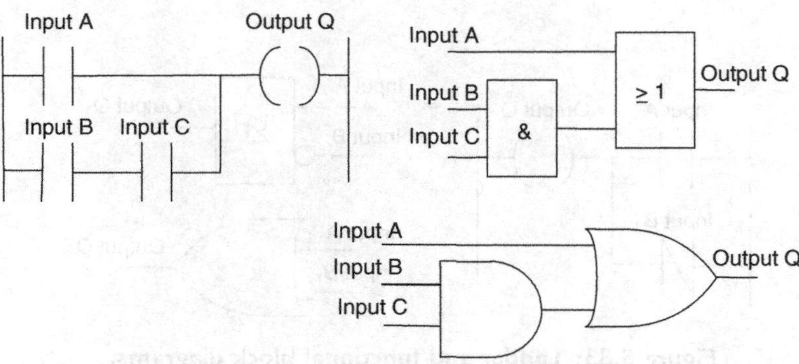

Figure 5.32: Ladder and functional block diagrams.

In Siemens notation it might be:

$$\text{I0.0} + \text{I0.1} \cdot \text{I0.2} = \text{Q2.0}$$

As a further illustration, consider the Boolean expression:

$$A + \bar{B} = Q$$

Figure 5.33 shows the ladder and functional block diagrams. Written in terms of Mitsubishi notation, the expression might be:

$$\text{X400} + \overline{\text{X401}} = \text{Y430}$$

and in Siemens notation:

$$\text{I0.0} + \overline{\text{I0.1}} = \text{Q2.0}$$

Consider the XOR gate and its assembly from NOT, AND, and OR gates, as shown in Figure 5.34.

The input to the bottom AND gate is:

$$A \text{ and } \bar{B}$$

and so its output is:

$$A \cdot \bar{B}$$

The input to the top AND gate is:

$$\bar{A} \text{ and } B$$

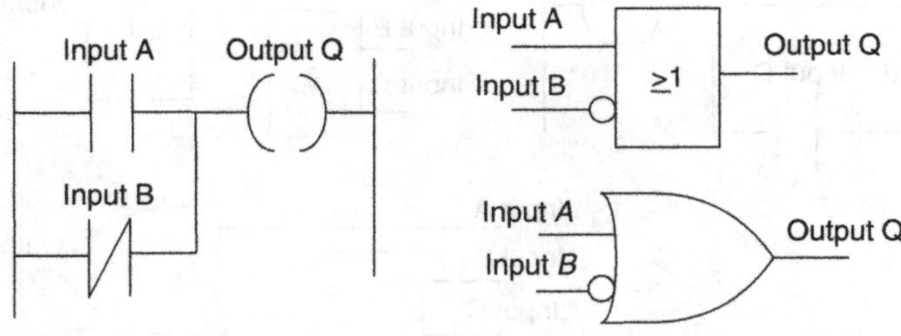

Figure 5.33: Ladder and functional block diagrams.

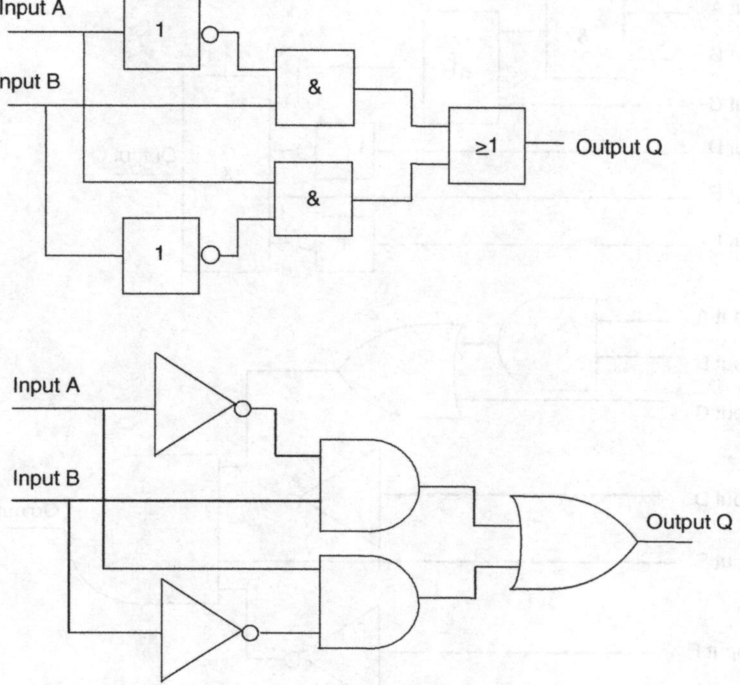

Figure 5.34: An XOR gate.

and so its output is:

$$\bar{A}\cdot B$$

Thus the Boolean expression for the output from the OR gate is:

$$A\cdot\bar{B} + \bar{A}\cdot B = Q$$

Consider a logic diagram with many inputs, as shown in Figure 5.35, and its representation by a Boolean expression and a ladder rung.

For inputs A and B we obtain an output from the upper AND gate of A·B. From the OR gate we obtain an output of A·B + C. From the lower AND gate we obtain an output Q of:

$$(A\cdot B + C)\cdot\bar{D}\cdot E\cdot\bar{F} = Q$$

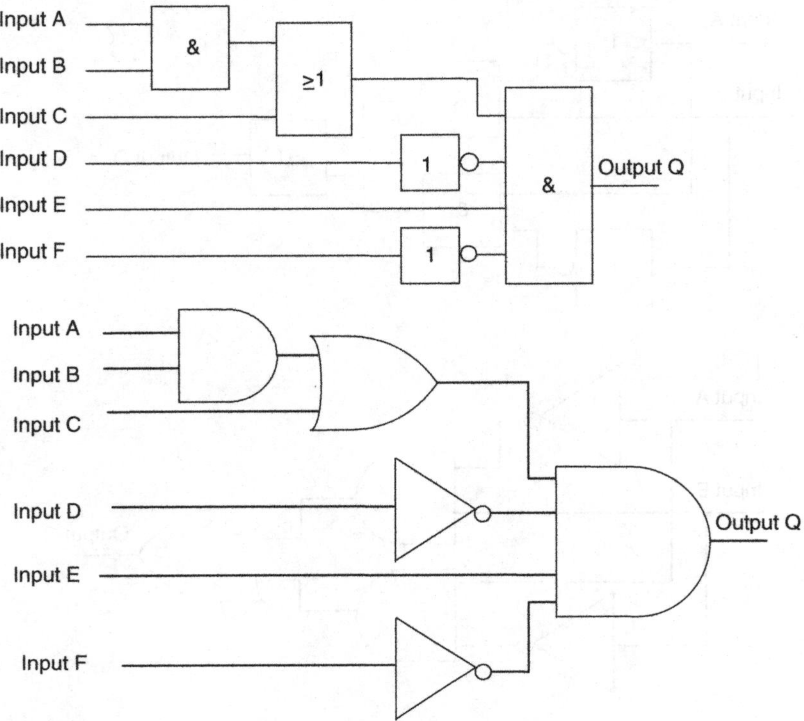

Figure 5.35: Logic diagram.

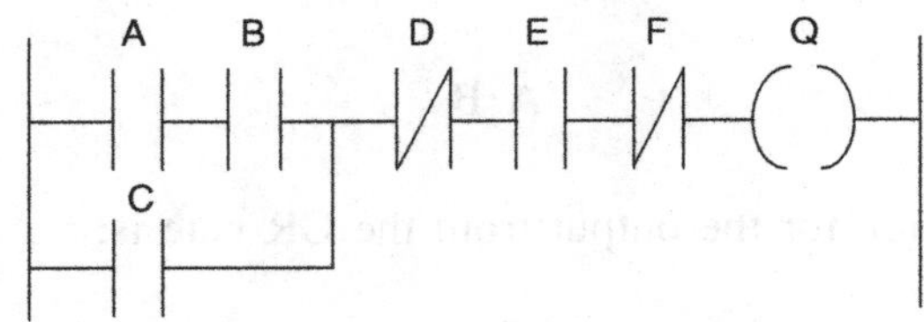

Figure 5.36: Ladder diagram for Figure 5.35.

The ladder diagram to represent this idea is shown in Figure 5.36.

5.7 Program Examples

The following tasks illustrate the application of the programming techniques given in this chapter.

A signal lamp is required to be switched on if a pump is running and the pressure is satisfactory, or if the lamp test switch is closed. For the inputs from the pump and the pressure sensors, we have an AND logic situation, since both are required if there is to be an output from the lamp. However, we have an OR logic situation with the test switch in that it is required to give an output of lamp on, regardless of whether there is a signal from the AND system. The function block diagram and the ladder diagram are thus of the form shown in Figure 5.37. Note that with the ladder diagram, we tell the PLC when it has reached the end of the program by the use of the END or RET instruction.

As another example, consider a valve that is to be operated to lift a load when a pump is running and either the lift switch or a switch indicating that the load has not already been lifted and is at the bottom of its lift channel is operated. We have an OR situation for the two switches and an AND situation involving the two switches and the pump. Figure 5.38 shows a possible program.

Next, consider a system where there has to be no output when any one of four sensors gives an output; otherwise there is to be an output. One way we could write a program for this situation is for each sensor to have contacts that are normally closed, so there is an output. When there is an input to the sensor, the contacts open and the output stops. Thus we have an AND logic situation. Figure 5.39 shows the functional block and ladder diagrams of a system that might be used.

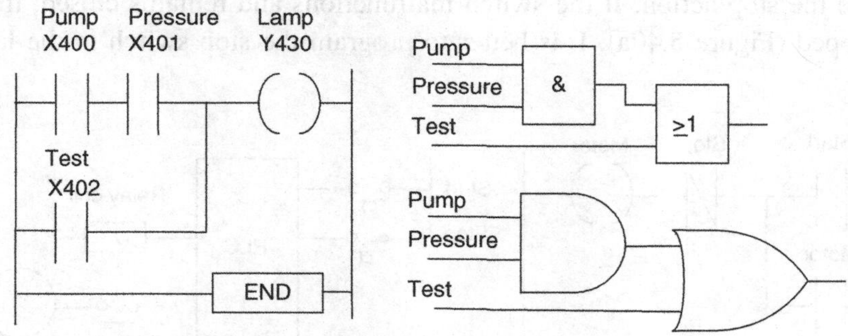

Figure 5.37: Signal lamp task.

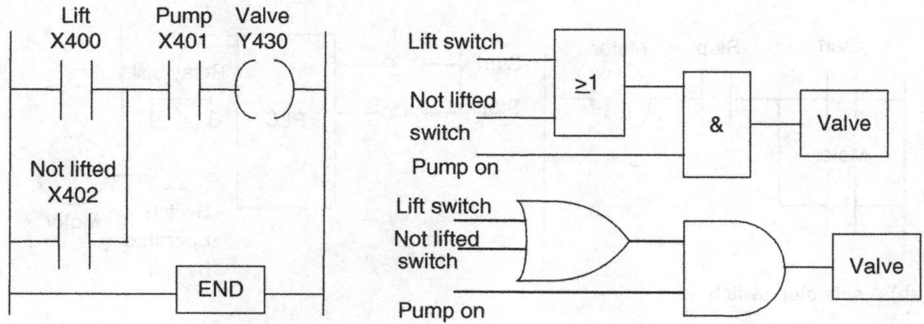

Figure 5.38: Valve operation program.

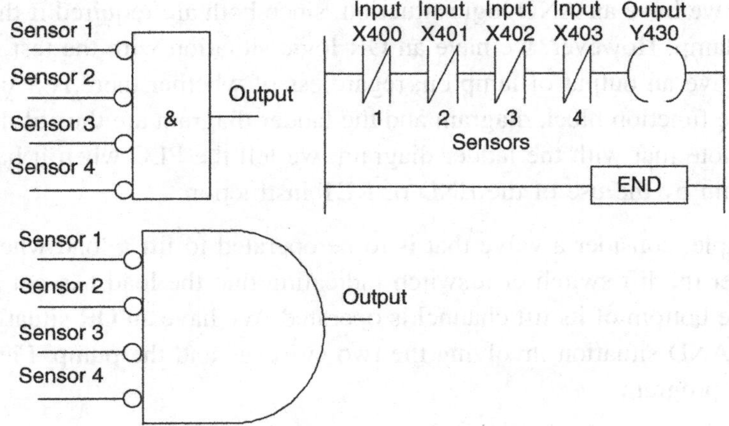

Figure 5.39: Output switched off by any one of four sensors being activated.

5.7.1 Location of Stop Switches

The location of stop switches with many applications has to be very carefully considered to ensure a safe system. A stop switch is *not* safe if it is normally closed and has to be opened to give the stop action. If the switch malfunctions and remains closed, the system cannot be stopped (Figure 5.40a). It is better to program the stop switch in the ladder

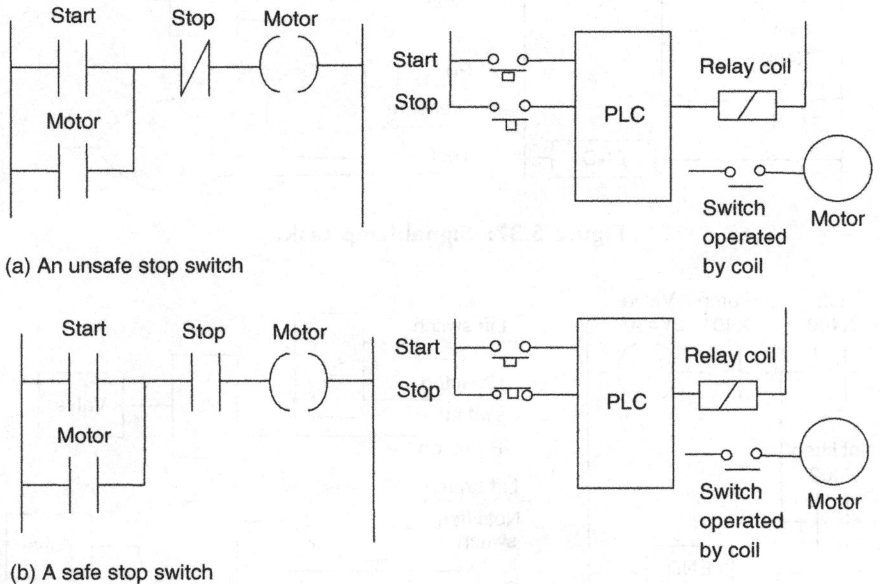

Figure 5.40: Motor stop switch location.

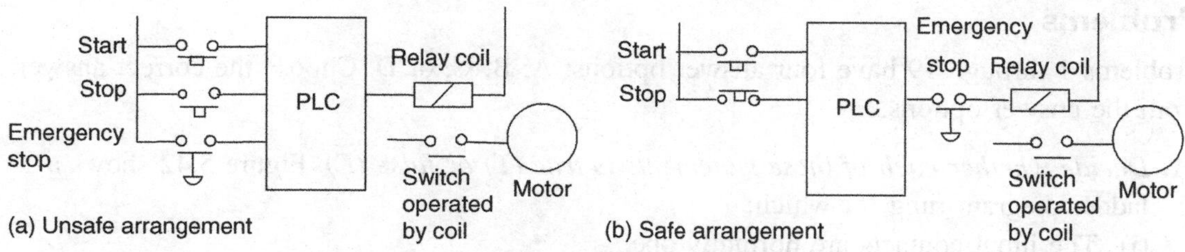

Figure 5.41: Location of emergency stop switch.

program as open in Figure 5.33b and use a stop switch that is normally closed and operating opens it. Thus there is an input signal to the system that closes the contacts in the program when it starts up.

Figure 5.41 shows where we can safely locate an emergency stop switch. If it is in the input to the PLC (Figure 5.41a), then if the PLC malfunctions it might not be possible to stop the motor. However, if the emergency stop switch is in the output, operating it will stop the motor and cause the start switch to become unlatched if the arrangement shown in Figure 5.41b is being used. The motor will thus not restart when the emergency stop button is released.

We must always have the situation that if a failure of the PLC occurs, the outputs must fall into a "fail-safe" state so that no harm can occur to anyone working in the plant.

Summary

Ladder programming (LAD) is a very common method of programming PLCs. Each rung on the ladder defines one operation in the control process and must start with an input or inputs and finish with at least one output. The program is scanned rung by rung, reading from left to right and finishing the cycle with the END rung. The program then restarts the cycle all over again. The inputs and outputs are all identified by their addresses. Rungs in ladder programs can be written to carry out the logic systems of AND, OR, NOT, NAND, NOR, and XOR with inputs. In latch programs, following an input, the output latches the input so that the output can continue even when the input has ceased. This is done by the output having a relay-type set of contacts that are activated when the output occurs and OR the input. Also, an input can be used to operate more than one output.

Function block diagrams (FBDs) can be used to program a PLC. Such programs are described as being a graphical language for depicting signal and data flows through blocks, which are reusable software elements. Logic gates are examples of such function blocks. Boolean algebra can be used to describe such programs.

Problems

Problems 1 through 19 have four answer options: A, B, C, or D. Choose the correct answer from the answer options.

1. *Decide whether each of these statements is true (T) or false (F).* Figure 5.42 shows a ladder diagram rung for which:
   (i) The input contacts are normally open.
   (ii) There is an output when there is an input to the contacts.
   A. (i) T (ii) T
   B. (i) T (ii) F
   C. (i) F (ii) T
   D. (i) F (ii) F

2. *Decide whether each of these statements is true (T) or false (F).* Figure 5.43 shows a ladder diagram rung for which:
   (i) The input contacts are normally open.
   (ii) There is an output when there is an input to the contacts.
   A. (i) T (ii) T
   B. (i) T (ii) F
   C. (i) F (ii) T
   D. (i) F (ii) F

3. *Decide whether each of these statements is true (T) or false (F).* Figure 5.44 shows a ladder diagram rung for which:
   (i) When only input 1 contacts are activated, there is an output.
   (ii) When only input 2 contacts are activated, there is an output.
   A. (i) T (ii) T
   B. (i) T (ii) F
   C. (i) F (ii) T
   D. (i) F (ii) F

Input Output

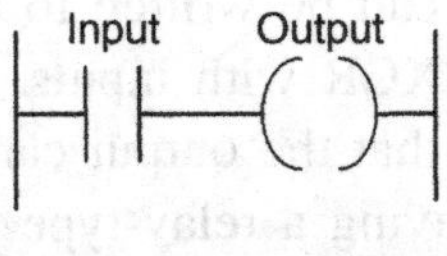

Figure 5.42: Diagram for Problem 1.

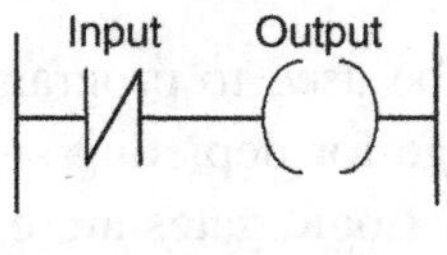

Input Output

Figure 5.43: Diagram for Problem 2.

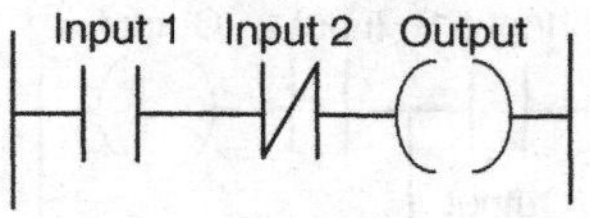

Input 1 Input 2 Output

Figure 5.44: Diagram for Problem 3.

4. *Decide whether each of these statements is true (T) or false (F).* Figure 5.45 shows a ladder diagram rung for which there is an output when:
   (i) Inputs 1 and 2 are both activated.
   (ii) Either input 1 or input 2 is activated.
   A. (i) T (ii) T
   B. (i) T (ii) F
   C. (i) F (ii) T
   D. (i) F (ii) F

5. *Decide whether each of these statements is true (T) or false (F).* Figure 5.46 shows a ladder diagram rung with an output when:
   (i) Inputs 1 and 2 are both activated.
   (ii) Input 1 or 2 is activated.
   A. (i) T (ii) T
   B. (i) T (ii) F
   C. (i) F (ii) T
   D. (i) F (ii) F

6. *Decide whether each of these statements is true (T) or false (F).* Figure 5.47 shows a ladder diagram rung for which there is an output when:
   (i) Input 1 is momentarily activated before reverting to its normally open state.
   (ii) Input 2 is activated.

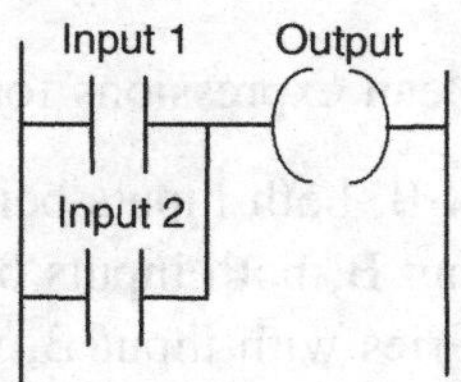

Figure 5.45: Diagram for Problem 4.

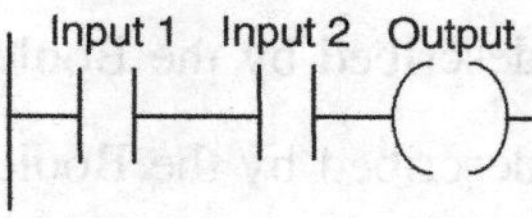

Input 1 Input 2 Output

Figure 5.46: Diagram for Problem 5.

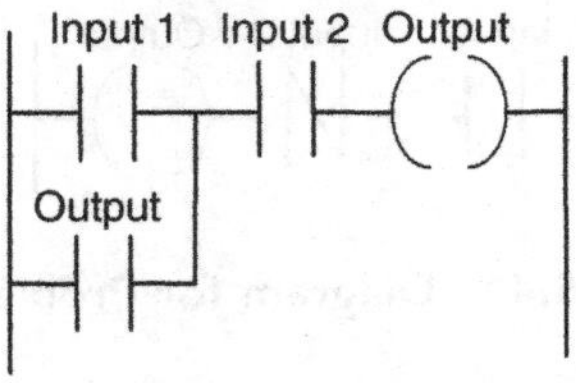

Figure 5.47: Diagram for Problem 6.

A. (i) T (ii) T
B. (i) T (ii) F
C. (i) F (ii) T
D. (i) F (ii) F

Problems 7 through 10 refer to the following logic gate systems:

A. AND
B. OR
C. NOR
D. NAND

7. Which form of logic gate system is given by a ladder diagram with a rung having two normally open sets of contacts in parallel?

8. Which form of logic gate system is given by a ladder diagram with a rung having two normally closed gates in parallel?

9. Which form of logic gate system is given by a ladder diagram with a rung having two normally closed gates in series?

10. Which form of logic gate system is given by a ladder diagram with a rung having two normally open gates in series?

Problems 11 through 14 concern Boolean expressions for inputs A and B.

A. Input A is in series with input B, both inputs being normally off.
B. Input A is in parallel with input B, both inputs being normally off.
C. Input A, normally off, is in series with input B, which is normally on.
D. Input A is in parallel with input B, both inputs being normally on.

11. Which arrangement of inputs is described by the Boolean relationship $A \cdot B$?

12. Which arrangement of inputs is described by the Boolean relationship $A + B$?

13. Which arrangement of inputs is described by the Boolean relationship $\bar{A} + \bar{B}$

14. Which arrangement of inputs is described by the Boolean relationship $A \cdot \bar{B}$

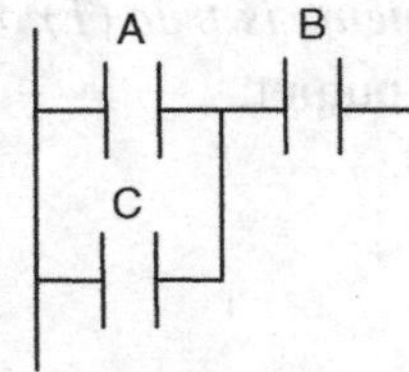

Figure 5.48: Diagram for Problem 15.

15. The arrangement of inputs in Figure 5.48 is described by the Boolean expression:
    A. A·B·C
    B. (A + C)·B
    C. (A + B)·C
    D. A·C + B

16. *Decide whether each of these statements is true (T) or false (F).* For the function block diagram in Figure 5.49, there is an output:
    (i) When A is 1.
    (ii) When B is 1.
    A. (i) T (ii) T
    B. (i) T (ii) F
    C. (i) F (ii) T
    D. (i) F (ii) F

17. *Decide whether each of these statements is true (T) or false (F).* For the function block diagram in Figure 5.50, there is an output:
    (i) When A is 1.
    (ii) When B is 1.
    A. (i) T (ii) T
    B. (i) T (ii) F
    C. (i) F (ii) T
    D. (i) F (ii) F

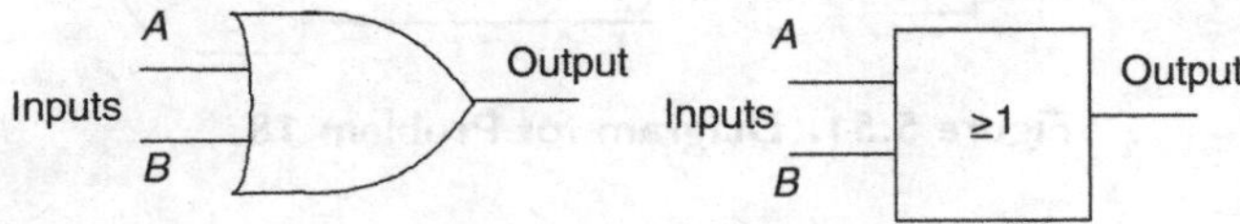

Figure 5.49: Diagram for Problem 16.

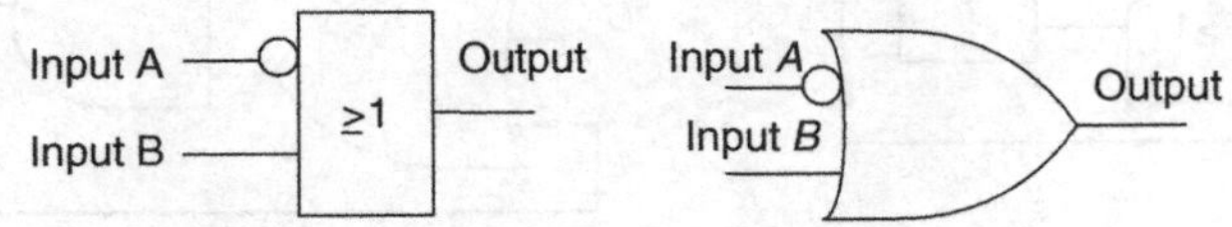

Figure 5.50: Diagram for Problem 17.

18. *Decide whether each of these statements is true (T) or false (F).* For the functional block diagram in Figure 5.51, there is an output:
   (i) When A is 1, B is 0 and C is 0.
   (ii) When A is 0, B is 1 and C is 1.
   A. (i) T (ii) T
   B. (i) T (ii) F
   C. (i) F (ii) T
   D. (i) F (ii) F

19. *Decide whether each of these statements is true (T) or false (F).* For the function block diagram in Figure 5.52, with A being a steady input condition and B a momentary input, there is an output:
   (i) When A is 1 and B is 0.
   (ii) When A is 0 and B is 1.
   A. (i) T (ii) T
   B. (i) T (ii) F
   C. (i) F (ii) T
   D. (i) F (ii) F

20. Figure 5.53 shows a ladder diagram. Which of the function block diagrams is its equivalent?

21. Figure 5.54 shows a function block diagram. Which of the ladder diagrams in Figure 5.54 is the equivalent?

22. Figure 5.55 shows a ladder diagram. Which of the diagrams showing inputs and output signals would occur with that ladder program?

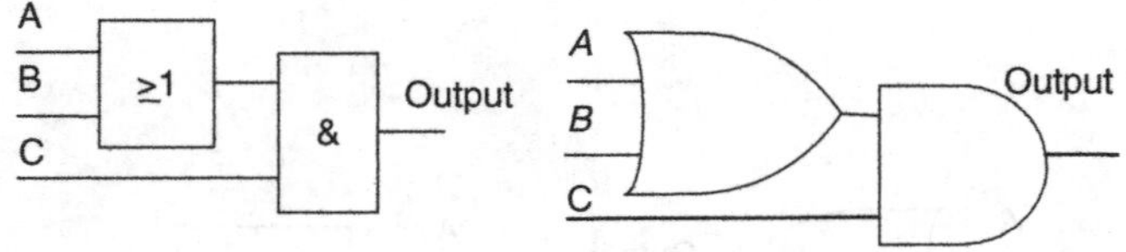

Figure 5.51: Diagram for Problem 18.

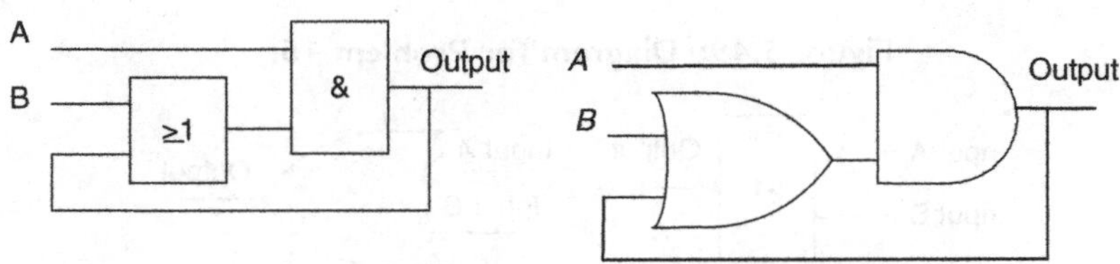

Figure 5.52: Diagram for Problem 19.

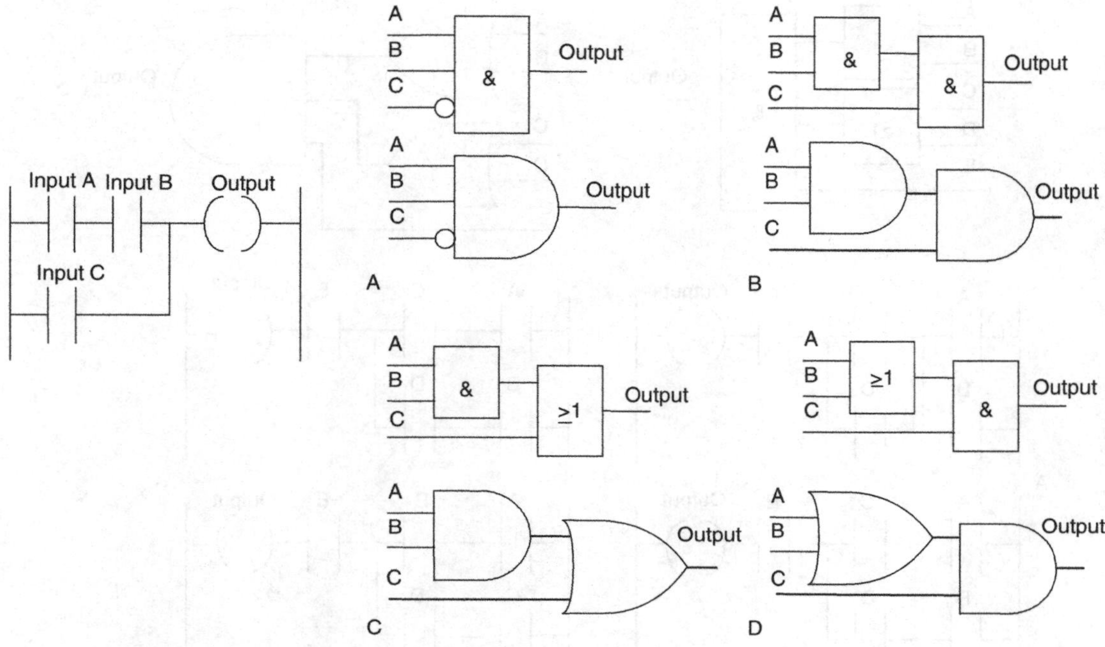

Figure 5.53: Diagram for Problem 20.

23. Figure 5.56 shows a ladder diagram. Which of the diagrams showing inputs and output signals would occur with that ladder program?

24. Figure 5.57 shows a ladder diagram. Which of the diagrams showing inputs and output signals would occur with that ladder program?

25. Draw the ladder rungs to represent:
    (a) Two switches are normally open and both have to be closed for a motor to operate.
    (b) Either of two, normally open, switches have to be closed for a coil to be energized and operate an actuator.
    (c) A motor is switched on by pressing a spring-return push-button start switch, and the motor remains on until another spring-return push-button stop switch is pressed.
    (d) A lamp is to be switched on if there is an input from sensor A or sensor B.
    (e) A light is to come on if there is no input to a sensor.
    (f) A solenoid valve is to be activated if sensor A gives an input.
    (g) For safety reasons, machines are often set up to ensure that a machine can only be operated by the operator pressing two switches simultaneously, one by the right hand and one by the left hand. This is to ensure that both the operator's hands will be on the switches and cannot be in the machine when it is operating. Draw a ladder rung for such a requirement.

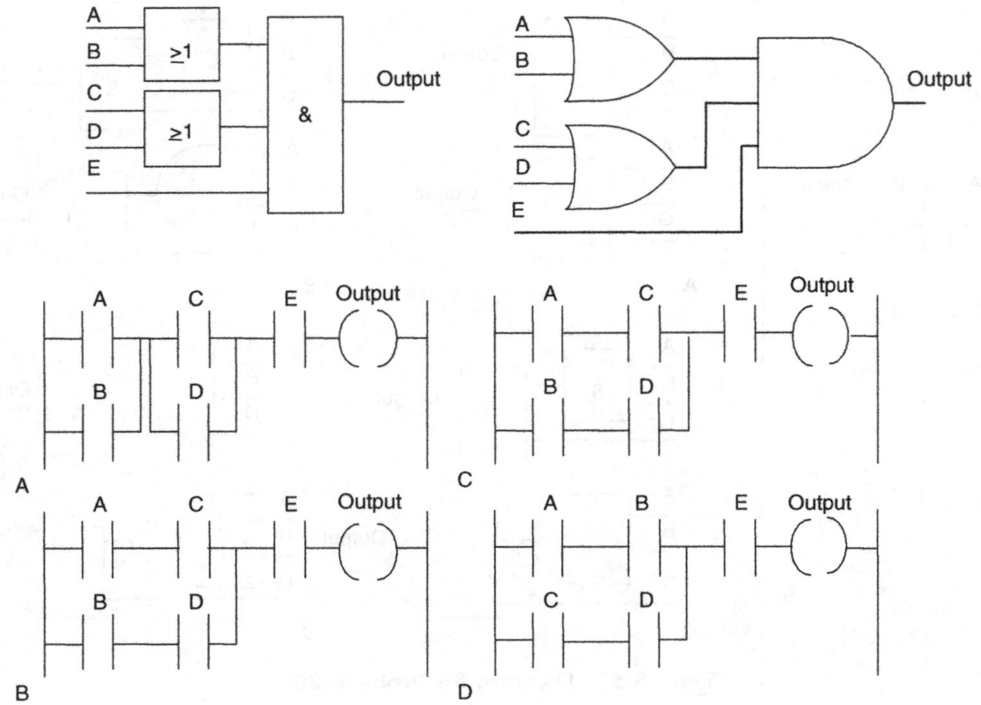

Figure 5.54: Diagram for Problem 21.

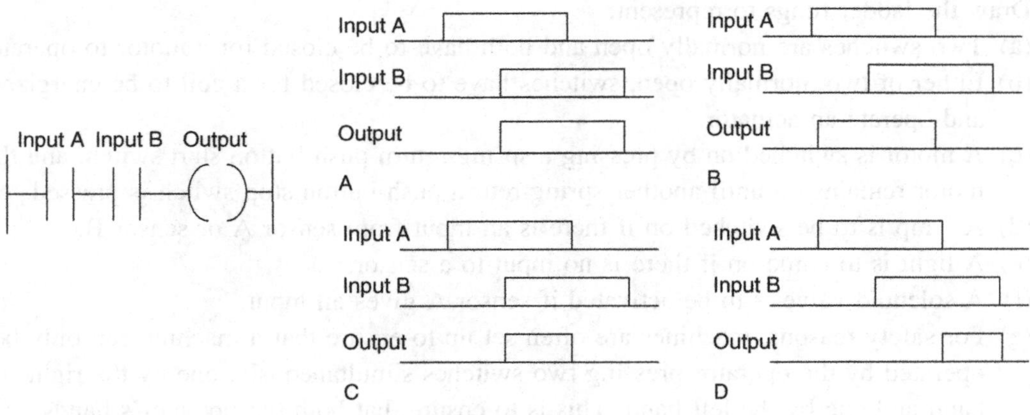

Figure 5.55: Diagram for Problem 22.

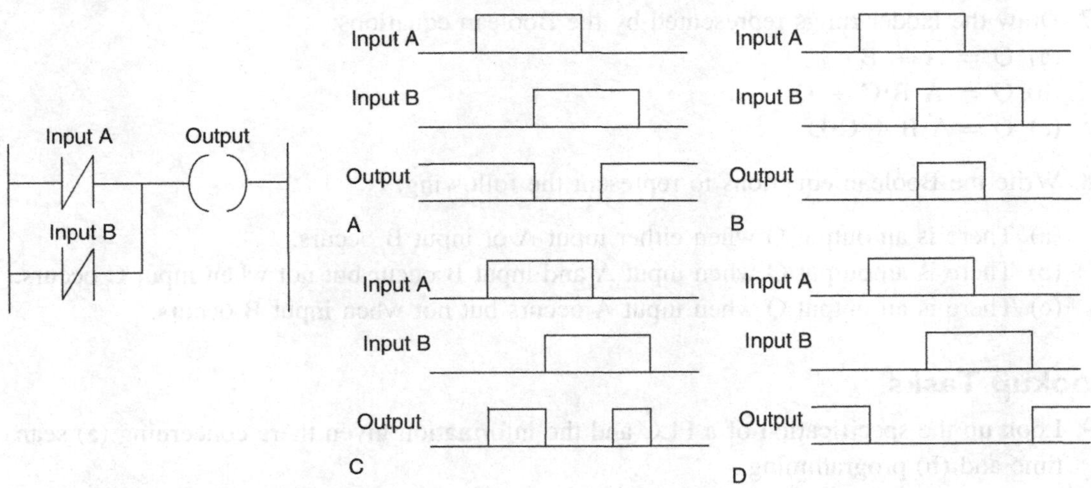

Figure 5.56: Diagram for Problem 23.

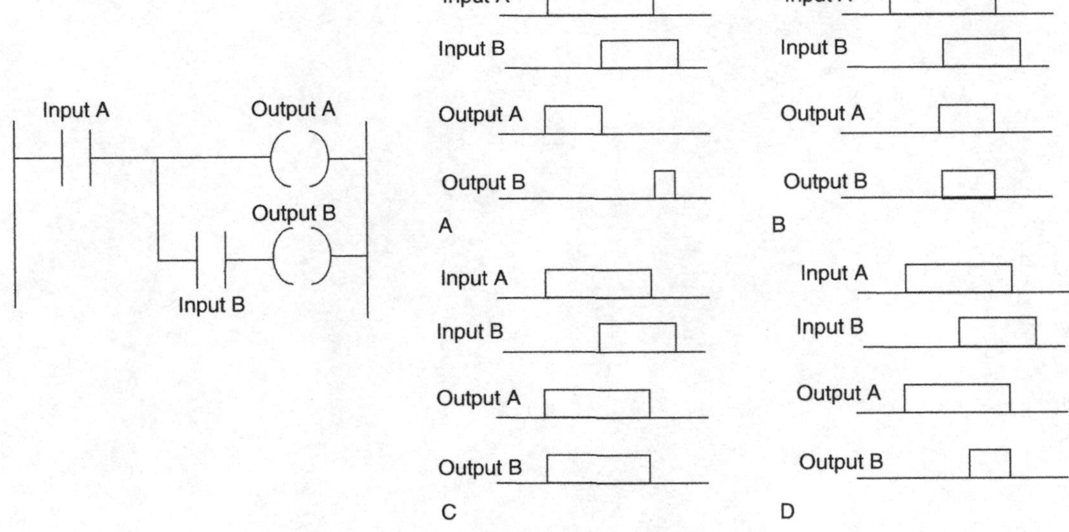

Figure 5.57: Diagram for Problem 24.

26. Draw the function block diagrams to represent:
    (a) There is to be a motor startup when either switch A or switch B is activated.
    (b) A motor is to be started when two normally open switches are activated and remain on, even if the first of the two switches goes off but not if the second switch goes off.
    (c) A pump is to be switched on if the pump start switch is on or a test switch is operated.

27. Draw the ladder rungs represented by the Boolean equations:
    (a) $Q = A + \bar{B}$
    (b) $Q = \bar{A} \cdot B \cdot C + D$
    (c) $Q = A \cdot B + C \cdot D$
28. Write the Boolean equations to represent the following:
    (a) There is an output Q when either input A or input B occurs.
    (b) There is an output Q when input A and input B occur but not when input C occurs.
    (c) There is an output Q when input A occurs but not when input B occurs.

Lookup Tasks

29. Look up the specification of a PLC and the information given there concerning (a) scan time and (b) programming.

CHAPTER 6

IL, SFC, and ST Programming Methods

This chapter continues from the previous chapter and discusses the other IEC 61131-3 programming languages, that is, instruction lists (ILs), sequential function charts (SFCs), and structured text (ST).

6.1 Instruction Lists

A programming method that can be considered to be the entering of a ladder program using text is the *instruction list* (IL). An instruction list gives programs as a series of instructions, with each instruction on a new line. Each instruction consists of an operator followed by one or more operands, that is, the subjects of the operator. Thus we might have:

```
LD   A
```

to indicate that the operand A is to be loaded, LD being the operator used to indicate loading. In terms of ladder diagrams, an operator may be regarded as a ladder element, and LD is equivalent to starting a rung with open contacts for input A. Another instruction might be:

```
OUT   Q
```

to indicate that there is to be an output to Q.

Mnemonic codes are used for operators, each code corresponding to an operator/ladder element. The codes used differ to some extent from manufacturer to manufacturer, though a standard under IEC 61131-3 has been proposed and is being widely adopted. Table 6.1 shows some of the codes used by manufacturers and the proposed standard for instructions used in this chapter (see later chapters for codes for other functions).

Instruction List is a low-level textual language that is simple to implement and used by a number of PLC manufacturers, mainly for small and medium-sized PLCs. It is particularly suitable for small, straightforward programs. Some manufactures do not support ILs but use only higher-level language of structured text (ST).

W. Bolton: Programmable Logic Controllers, Sixth Edition. http://dx.doi.org/10.1016/B978-0-12-802929-9.00006-6

Table 6.1: Instruction Code Mnemonics

IEC 61131-3	Mitsubishi	OMRON	Siemens	Operation	Ladder Diagram
LD	LD	LD	A	Load operand into result register.	Start a rung with open contacts.
LDN	LDI	LD NOT	AN	Load negative operand into result register.	Start a rung with closed contacts.
AND	AND	AND	A	Boolean AND.	Series element with open contacts.
ANDN	ANI	AND NOT	AN	Boolean AND with negative operand.	Series element with closed contacts.
OR	OR	OR	O	Boolean OR.	Parallel element with open contacts.
ORN	ORI	OR NOT	ON	Boolean OR with negative operand.	Parallel element with closed contacts.
ST	OUT	OUT	=	Store result register into operand.	An output.

As an illustration of the use of IEC 61131-3 operators, consider the following:

```
LD     A     (*Load A*)
AND    B     (*AND B*)
ST     Q     (*Store result in Q, i.e. output to Q*)
```

In the first line of the program, LD is the operator, A the operand, and the words at the ends of program lines and in parentheses shown and preceded and followed by * are comments added to explain what the operation is and are not part of the program operation instructions to the PLC. LD A is thus the instruction to load A into the memory register. It can then later be called on for further operations. The next line of the program has the Boolean operation AND performed with A and B. The last line has the result stored in Q, that is, output to Q.

Labels can be used to identify various entry points to a program, useful, as we will find later, for jumps in programs; these precede the instruction and are separated from it by a colon. Thus we might have:

```
PUMP_OK:    LD    C    (*Load C*)
```

with the instruction earlier in the program to jump to PUMP_OK if a particular condition is realized.

With the IEC 61131-3 operators, an N after the operator is used to negate its value. For example, if we have:

```
LD      A    (*Load A*)
ANDN    B    (*AND NOT B*)
```

the ANDN operator inverts the value of ladder contacts and ANDs the result.

6.1.1 Ladder Programs and Instruction Lists

When looked at in terms of ladder diagrams, whenever a rung is started, it must use a "start a rung" code. This might be LD, or perhaps A or L, to indicate that the rung is starting with open contacts, or LDI, or perhaps LDN, LD NOT, AN, or LN, to indicate it is starting with closed contacts. All rungs must end with an output or store result code. This might be OUT or = or ST. The following shows how individual rungs on a ladder are entered using the Mitsubishi mnemonics for the AND gate, shown in Figure 6.1a.

The rung starts with LD because it is starting with open contacts. For Figure 6.1a, since the address of the input is X400, the instruction is LD X400. This is followed by another open contacts input, and so the next program line involves the instruction AND with the address of the element; thus the instruction is AND X401. The rung terminates with an output, so the instruction OUT is used with the address of the output, that is, OUT Y430. The single rung of a ladder would thus be entered as:

```
LD    X400
AND   X401
OUT   Y430
```

For the same rung with Siemens notation (Figure 6.1b), we have:

```
A    I0.1
A    I0.2
=    Q2.0
```

Consider another example: an OR gate. Figure 6.2a shows the gate with Mitsubishi notation.

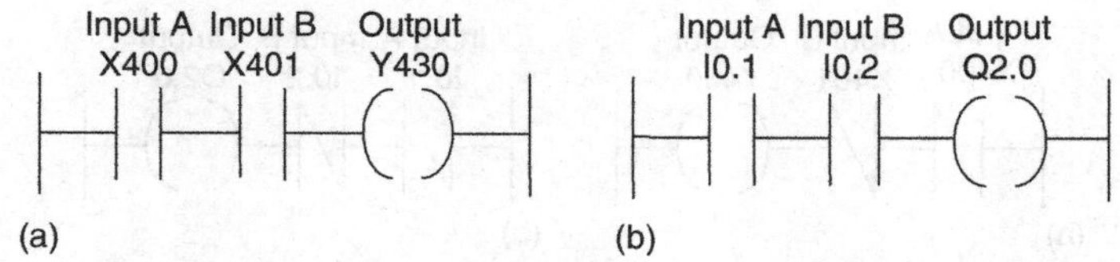

Figure 6.1: AND gate: (a) Mitsubishi, and (b) Siemens.

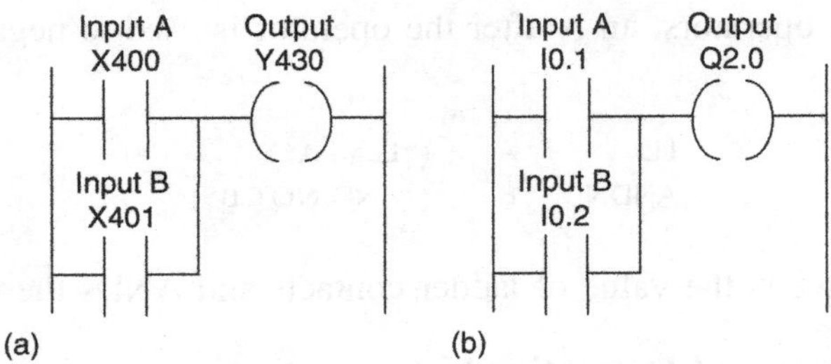

Figure 6.2: OR gate: (a) Mitsubishi, and (b) Siemens.

The instruction for the rung in Figure 6.2a starts with an open contact and is LD X400. The next item is the parallel OR set of contacts X401. Thus the next instruction is OR X401. The last step is the output, hence OUT Y430. The instruction list would thus be:

```
LD    X400
OR    X401
OUT   Y430
```

Figure 6.2b shows the Siemens version of the OR gate. The following is the Siemens instruction list:

```
A     I0.1
O     I0.2
=     Q2.0
```

Figure 6.3a shows the ladder system for a NOR gate in Mitsubishi notation.

The rung in Figure 6.3a starts with normally closed contacts, so the instruction is LDI. When added to Mitsubishi instruction, I is used to indicate the inverse of the instruction. The next step is a series of normally closed contacts and so the instruction is ANI, again the I being used to make an AND instruction the inverse. I is also the instruction for a NOT gate. The instructions for the NOR gate rung of the ladder would thus be entered as:

```
LDI   X400
ANI   X401
OUT   Y430
```

Input A X400 Input B X401 Output Y430

(a)

Input A I0.1 Input B I0.2 Output Q2.0

(b)

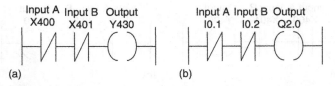

Figure 6.3: NOR gate: (a) Mitsubishi, and (b) Siemens.

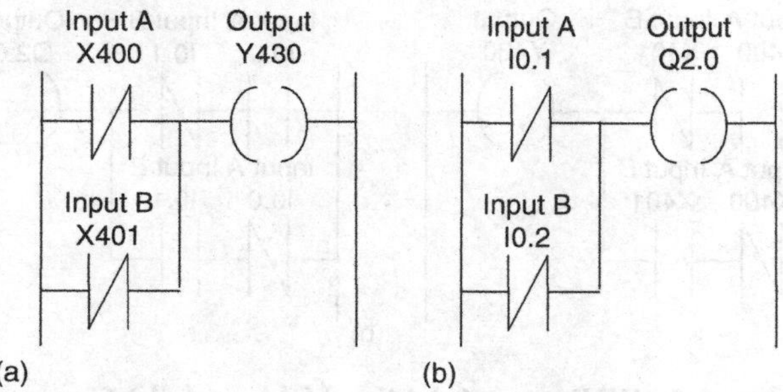

Figure 6.4: NAND gate: (a) Mitsubishi, and (b) Siemens.

Figure 6.3b shows the NOR gate with Siemens notation. Note that N added to an instruction is used to make the inverse. The instruction list then becomes:

```
LN   I0.1
AN   I0.2
=    Q2.0
```

Consider the rung shown in Figure 6.4a in Mitsubishi notation, a NAND gate.

Figure 6.4a starts with the normally closed contacts X400 and so starts with the instruction LDI X400. The next instruction is for a parallel set of normally closed contacts; thus the instruction is ORI X401. The last step is the output, hence OUT Y430. The instruction list is thus:

```
LDI   X400
ORI   X401
OUT   Y430
```

Figure 6.4b shows the NAND gate in Siemens notation. The instruction list is then:

```
AN   I0.1
ON   I0.2
=    Q2.0
```

6.1.2 Branch Codes

The EXCLUSIVE OR (XOR) gate shown in Figure 6.5 has two parallel arms with an AND situation in each arm.

Figure 6.5a shows Mitsubishi notation. With such a situation, Mitsubishi uses an ORB instruction to indicate "OR together parallel branches." The first instruction is for a normally open pair of contacts X400. The next instruction is for a series set of normally closed contacts X401, hence ANI X401. After reading the first two instructions, the third instruction starts a new line. It is recognized as a new line because it starts with LDI, all

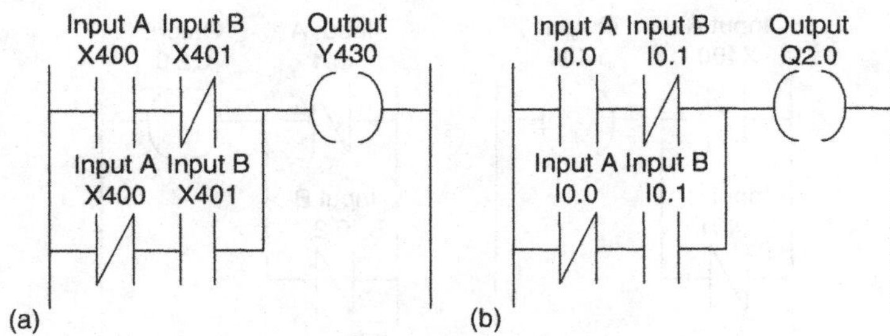

Figure 6.5: XOR gate: (a) Mitsubishi, and (b) Siemens.

new lines starting with LD or LDI. But the first line has not been ended by an output. The PLC thus recognizes that a parallel line is involved for the second line and reads together the listed elements until the ORB instruction is reached. The mnemonic ORB (OR branches/blocks together) indicates to the PLC that it should OR the results of the first and second instructions with that of the new branch with the third and fourth instructions. The list concludes with the output OUT Y430. The instruction list would thus be entered as:

```
LD    X400
ANI   X401
LDI   X400
AND   X401
ORB
OUT   Y430
```

Figure 6.5b shows the Siemens version of an XOR gate. Brackets are used to indicate that certain instructions are to be carried out as a block. They are used in the same way as brackets in any mathematical equation. For example, (2 + 3) / 4 means that the 2 and 3 must be added before dividing by 4. Thus with the Siemens instruction list we have in step 0 the instruction A(. The brackets close in step 3. This means that the A in step 0 is applied only after the instructions in steps 1 and 2 have been applied.

Step	Instruction	
0	A(	
1	A	I0.0
2	AN	I0.1
3	)	
4	O(	
5	AN	I0.0
6	A	I0.1
7	)	
8	=	Q2.0

The IEC 61131-3 standard for such programming is to use brackets in the way used in the previous Siemens example, that is, in the same way brackets are used in normal arithmetic. This enables instructions contained within brackets to be deferred until the bracket is completed. Thus the IEC instruction list program:

```
LD X
ADD( B
MUL( C
ADD D
)
)
```

Gives X + (B × (C + D)).

Figure 6.6 shows a circuit that can be considered as two branched AND blocks. Figure 6.6a shows the circuit in Mitsubishi notation. The instruction used here is ANB. The instruction list is thus:

Step	Instruction	
0	LD	X400
1	OR	X402
2	LD	X401
3	OR	X403
4	ANB	
5	OUT	Y430

Figure 6.6b shows the same circuit in Siemens notation. Such a program is written as an instruction list using brackets. The A instruction in step 0 applies to the result of steps 1 and 2. The A instruction in step 4 applies to the result of steps 5 and 6. The program instruction list is thus:

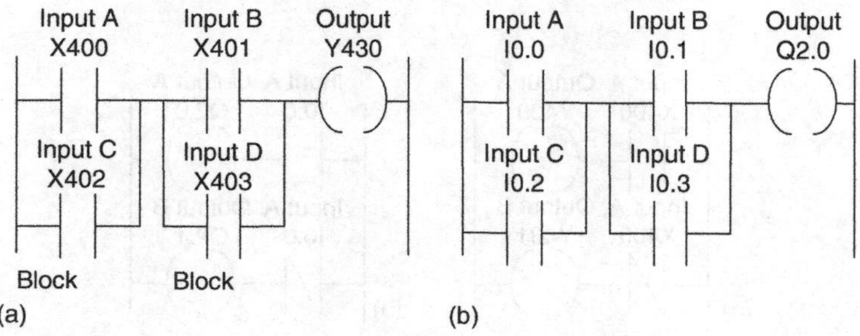

Figure 6.6: Two branched AND gates: (a) Mitsubishi, and (b) Siemens.

Step	Instruction	
0	A(	
1	A	I0.0
2	O	I0.2
3	)	
4	A(	
5	A	I0.1
6	O	I0.3
7	)	
8	=	Q2.0

6.1.3 More Than One Rung

Figure 6.7a shows a ladder, in Mitsubishi notation, with two rungs. In writing the instruction list we just write the instructions for each line in turn. The instruction LD or LDI indicates to the PLC that a new rung is starting. The instruction list is thus:

```
LD     X400
OUT    Y430
LDI    X400
OUT    Y431
```

The system is one where when X400 is not activated, there is an output from Y431 but not Y430. When X400 is activated, there is then an output from Y430 but not Y431.

Figure 6.7b shows the same program in Siemens notation. The = instruction indicates the end of a line. The A or AN instruction does not necessarily indicate the beginning of a rung since the same instruction is used for AND and AND NOT. The instruction list is then:

```
A      I0.0
=      Q2.0
AN     I0.0
=      Q2.1
```

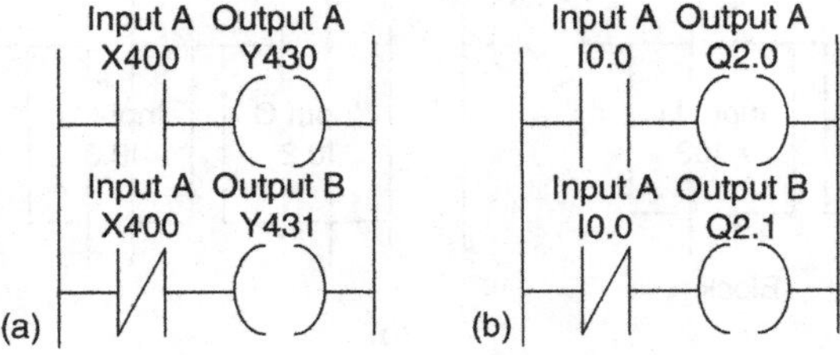

Figure 6.7: Toggle circuit: (a) Mitsubishi, and (b) Siemens.

6.1.4 Programming Examples

The following tasks are intended to illustrate the application of the programming techniques given in this section and are the examples for which ladder diagrams and function block diagrams were derived in Section 5.7. (See that section for an explanation of the ladder diagrams; here we show the instruction lists relating to the programs.)

A signal lamp is required to be switched on if a pump is running and the pressure is satisfactory or if the lamp test switch is closed. Figure 6.8 shows the ladder program and the related instruction list.

For a valve that is to be operated to lift a load when a pump is running and either the lift switch operated or a switch operated indicating that the load has not already been lifted and is at the bottom of its lift channel, Figure 6.9 shows the ladder program and the related instruction list.

For a system in which there has to be no output when any one of four sensors gives an output and otherwise there is to be an output, Figure 6.10 shows the ladder program and the instruction list.

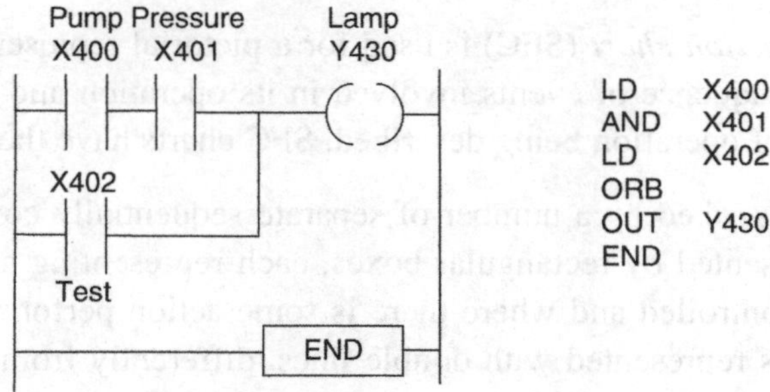

Figure 6.8: Signal lamp task.

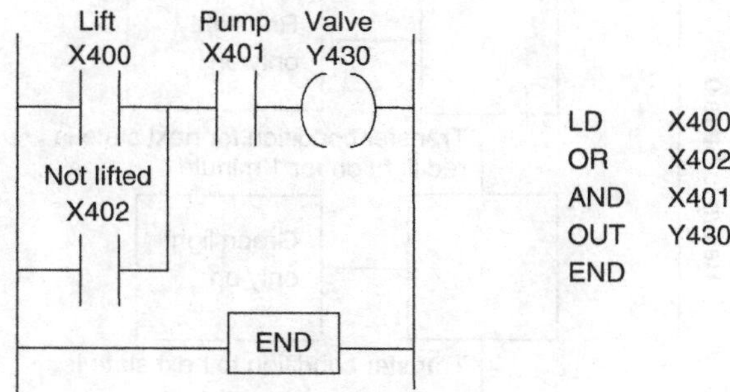

Figure 6.9: Valve operation program.

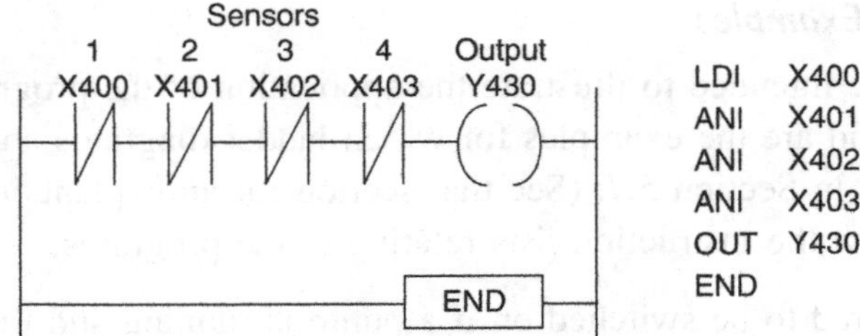

Figure 6.10: Output switched off by any one of four sensors being activated.

6.2 Sequential Function Charts

If we wanted to describe a traffic lamp sequence of red-green, one way we could do this would be to represent it as a sequence of functions or states such as red light state and green light state and the inputs and outputs to each state. Figure 6.11 illustrates this. State 0 has an input that is triggered after the green light has been on for 1 minute and an output of red light on, i.e. the transfer condition from the red light is a time of 1 minute. State 1 has an input that is triggered after the red light has been on for 1 minute and an output of green light on, i.e. the transfer condition from the green light is a time of 1 minute. When the green light has been on for 1 minute, there is a transfer back to State 0.

The term *sequential function chart* (SFC) is used for a pictorial representation of a system's operation to show the sequence of events involved in its operation and Figure 6.11 is an illustration of the type of operation being described. SFC charts have the following elements:

1. The operation is described by a number of separate sequentially connected states or steps that are represented by rectangular boxes, each representing a particular state of the system being controlled and where there is some action performed. The initial start step in a program is represented with double lines, differently from the other steps. Figure 6.12 shows a start step and later steps.

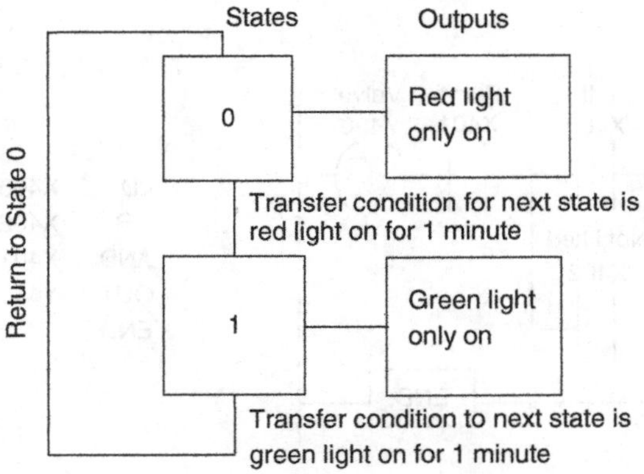

Figure 6.11: Sequence for traffic lights.

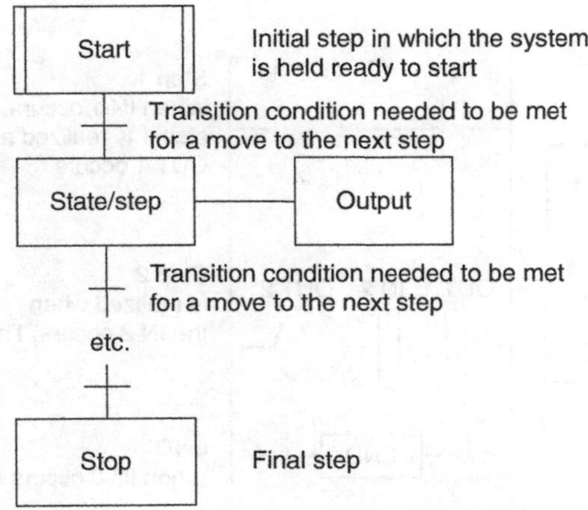

Figure 6.12: A state/step and its transition.

2. Each connecting line between states has a horizontal bar representing the transition condition that has to be realized before the system can move from one state to the next. Two steps can never be directly connected; they must always be separated by a transition. Two transitions can never directly follow from one to another; they must always be separated by a step.
3. The program checks the transition conditions so that when realized the next state following the transition is moved to.
4. The process thus continues from one state to the next until the complete machine cycle is completed.
5. Outputs/actions at any state/step are represented by horizontally linked boxes and occur when that state has been realized. Thus we might have items such as Wait 1 minute or Open valve 1 or Close valve 1.

As an illustration, Figure 6.13 shows part of an SFC and its equivalent ladder diagram. The program starts when IN 1 is realized and this gives step 1 output. When this output has completed, the next transition occurs and step 2 occurs with the resulting output 2. When this is completed, the transfer condition occurs that allows for the move to the End step.

As an illustration of the principles of SFC, consider the situation with, say, part of the washing cycle of a domestic washing machine where the drum is to be filled with water, and then when the drum is full, a heater has to be switched on and remain on until the temperature reaches the required level. Then the drum is to be rotated for a specified time. We have a sequence of states that can be represented in the manner shown in Figure 6.14.

The transition defines the conditions that must occur in order to go to the next step in the program, and so they have to occur between each step in a program. Until the conditions are

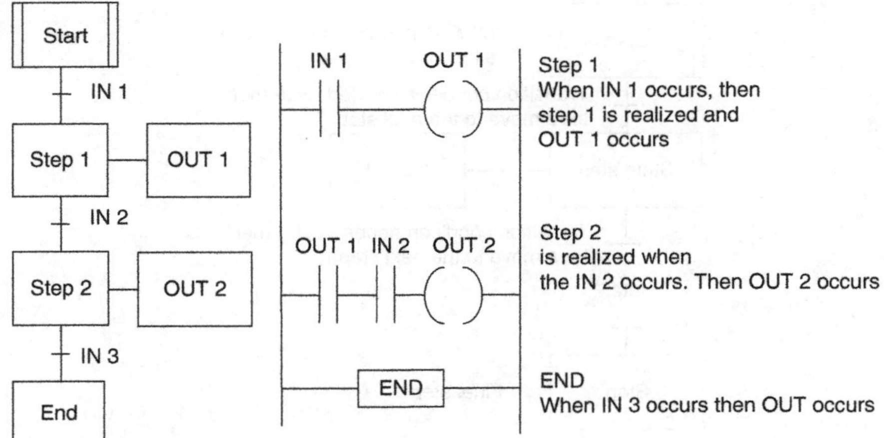

Figure 6.13: SFC and equivalent ladder diagram.

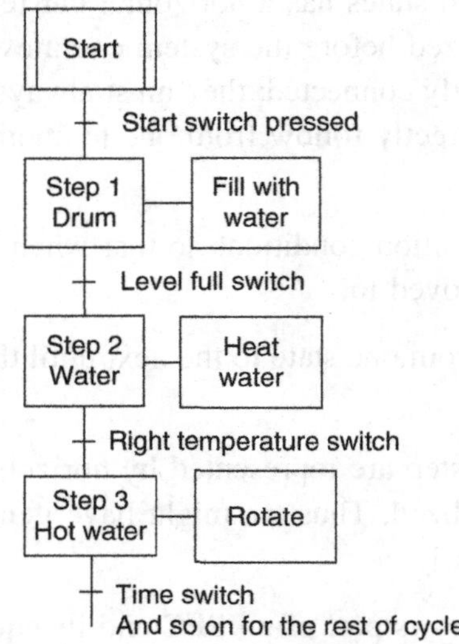

Figure 6.14: Washing machine.

realized the program continues to execute the current step. Thus we might have for a transition:

If true go to the next step. If false continue the step above.

Other possibilities are conditions such as:

Temperature > 50
Pump = On
Valve_Open AND Compressor_On
Product = 10

that have to be realized before progression can occur to the next step in the program. When programming, transitions can be entered using a Boolean expression in structured text (see Chapter 6.3) to check whether a condition is true 1 or false 0. Alternatively a subroutine can be called up to check for the required condition and when realized give a true 1 response, otherwise false 0. At the end of the subroutine an End of Transition (EOT) instruction can then be used to set the state of the transition to the Boolean value realized by the subroutine.

Actions are added to steps to indicate the different functions that the step performs. Thus we might have switching a motor on or perhaps opening a valve or calling up a subroutine.

The Sequential Function Chart language is a powerful graphical technique for describing the sequential behavior of a program. Graphical languages have been used for a number of years, Grafset being a European graphical language. The IEC 61131-1 standard resembles many of the features of Grafset.

6.2.1 Branching and Convergence

Selective branching is illustrated in Figure 6.15 and allows for different states to be realized, depending on the transfer condition that occurs. Transitions are needed below the horizontal lines in order to indicate the conditions that have to be met to progress from the step above the horizontal line to a step below it.

Parallel branching (Figure 6.16), represented by a pair of horizontal lines, allows for two or more different states to be realized and proceed simultaneously. A transition is required outside the pair of horizontal lines in order to indicate the condition that has to be met in order for the group of steps that follow to be simultaneously realized.

Figures 6.17 and 6.18 show how convergence is represented by an SFC. In Figure 6.17 the sequence can go from state 2 to state 4 if IN 4 occurs or from state 3 to state 4

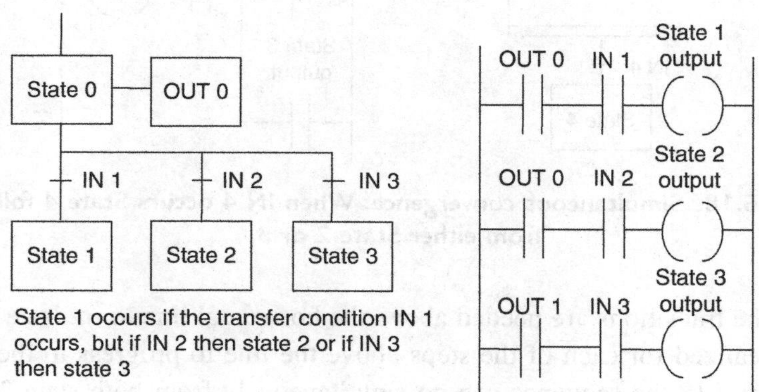

Figure 6.15: Selective branching: The state that follows State 0 will depend on whether transition IN1, IN2, or IN3 occurs.

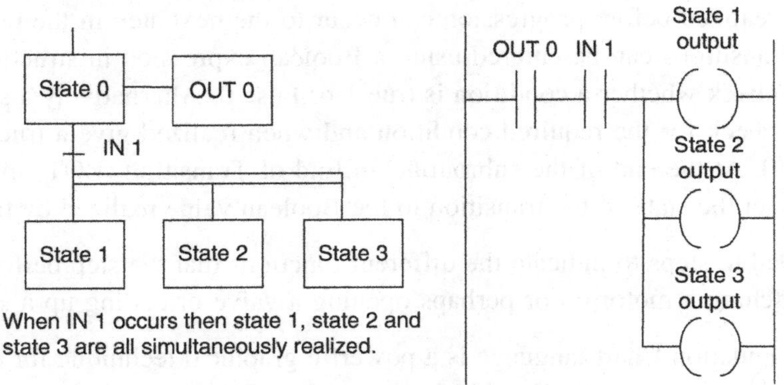

Figure 6.16: Parallel branching states 1, 2, and 3 occur simultaneously when transition IN 1 occurs.

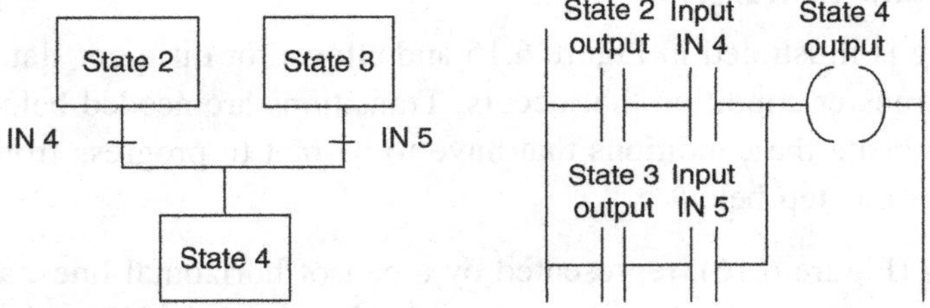

Figure 6.17: Convergence: State 4 follows when either IN 4 or IN5 occurs.

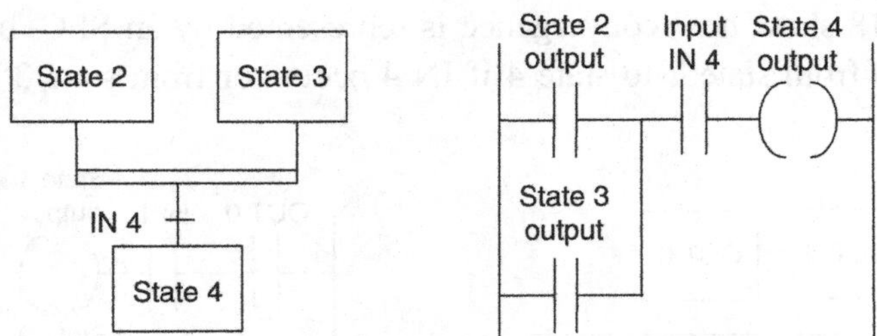

Figure 6.18: Simultaneous convergence: When IN 4 occurs State 4 follows from either State 2 or 3.

if IN 5 occurs. The transitions are needed above the horizontal line to indicate the conditions that have to be realized for each of the steps above the line to progress to the step below the line. In Figure 6.18 the sequence can go simultaneously from both state 2 and state 3 to state 4 if IN 4 occurs. The transition is below the pair of horizontal lines and indicates the condition that has to be realized for progression to the step that follows.

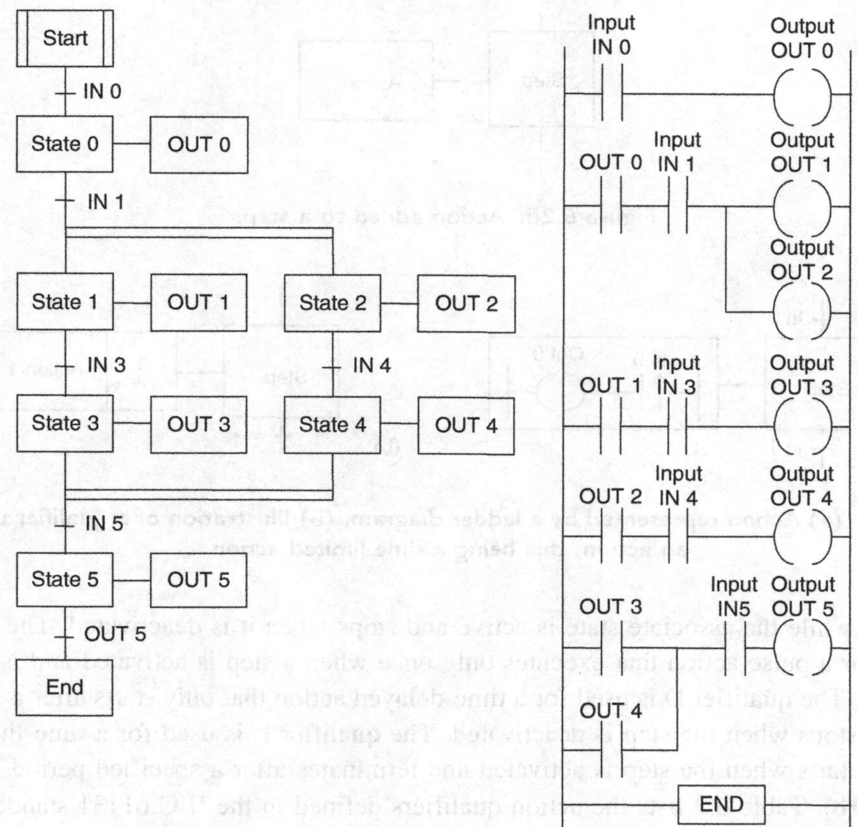

Figure 6.19: Part of an SFC and its equivalent ladder program.

As an illustration of the use of the preceding, Figure 6.19 shows part of a program represented by both its SFC and ladder programs.

6.2.2 Actions

With steps, there is an action or actions that have to be performed. Such actions, such as the outputs in the preceding example, are depicted as rectangular boxes attached to the state (Figure 6.20). The behavior of the action can be given using a ladder diagram, a function block diagram, an instruction list, or structured text. Thus, where a ladder diagram is used, the behavior of the action is shown by the ladder diagram being enclosed within the action box. The action is then activated when there is a power flow into the action box. Figure 6.21a illustrates this concept.

Action boxes can be preceded by qualifiers to specify the conditions to exist for the action. In the absence of a qualifier or the qualifier N, the action is not stored and is executed

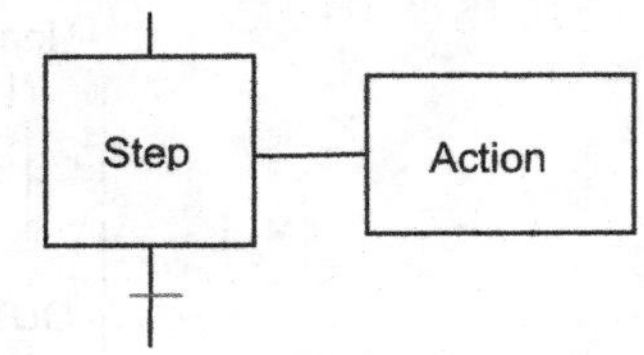

Figure 6.20: Action added to a step.

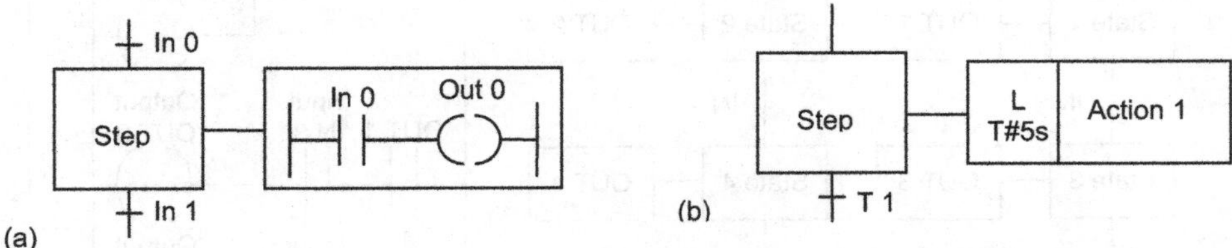

Figure 6.21: (a) Action represented by a ladder diagram, (b) Illustration of a qualifier used with an action, this being a time-limited action.

continually while the associate state is active and stops when it is deactivated. The qualifier P is used for a pulse action that executes only once when a step is activated and is then deactivated. The qualifier D is used for a time-delayed action that only starts after a specified period and stops when the step is deactivated. The qualifier L is used for a time-limited action that starts when the step is activated and terminates after a specified period (Figure 6.21b). Table 6.2 lists the action qualifiers defined in the IEC 61131 standard.

Table 6.2: Action Qualifiers

Qualifier	Description
None	Non-stored, the default, same as N.
N	Non-stored, executes while the associated step is active and then stops.
R	Resets a stored action.
S	Stored. Stays active until a Reset action turns off this action.
L	Time limited action. Terminates after a given time. A time period must be specified.
D	Time delayed action. Starts after a given time which must be specified.
P	A pulse action that occurs once when the step is activated and once when it is deactivated.
P1	A pulse action that only occurs when a step is activated.
P0	A pulse action that only occurs when a step is deactivated.
SD	The action starts after a given time, even if the associated step is deactivated before that time has elapsed.
DS	The action starts a specific time after the step is activated and the step is still active and stops when the step is deactivated.
SL	The action starts when the step is activated and is time limited, executing for a given period.

When a transition condition is realized and one step is left and the next started, it is necessary to consider what state the action of that step needs to be left in. There may be a need to turn off an activated device that the step had turned on. This can be done in programming the action for that device to carry out such a task. If it is to be returned to its initial state then program it to go to the initial step position.

6.2.3 Programming a PLC

Software is supplied by PLC manufacturers to enable programs to be entered for use by a PLC. As an indication, the following outline is the procedure that would be used with Allen-Bradley software to program a sequential function chart. With the SFC toolbar on the screen, the button for the item wanted is clicked and then dragged to the required location on the SFC chart. Thus the element for a step followed by a transition might be so dragged. To connect two elements, for example two steps, together, a pin on one of the elements is clicked and then the pin on the other element is clicked. A valid connection point is shown by a green dot appearing. A simultaneous button on the toolbar can be clicked and dragged to add a simultaneous branch to the program, likewise a selection branch. To program a transition, the text area of the transition icon is double clicked and if a Boolean expression is required it is just typed in. To call a subroutine, the transition icon is right clicked, Set JSR selected and then the routine selected, then OK clicked. To add an action to a step, the step is right clicked and Add Action selected. The button in the action element is clicked to a series of tabs which are used to specify the qualifier for the action, e.g. L for a time qualifier, the action order and a tag. The text entry of the action element is right clicked and Set JSR selected for subroutines to be added.

6.3 Structured Text

Structured text is a programming language that strongly resembles the programming language Pascal. Programs are written as a series of statements separated by semicolons. The statements use predefined statements and subroutines to change variables, these being defined values, internally stored values, or inputs and outputs.

Assignment statements are used to indicate how the value of a variable is to be changed, for example

```
Light: = SwitchA;
```

is used to indicate that a light, the variable, is to have its "value" changed, that is, switched on or off, when switch A changes its "value," that is, is on or off. The general format of an assignment statement is:

```
X:= Y;
```

where Y represents an expression that produces a new value for the variable X and := is the assignment symbol. The variable retains the assigned value until another assignment changes the value. Other examples are:

```
Light:– SwitchA OR SwitchB;
```

to indicate that the light is switched on by either switch A OR switch B. Using the AND function, we might have:

```
Start:– Steam AND Pump;
```

to indicate that start occurs when steam AND the pump are on.

Table 6.3 shows some of the operators, such as the OR and AND in the preceding statements, that are used in structured text programs and their relative precedence when an expression is being evaluated. Parentheses (brackets) are used to group expressions within expressions to ensure that the expression is executed in the required sequence. For example:

```
InputA := 6;
InputB := 4;
InputC := 2;
OutputQ := InputA/3 + InputB/(3 - InputC);
```

Table 6.3: Structured Text Operators

Operator	Description	Precedence
(...)	Parenthesized (bracketed) expression	Highest
Function(...)	List of parameters of a function	
**	Raising to a power	
-, NOT	Negation, Boolean NOT	
*, /, MOD	Multiplication, division, modulus operation	
+, -	Addition, subtraction	
<, >, <=, >=	Less than, greater than, less than or equal to, greater than or equal to	
=, <>	Equality, inequality	
AND, &	Boolean AND	
XOR	Boolean XOR	
OR	Boolean OR	Lowest

has (3 – InputC) evaluated before its value is used as a divisor, so the second part of the OutputQ statement is 4/(3 – 2) = 4. Division has precedence over addition, so the first part of the statement is evaluated before the addition, that is, 6/3. So we have for OutputQ the value 2 + 4 = 6.

Structured text is not case sensitive; thus lowercase or capital letters can be used as is felt necessary to aid clarity. Likewise, spaces are not necessary but can be used to aid clarity; likewise indenting lines. All the identities of directly represented variables start with the % character and are followed by a one- or two-letter code to identify whether the memory location is associated with inputs, outputs, or internal memory and whether it is bits, bytes, or words, such as

```
%IX100 (*Input memory bit 100*)

%ID200 (*Input memory word 200*)

%QX100 (*Output memory bit 100*)
```

The first letter is I for input memory location, Q for output memory location, or M for internal memory. The second letter is X for bit, B for byte (8 bit), W for word (16 bits), D for double word (32 bits), or L for long word (64 bits).

AT is used to fix the memory location for a variable. Thus we might have:

```
Input1 AT %IX100; (*Input1 is located at input memory bit 100*)
```

6.3.1 Conditional Statements

The IF statement:

```
IF fluid_temp THEN
```

is used to indicate that if the fluid temp variable is ON, that is, 1, the actions following that line in the program are to occur. The IF statement:

```
IF NOT fluid_temp THEN
```

is used to indicate that if the fluid temp variable is NOT 1, the actions following that line in the program are to occur. The IF statement:

```
IF fluid_temp1 OR fluid_temp2 THEN
```

is used to indicate that if the fluid temp variable 1, the fluid temp variable 2 is ON, that is, 1, the actions following that line in the program are to occur.

IF ... THEN ... ELSE is used when selected statements are to be executed when certain conditions occur. For example:

```
IF (Limit_switch1 AND Workpiece_Present) THEN
    Gate1 := Open;
    Gate2 := Close;
ELSE
    Gate1 := Close;
    Gate2 := Open;
END_IF;
```

Note that the end of the IF statement has to be indicated. Another example, using PLC addresses, is:

```
IF (I:000/00 = 1) THEN
    O:001/00 := 1;
ELSE
    O:000/01 := 0;
END_IF;
```

So, if there is an input to I:000/00 to make it 1, output O:001/00 is 1; otherwise it's 0.

CASE is used to give the condition that selected statements are to be executed if a particular integer value occurs *else* some other selected statements. For example, for temperature control we might have:

```
CASE (Temperature) OF
    0 ... 40 :      Furnace_switch := On;
    40 ... 100:     Furnace_switch := Off;
ELSE
                    Furnace_switch := Off;
END_CASE;
```

Note, as with all conditional statements, the end of the CASE statement has to be indicated. Another example might be, for the operation of a motor with fans being required to operate at different speeds based on the operation of particular switch positions:

```
CASE speed_setting OF
     1:        speed := 5;
     2:        speed := 10;
     3:        speed := 15; fan 1 := ON;
     4:        speed := 20; fan 2 := ON;
ELSE
               Speed :=0; speed fault := TRUE;
END_CASE
```

6.3.2 Iteration Statements

These are used where it is necessary to repeat one or more statements a number of times, depending on the state of some variable. The FOR ... DO iteration statement allows a set of statements to be repeated depending on the value of the iteration integer variable. For example:

```
FOR Input := 10 to 0 BY -1
    DO
    Output := Input;
END_FOR;
```

has the output decreasing by 1 each time the input, dropping from 10 to 0, decreasing by 1.

WHILE ... DO allows one or more statements to be executed while a particular Boolean expression remains true, such as:

```
OutputQ := 0;
WHILE InputA AND InputB
    DO
    OutputQ =: OutputQ + 1;
END_WHILE;
```

REPEAT ... UNTIL allows one or more statements to be executed and repeated while a particular Boolean expression remains true.

```
OutputQ := 0
REPEAT
    OutputQ := OutputQ + 1;
UNTIL (Input1 = Off) OR (OutputQ > 5)
END_REPEAT;
```

6.3.3 Structured Text Programs

Programs have first to define the data types required to represent data, such as:

```
TYPE Motor: (Stopped, Running);
END_TYPE;

TYPE Valve: (Open, shut);
END_TYPE;
TYPE Pressure: REAL; (*The pressure is an analogue value*)
END_TYPE;
```

the variables, that is, signals from sensors and output signals to be used in a program, such as:

```
VAR_IN (*Inputs*)
    PumpFault : BOOL; (*Pump operating fault is a Boolean variable*)
END_VAR;
VAR_OUT (*Outputs*)
    Motor_speed : REAL;
END_VAR;

VAR_IN
    Value: INT; (*The value is an integer*)
END_VAR;
```

```
VAR
    Input1 AT %IX100; (*Input1 is located at input memory bit 100*)
END_VAR;
```

and any initial values to be given to variables, such as:

```
VAR
    Temp : REAL =100; (*Initial value is an analogue number 100*)
END_VAR;
```

before getting down to the instruction statements.

The following is an example of a function block that might appear in a larger program and is concerned with testing voltages:

```
FUNCTION_BLOCK TEST_VOLTAGE
    VAR_INPUT
      VOLTS1, VOLTS2, VOLTS3
    END_VAR
    VAR_OUTPUT
      OVERVOLTS : BOOL;
    END_VAR
    IF VOLTS1 > 12 THEN
      OVERVOLTS :=TRUE; RETURN;
    END_IF;
    IF VOLTS2 > 12 THEN
      OVERVOLTS :=TRUE; RETURN;
    END_IF;
    IF VOLTS3 > 12 THEN
      OVERVOLTS :=TRUE;
    END_IF;
END_FUNCTION_BLOCK;
```

Sensor 1 Valve 1

```
Valve_1 := Sensor_1

IF Sensor_1 THEN
    Valve_1 := 1;
END_IF
```

Figure 6.22: A ladder program rung and two alternative STC equivalents.

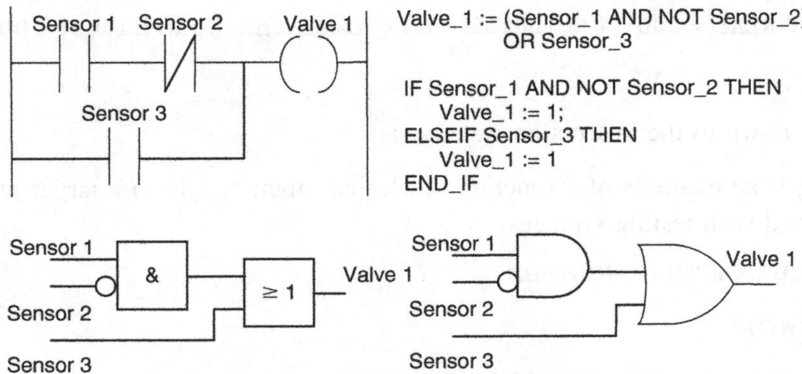

Figure 6.23: A ladder program rung, its function box equivalent, and two STC equivalents.

If the testing of volts 1, volts 2, or volts 3 indicates that any one of them is more than 12, the output OVERVOLTS is set to true and the RETURN statement called to terminate the execution of the function block. In the rest of the program, when OVERVOLTS is set to true, the program will initiate some action.

6.3.4 Comparison with Ladder Programs

Figure 6.22 shows a ladder rung and its equivalent expressions in structured text; Figure 6.23 shows another ladder rung and equivalents in function box and STC.

Summary

A programming method that can be considered to be the entering of a ladder program using text is the *instruction list (IL)*. An IL gives programs as a series of instructions, each instruction being on a new line. Each instruction consists of an operator followed by one or more operands, that is, the subjects of the operator. Mnemonic codes are used, each code corresponding to an operator/ladder element.

The *sequential function chart* (SFC) programming method is used for a pictorial representation of a system's operation to show the sequence of the events involved in its operation. The operation is described by a number of separate sequentially connected states or steps that are represented by rectangular boxes, each representing a particular state of the system being controlled. Each connecting line between states has a horizontal bar representing the transition condition that has to be realized before the system can move from one state to the next. When the transfer conditions to the next state are realized, the next state or step in the program occurs. The process thus continues from one state to the next until the entire machine cycle is completed. Outputs/actions at any state are represented by horizontally linked boxes and occur when that state has been realized.

With the structured text (ST) programming method, programs are written as a series of statements separated by semicolons. The statements use predefined statements and subroutines to change variables, these being defined values, internally stored values, or inputs and outputs. Assignment statements are used to indicate how the value of a variable is be changed, such as X := Y. Structured text is not case sensitive and spaces are not necessary but can be used to aid clarity. IF ... THEN ... ELSE is used when selected statements are to be executed when certain conditions occur. CASE is used to give the condition that selected statements are to be executed if a particular integer value occurs *else* some other selected statements. FOR ... DO ... allows a set of statements to be repeated depending on the value of the iteration integer variable. WHILE ... DO .. allows one or more statements to be executed while a particular Boolean expression remains true. REPEAT ... UNTIL ... allows one or more statements to be executed and repeated while a particular Boolean expression remains true.

Problems

Problems 1 through 24 have four answer options: A, B, C, or D. Choose the correct answer from the answer options.

1. Decide whether each of these statements is true (T) or false (F). The instruction list:

```
LD      X401
AND     X402
OUT     Y430
```

describes a ladder diagram rung for which there is an output when:
(i) Input X401 is activated but X402 is not.
(ii) Input X401 and input X402 are both activated.
A. (i) T (ii) T
B. (i) T (ii) F
C. (i) F (ii) T
D. (i) F (ii) F

2. *Decide whether each of these statements is true (T) or false (F).* The instruction list:

```
LD   X401
OR   X402
OUT  Y430
```

describes a ladder diagram rung for which there is an output when:
(i) Input X401 is activated but X402 is not.
(ii) Input X402 is activated but X401 is not.
A. (i) T (ii) T
B. (i) T (ii) F
C. (i) F (ii) T
D. (i) F (ii) F

3. *Decide whether each of these statements is true (T) or false (F).* The instruction list:

```
LD   X401
ANI  X402
OUT  Y430
```

describes a ladder diagram rung for which there is an output when:
(i) Input X401 is activated but X402 is not.
(ii) Input X401 and input X402 are both activated.
A. (i) T (ii) T
B. (i) T (ii) F
C. (i) F (ii) T
D. (i) F (ii) F

4. *Decide whether each of these statements is true (T) or false (F).* The instruction list:

```
LDI  X401
ANI  X402
OUT  Y430
```

describes a ladder diagram rung for which there is an output when:
(i) Input X401 is activated but X402 is not.
(ii) Input X401 and input X402 are both activated.
A. (i) T (ii) T
B. (i) T (ii) F
C. (i) F (ii) T
D. (i) F (ii) F

5. *Decide whether each of these statements is true (T) or false (F).* The instruction list:

```
LD    X401
OR    Y430
ANI   X402
OUT   Y430
```

describes a ladder diagram rung for which there is:

(i) An output when input X401 is momentarily activated.

(ii) No output when X402 is activated.

A. (i) T (ii) T

B. (i) T (ii) F

C. (i) F (ii) T

D. (i) F (ii) F

6. *Decide whether each of these statements is true (T) or false (F).* The instruction list:

```
A   I0.1
A   I0.2
=   Q2.0
```

describes a ladder diagram rung for which there is an output when:

(i) Input I0.1 is activated but I0.2 is not.

(ii) Input I0.1 and input I0.2 are both activated.

A. (i) T (ii) T

B. (i) T (ii) F

C. (i) F (ii) T

D. (i) F (ii) F

7. *Decide whether each of these statements is true (T) or false (F).* The instruction list:

```
A   I0.1
O   I0.2
=   Q2.0
```

describes a ladder diagram rung for which there is an output when:

(i) Input I0.1 is activated but I0.2 is not.

(ii) Input I0.2 is activated but I0.1 is not.

A. (i) T (ii) T

B. (i) T (ii) F

C. (i) F (ii) T

D. (i) F (ii) F

8. *Decide whether each of these statements is true (T) or false (F).* The instruction list:

```
A    I0.1
AN   I0.2
=    Q2.0
```

describes a ladder diagram rung for which there is an output when:
(i) Input I0.1 is activated but I0.2 is not.
(ii) Input I0.1 and input I0.2 are both activated.
A. (i) T (ii) T
B. (i) T (ii) F
C. (i) F (ii) T
D. (i) F (ii) F

9. *Decide whether each of these statements is true (T) or false (F).* The instruction list:

```
AN   I0.1
AN   I0.2
=    Q2.0
```

describes a ladder diagram rung for which there is an output when:
(i) Input I0.1 is activated but I0.2 is not.
(ii) Input I0.1 and input I0.2 are both activated.
A. (i) T (ii) T
B. (i) T (ii) F
C. (i) F (ii) T
D. (i) F (ii) F

10. *Decide whether each of these statements is true (T) or false (F).* The instruction list:

```
A    I0.1
O    Q2.0
AN   I0.2
=    Q2.0
```

describes a ladder diagram rung for which there is:
(i) An output when input I0.1 is momentarily activated.
(ii) No output when I0.2 is activated.
A. (i) T (ii) T
B. (i) T (ii) F
C. (i) F (ii) T
D. (i) F (ii) F

11. *Decide whether each of these statements is true (T) or false (F).* The instruction list:

```
LD   X401
OUT  Y430
LDI  X401
OUT  Y431
```

describes a program for which:
(i) When X401 is activated, there is an output from Y430 but not Y431.
(ii) When X401 is not activated, there is an output from Y431 but not Y430.
A. (i) T (ii) T
B. (i) T (ii) F
C. (i) F (ii) T
D. (i) F (ii) F

12. *Decide whether each of these statements is true (T) or false (F).* The instruction list:

```
LD   X400
OR   X401
OR   X402
AND  X403
OUT  Y431
```

describes a program for which there will be an output from Y431 when:
(i) Just X400 or X401 or X402 is activated.
(ii) Just X400 and X403 are activated.
A. (i) T (ii) T
B. (i) T (ii) F
C. (i) F (ii) T
D. (i) F (ii) F

13. *Decide whether each of these statements is true (T) or false (F).* The instruction list:

```
LD   X400
AND  X401
OR   X402
OUT  Y430
```

describes a program for which there will be an output from Y430 when:
(i) Just X400 or X402 is activated.
(ii) Just X400 and X401 are activated.
A. (i) T (ii) T
B. (i) T (ii) F
C. (i) F (ii) T
D. (i) F (ii) F

14. *Decide whether each of these statements is true (T) or false (F).* For the sequential function chart shown in Figure 6.24:
    (i) State 1 is realized when condition X1 is realized.
    (ii) Output 1 occurs when condition X2 is realized.
    A. (i) T (ii) T
    B. (i) T (ii) F
    C. (i) F (ii) T
    D. (i) F (ii) F

15. *Decide whether each of these statements is true (T) or false (F).*
    For the sequential function chart shown in Figure 6.25, if State 1 is active:
    (i) State 2 is realized when condition X2 is realized.
    (ii) State 3 occurs when condition X3 is realized.
    A. (i) T (ii) T
    B. (i) T (ii) F
    C. (i) F (ii) T
    D. (i) F (ii) F

16. For the ladder program described in Figure 6.26a, which of the sequential function charts in Figure 6.26b will represent it?
    Problems 17, 18, and 19 concern the sequential function chart shown in Figure 6.27.

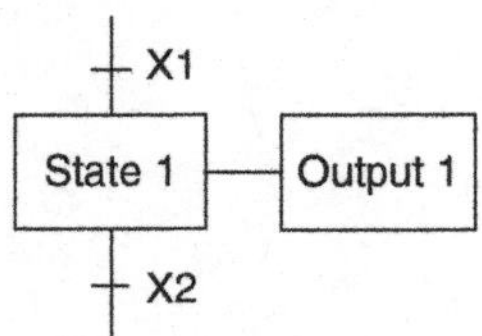

Figure 6.24: Diagram for Problem 14.

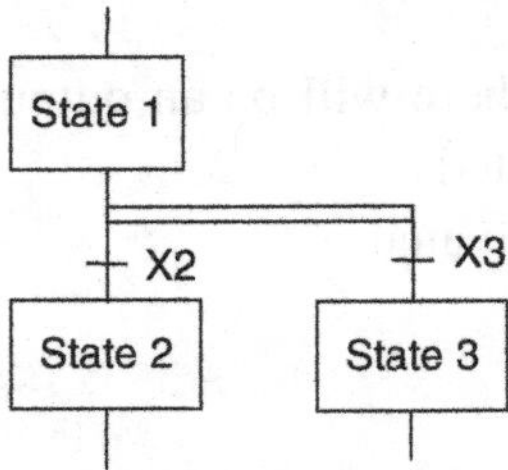

Figure 6.25: Diagram for Problem 15.

Input X1
Output Y1
Input X2
Input X3
Output Y2
(a)

X1
State 1
Y1
X2
State 2
Y2
X3
A

X1
State 1
Y1
X2
X3
State 2
Y2
B

X1
X2
X3
State 1
Y1
State 2
Y2
C
(b)

X2
X1
X3
State 1
Y1
State 2
Y2
D

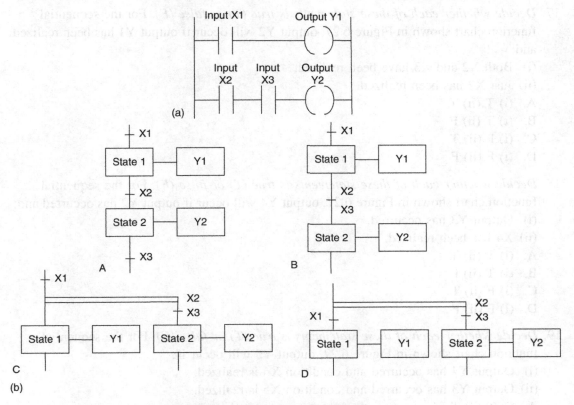

Figure 6.26: Diagram for Problem 16.

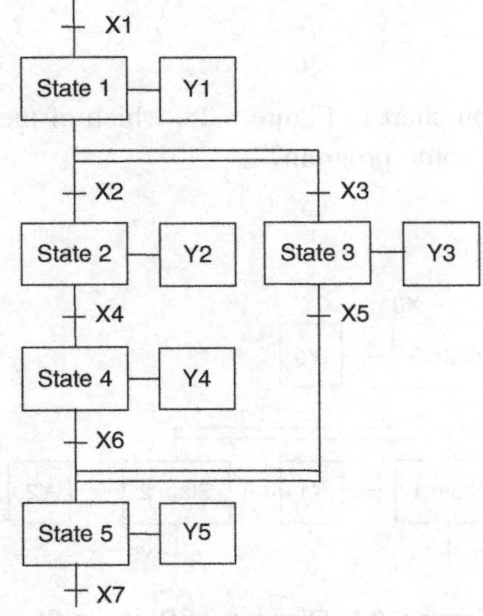

Figure 6.27: Diagram for Problems 17, 18, and 19.

17. *Decide whether each of these statements is true (T) or false (F).* For the sequential function chart shown in Figure 6.27, output Y2 will occur if output Y1 has been realized and:
    (i) Both X2 and X3 have been realized.
    (ii) Just X2 has been realized.
    A. (i) T (ii) T
    B. (i) T (ii) F
    C. (i) F (ii) T
    D. (i) F (ii) F

18. *Decide whether each of these statements is true (T) or false (F).* For the sequential function chart shown in Figure 6.27, output Y4 will occur if output Y2 has occurred and:
    (i) Output Y3 has occurred.
    (ii) X4 has been realized.
    A. (i) T (ii) T
    B. (i) T (ii) F
    C. (i) F (ii) T
    D. (i) F (ii) F

19. *Decide whether each of these statements is true (T) or false (F).* For the sequential function chart shown in Figure 6.27, output Y5 will occur if:
    (i) Output Y4 has occurred and condition X6 is realized.
    (ii) Output Y3 has occurred and condition X5 is realized.
    A. (i) T (ii) T
    B. (i) T (ii) F
    C. (i) F (ii) T
    D. (i) F (ii) F

20. For the sequential function chart in Figure 6.28, which of the ladder programs in Figure 6.29 describes the same program?

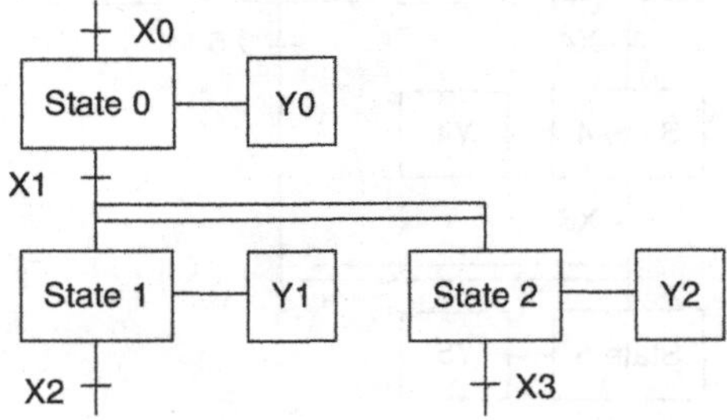

Figure 6.28: Diagram of Problem 20.

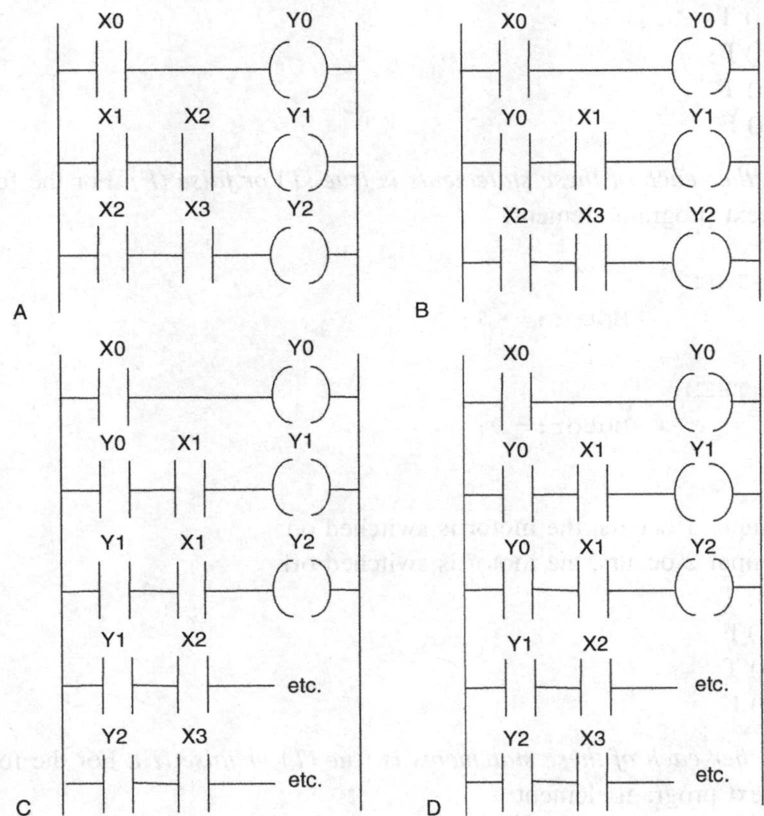

Figure 6.29: Diagram for Problem 20.

21. *Decide whether each of these statements is true (T) or false (F).* For the following structured text program element:

```
VAR
            i: INT;
END_VAR;
i :- 0;
REPEAT
            i :- i + 1;
UNTIL i:= 5;
END_REPEAT;
```

(i) The variable *i* can only have the 0 or 1 values.
(ii) Each time the program repeats, *i* has its value increased by 1.

A. (i) T (ii) T
B. (i) T (ii) F
C. (i) F (ii) T
D. (i) F (ii) F

22. *Decide whether each of these statements is true (T) or false (F).* For the following structured text program element:

```
IF Input1 THEN
                    Motor:=- 1;
END_IF;
IF Input2 THEN
                    Motor:= 0;
END_IF;
```

(i) When input 1 occurs, the motor is switched on.
(ii) When input 2 occurs, the motor is switched off.
A. (i) T (ii) T
B. (i) T (ii) F
C. (i) F (ii) T
D. (i) F (ii) F

23. *Decide whether each of these statements is true (T) or false (F).* For the following structured text program element:

```
IF (Limit_switch_1 AND Workpiece_Present) THEN
                    Gate_1 :- Open;
                    Gate_2 :- Closed;
ELSE
                    Gate_1 :- Closed;
                    Gate_2 :- Open;
END_IF;
```

(i) If only the workpiece is present, gate 1 is open and gate 2 is closed.
(ii) If only the limit switch is activated, gate 1 is closed and gate 2 is open.
A. (i) T (ii) T
B. (i) T (ii) F
C. (i) F (ii) T
D. (i) F (ii) F

24. *Decide whether each of these statements is true (T) or false (F).* For the following structured text program element:

```
VAR
                    Start_Up AT %IX120;
END_VAR
```

(i) Start_Up can be found at input memory location bit 120.
(ii) Start_Up has the value 120 bits.
A. (i) T (ii) T
B. (i) T (ii) F
C. (i) F (ii) T
D. (i) F (ii) F

25. Write a sequential function chart program for following the operation of a start switch, after which a tank is filled by opening valve 1 until a level switch 1 is triggered, then the tank is drained by opening drain valve 2 until level switch 2 is triggered, then the sequence is repeated.

26. Write a structured text program for the following: a tank is filled by opening valve 1, as long as level switch 1 is not triggered and the drain valve is closed.

27. Write a structured text program to set the temperature of an enclosure by switches to the values 40, 50, 60, and 70, and switch on fan 1 when the temperature is 60 and fan 2 when it is 70.

CHAPTER 7

Internal Relays

This chapter continues from the previous chapters on programming and introduces *internal relays*. A variety of other terms are often used to describe these elements, such as *auxiliary relays*, *markers*, *flags*, *coils*, and *bit storage*. These are one of the elements included among the special built-in functions with PLCs and are very widely used in programming. A small PLC might have a hundred or more internal relays, some of them battery backed so that they can be used in situations where it is necessary to ensure safe shutdown of a plant in the event of power failure. Later chapters consider other common built-in elements.

7.1 Internal Relays

In PLCs there are elements that are used to hold data, that is, bits, and behave like relays, being able to be switched on or off and to switch other devices on or off. Hence the term *internal relay*. Such internal relays do not exist as real-world switching devices but are merely bits in the storage memory that behave in the same way as relays. For programming, they can be treated in the same way as an external relay output and input. Thus inputs to external switches can be used to give an output from an internal relay. This then results in the internal relay contacts being used, in conjunction with other external input switches, to give an output, such as activating a motor. Thus we might have (Figure 7.1):

On one rung of the program:

> Inputs to external inputs activate the internal relay output.

On a later rung of the program:

> As a consequence of the internal relay output, internal relay contacts are activated and so control some output.

In using an internal relay, it has to be activated on one rung of a program and then its output used to operate switching contacts on another rung, or rungs, of the program. Internal relays can be programmed with as many sets of associated contacts as desired.

To distinguish internal relay outputs from external relay outputs, they are given different types of addresses. Different manufacturers tend to use different terms for internal relays and

W. Bolton: Programmable Logic Controllers, Sixth Edition. http://dx.doi.org/10.1016/B978-0-12-802929-9.00007-8

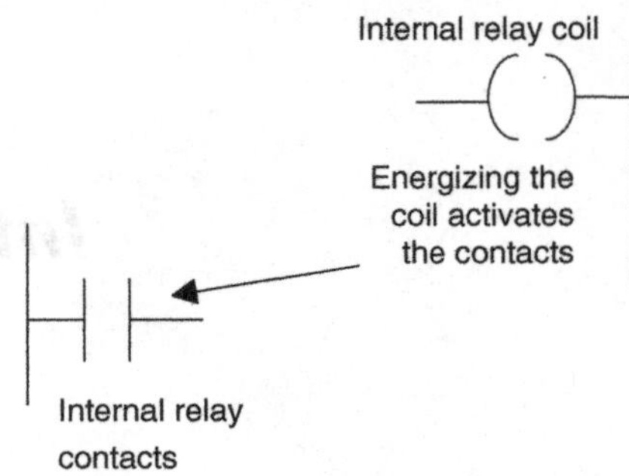

Figure 7.1: Internal relay.

have different ways of expressing their addresses. For example, Mitsubishi uses the term *auxiliary relay* or *marker* and the notation M100, M101, and so on. Siemens uses the term *flag* and the notation F0.0, F0.1, and so on. Telemecanique uses the term *bit* and the notation B0, B1, and so on. Toshiba uses the term *internal relay* and the notation R000, R001, and so on. Allen-Bradley uses the term *bit storage* and notation in the PLC-5 of the form B3/001, B3/002, and so on.

7.2 Ladder Programs

With ladder programs, an internal relay output is represented using the symbol for an output device, namely ( ), with an address that indicates that it is an internal relay. Thus, with a Mitsubishi PLC, we might have the address M100, the M indicating that it is an internal relay or marker rather than an external device. The internal relay switching contacts are designated with the symbol for an input device, namely | |, and given the same address as the internal relay output, such as M100.

7.2.1 Programs with Multiple Input Conditions

As an illustration of the use that can be made of internal relays, consider the following situation. A system is to be activated when two different sets of input conditions are realized. We might just program this as an AND logic gate system; however, if a number of inputs have to be checked in order that each of the input conditions can be realized, it may be simpler to use an internal relay. The first input conditions then are used to give an output to an internal relay. This relay has associated contacts that then become part of the input conditions with the second input.

Figure 7.2 shows a ladder program for such a task. For the first rung, when input In 1 or input In 3 is closed and input In 2 closed, internal relay IR 1 is activated. This results in the contacts for IR 1 closing. If input In 4 is then activated, there is an output from output Out 1. Such a task might be involved in the automatic lifting of a barrier when someone approaches from either side. Input In 1 and input In 3 are inputs from photoelectric sensors that detect the presence of a person

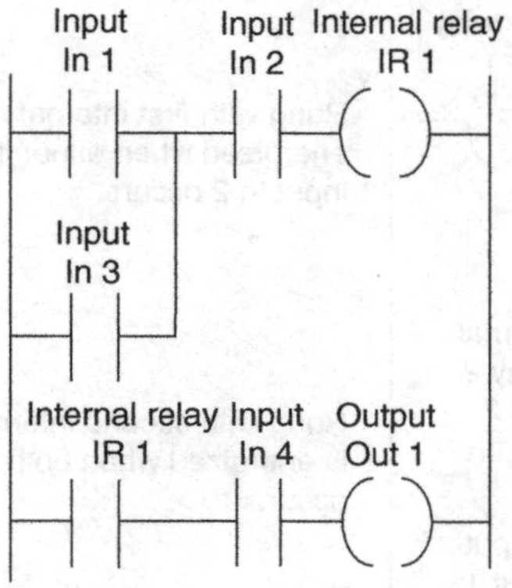

Figure 7.2: Internal relay.

approaching or leaving from either side of the barrier, input In 1 being activated from one side of it and input In 3 from the other. Input In 2 is an enabling switch to enable the system to be closed down. Thus when input In 1 or input In 3, and input In 2, are activated, there is an output from internal relay 1. This will close the internal relay contacts. If input In 4, perhaps a limit switch, detects that the barrier is closed, then it is activated and closes. The result is then an output from Out 1, a motor that lifts the barrier. If the limit switch detects that the barrier is already open, the person having passed through it, then it opens and so output Out 1 is no longer energized and a counterweight might then close the barrier. The internal relay has enabled two parts of the program to be linked, one part being the detection of the presence of a person and the second part the detection of whether the barrier is already up or down. Figure 7.3a shows how Figure 7.2 would appear in Mitsubishi notation and Figure 7.3b shows how it would appear in Siemens notation.

Figure 7.4 is another example of a ladder program involving internal relays. Output 1 is controlled by two input arrangements. The first rung shows the internal relay IR 1, which is

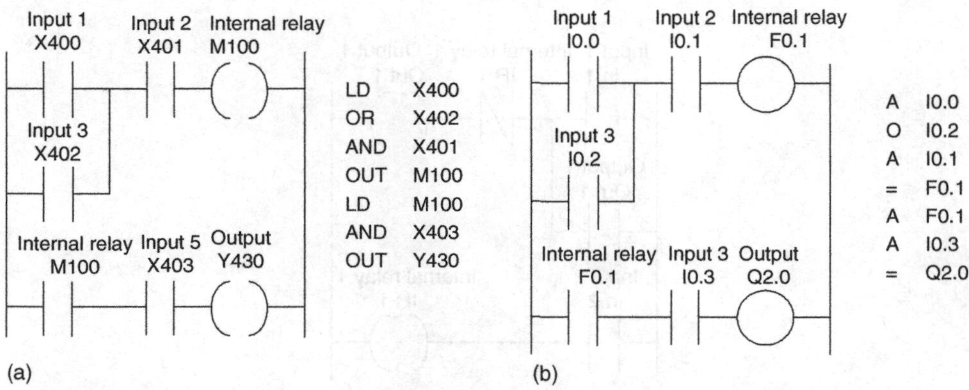

Figure 7.3: Figure 7.2 (a) in Mitsubishi notation, and (b) in Siemens notation.

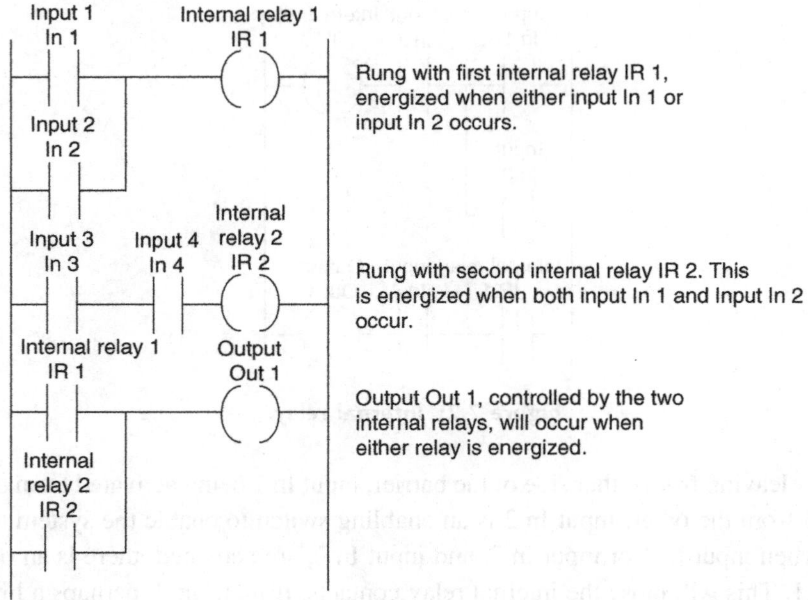

Figure 7.4: Use of two internal relays.

energized if input In 1 or In 2 is activated and closed. The second rung shows internal relay IR 2, which is energized if inputs In 3 and In 4 are both energized. The third rung shows that output Out 1 is energized if internal relay IR 1 or IR 2 is activated. Thus there is an output from the system if either of two sets of input conditions is realized.

7.2.2 Latching Programs

Another use of internal relays is for resetting a latch circuit. Figure 7.5 shows an example of such a ladder program.

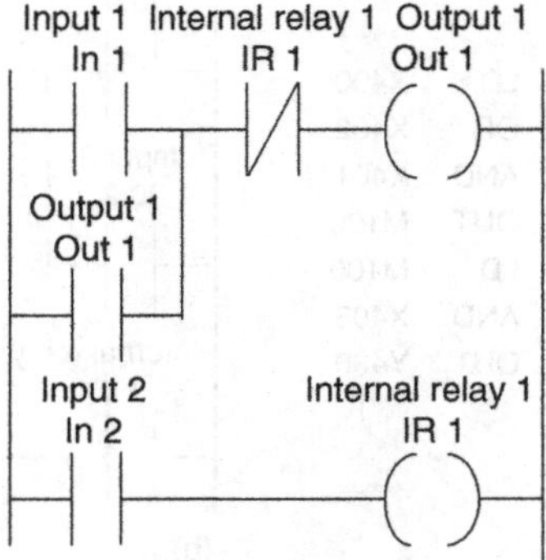

Figure 7.5: Resetting latch.

When the input In 1 contacts are momentarily closed, there is an output at Out 1. This closes the contacts for Out 1 and so maintains the output, even when input In 1 opens. When input In 2 is closed, the internal relay IR 1 is energized and so opens the IR 1 contacts, which are normally closed. Thus the output Out 1 is switched off and so the output is unlatched.

Consider a situation requiring latch circuits where there is an automatic machine that can be started or stopped using push-button switches. A latch circuit is used to start and stop the power being applied to the machine. The machine has several outputs that can be turned on if the power has been turned on and are off if the power is off. It would be possible to devise a ladder diagram that has individually latched controls for each such output. However, a simpler method is to use an internal relay. Figure 7.6 shows such a ladder diagram. The first rung has the latch for keeping the internal relay IR 1 on when the start switch gives a momentary input. The second rung will then switch the power on. The third rung will also switch on and give output Out 2 if the input 2 contacts are closed. The third rung will also switch on and give output Out 3 if the input 3 contacts are closed. Thus all the outputs can be switched on when the start push button is activated. All the outputs will be switched off if the stop switch is opened. Thus all the outputs are latched by IR 1.

7.2.3 Response Time

The time taken between an input occurring and an output changing depends on such factors as the electrical response time of the input circuit, the mechanical response of the output

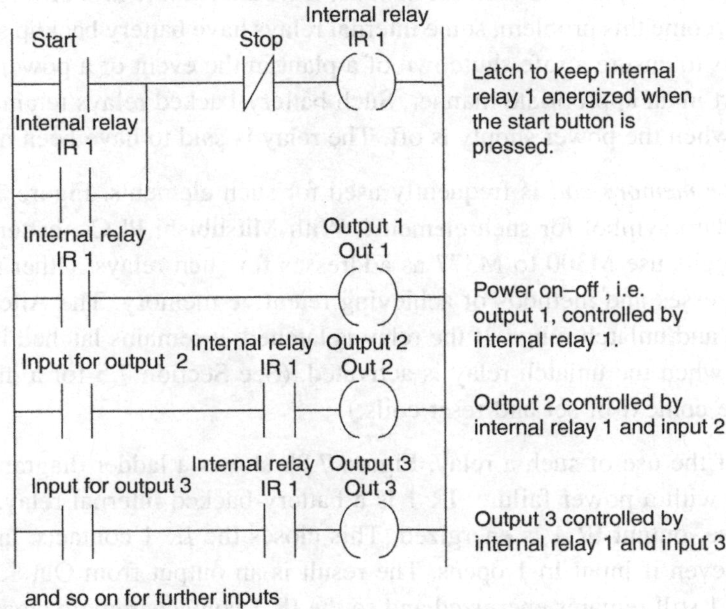

Figure 7.6: Starting of multiple outputs.

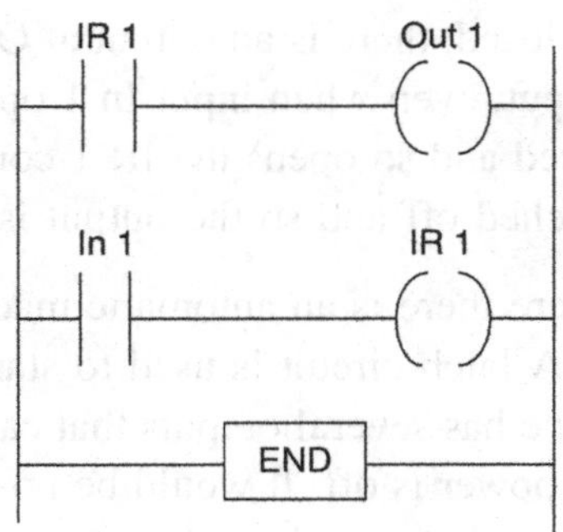
IR 1
Out 1
IR 1 not energized as a result of input to In 1 in the first scan until the second scan of the program.
In 1
IR 1
END

Figure 7.7: Response time lag arising from scan time.

device, and the scan time of the program. A ladder program is read from left to right and from top to bottom. Thus if an output device, such as an internal relay, is set in one scan cycle and the output has to be fed back to earlier in the program, it will require a second scan of the program before it can be activated. Figure 7.7 illustrates this concept.

7.3 Battery-Backed Relays

If the power supply is cut off from a PLC while it is being used, all the output relays and internal relays will be turned off. Thus when the power is restored, all the contacts associated with those relays will be set differently from when the power was on. Therefore, if the PLC was in the middle of some sequence of control actions, it would resume at a different point in the sequence. To overcome this problem, some internal relays have battery backup so that they can be used in circuits to ensure a safe shutdown of a plant in the event of a power failure and so enable it to restart in an appropriate manner. Such battery-backed relays retain their state of activation, even when the power supply is off. The relay is said to have been made *retentive*.

The term *retentive memory coil* is frequently used for such elements. Figure 7.8a shows the IEC 1131-3 standard symbol for such elements. With Mitsubishi PLCs, battery-backed internal relay circuits use M300 to M377 as addresses for such relays. Other manufacturers use different addresses and methods of achieving retentive memory. The Allen-Bradley PLC-5 uses latch and unlatch rungs. If the relay is latched, it remains latched if power is lost and is unlatched when the unlatch relay is activated. (See Section 7.5 for a discussion of such relays in the context of set and reset coils.)

As an example of the use of such a relay, Figure 7.8b shows a ladder diagram for a system designed to cope with a power failure. IR 1 is a battery-backed internal relay. When input In 1 contacts close, output IR 1 is energized. This closes the IR 1 contacts, latching so that IR 1 remains on even if input In 1 opens. The result is an output from Out 1. If there is a power failure, IR 1 still remains energized and so the IR 1 contacts remain closed and there is an output from Out 1.

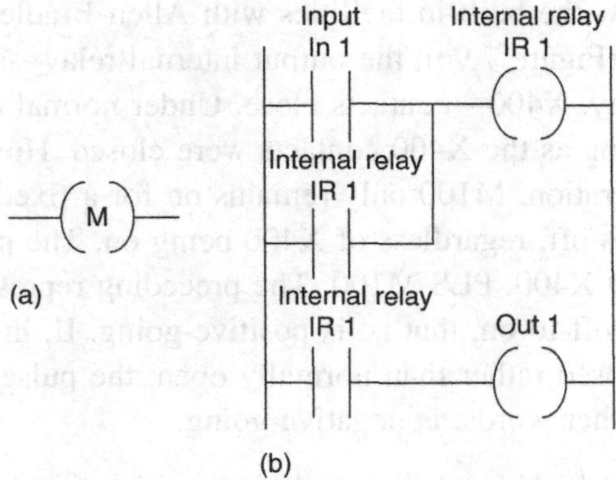

Figure 7.8: (a) Retentive memory coil, and (b) battery-backed relay program.

7.4 One-Shot Operation

One of the functions provided by some PLC manufacturers is the ability to program an internal relay so that its contacts are activated for just one cycle, that is, one scan through the ladder program. Hence when operated, the internal relay provides a fixed duration pulse at its contacts. This function is often termed *one-shot*. Though some PLCs have such a function as part of their programs, such a function can also easily be developed with just two rungs of a ladder program. Figure 7.9 shows such a pair of rungs.

For Figure 7.9a, when the trigger input occurs, it gives a trigger output in rung 1. In rung 2 it gives a cycle control output on an internal relay. Because rung 2 occurs after rung 1, the effect of the cycle control is not felt until the next cycle of the PLC program, when it opens the cycle control contacts in rung 1 and stops the trigger output. The trigger output then remains off, despite there being a trigger input. The trigger output can only occur again when the trigger output is switched off and then switched on again.

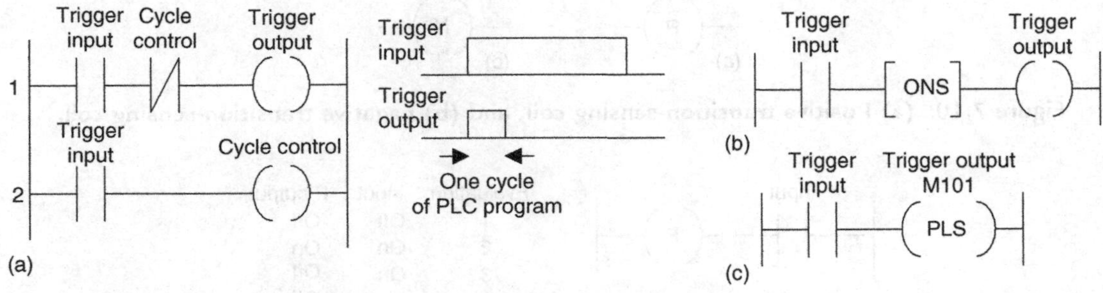

Figure 7.9: One-shot (a) program, (b) facility in an Allen-Bradley PLC, and (c) facility in a Mitsubishi PLC.

Figures 7.9b and 7.9c show the built-in facilities with Allen-Bradley and Mitsubishi PLCs. With the Mitsubishi PLC (Figure 7.9c), the output internal relay—say, M100—is activated when the trigger input—say, X400—contacts close. Under normal circumstances, M100 would remain on for as long as the X400 contacts were closed. However, if M100 has been programmed for pulse operation, M100 only remains on for a fixed period of time—one program cycle. It then goes off, regardless of X400 being on. The programming instructions that would be used are LD X400, PLS M100. The preceding represents pulse operation when the input goes from off to on, that is, is positive-going. If, in Figure 7.9c, the trigger input is made normally closed rather than normally open, the pulse occurs when the input goes from on to off—in other words, is negative-going.

The IEC 61131-3 gives standards for the symbols for positive transition-sensing and negative transition-sensing coils (Figure 7.10).

With the positive transition-sensing coil, if the power flow to it changes from off to on, the output is set on for one ladder rung evaluation. With the negative transition-sensing coil, if the power to it changes from off to on, the output is set on for one ladder rung evaluation. Thus, for the ladder rung of Figure 7.11, with the input off there is no output. When the input switches on, there is an output from the coil. However, the next and successive cycles of the program do not give outputs from the coil even though the switch remains on. The coil only gives an output the first time the switch is on.

7.5 Set and Reset

Another function that is often available is the ability to set and reset an internal relay. The set instruction causes the relay to self-hold, that is, latch. It then remains in that condition until the reset instruction is received. The term *flip-flop* is often used. Figure 7.12 shows the IEC 61131-3 standards for such coils. The SET coil is switched on when power is supplied to it and remains set until it is RESET. The RESET coil is reset to the off state when power is supplied to it and remains off until it is SET.

Figure 7.10: (a) Positive transition-sensing coil, and (b) negative transition-sensing coil.

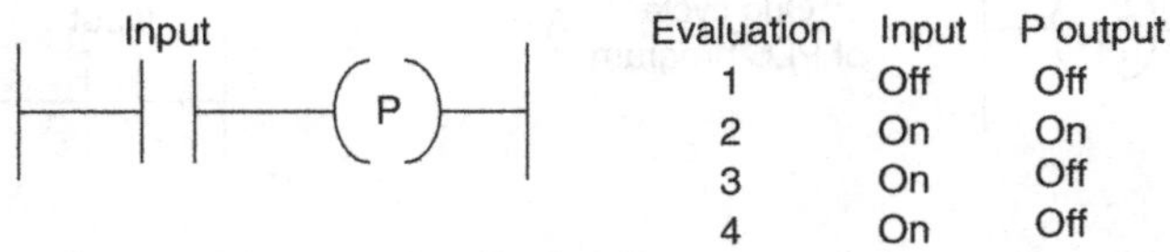

Input
P
Evaluation Input P output
1 Off Off
2 On On
3 On Off
4 On Off

Figure 7.11: Ladder rung with a positive-transition sensing coil.

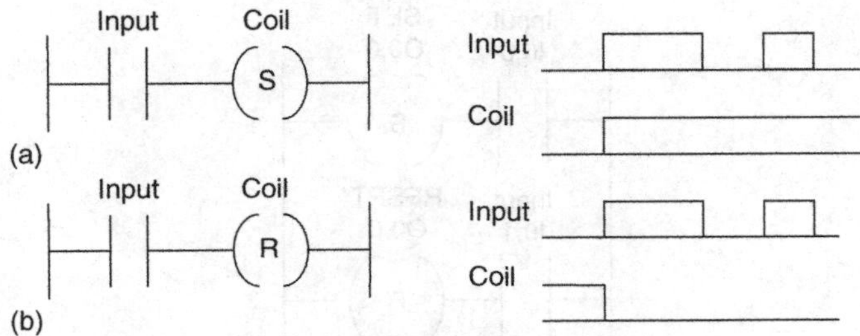

Figure 7.12: (a) SET and (b) RESET coils.

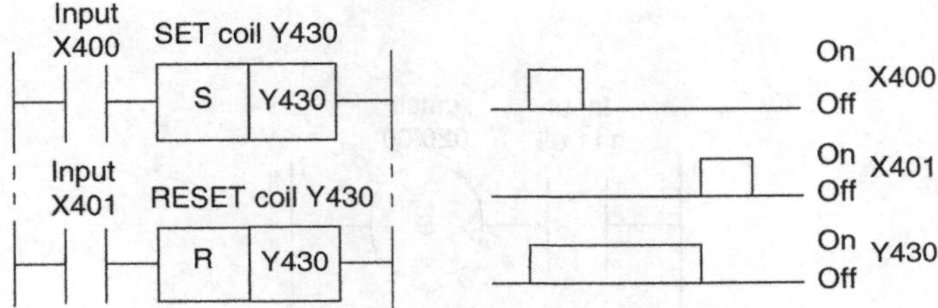

Figure 7.13: SET and RESET.

Figure 7.13 shows an example of a ladder diagram involving such a function. Activation of the first input, X400, causes the output Y430 to be turned on and set, that is, latched. Thus if the first input is turned off, the output remains on. Activation of the second input, X401, causes the output Y430 to be reset, that is, turned off and latched off. Thus the output Y430 is on for the time between X400 being momentarily switched on and X401 being momentarily switched on. Between the two rungs indicated for the set and reset operations, there could be other rungs for other activities to be carried out, with the set rung switching on an output at the beginning of the sequence and off at the end.

The programming instructions for the ladder rungs in the program for Figure 7.13 are:

```
LD    X400
S     Y430
Other program rungs are:
LD    X401
R     Y430
```

With a Telemecanique PLC, the ladder diagram would be as shown in Figure 7.14 and the programming instructions would be:

```
L     I0,0
S     O0,0
L     I0,1
R     O0,0
```

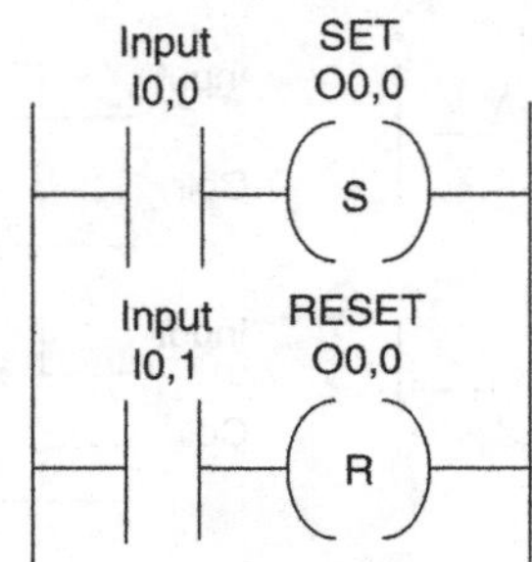

Figure 7.14: SET and RESET (Telemecanique PLC).

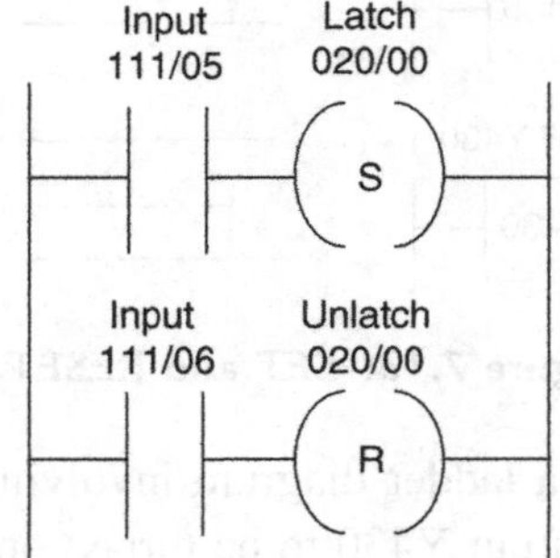

Figure 7.15: Latch and unlatch (Allen-Bradley PLC).

With an Allen-Bradley PLC, the terms *latch* and *unlatch* are used. Figure 7.15 shows the ladder diagram.

The SET and RESET coil symbols are often combined in a single box symbol. Figure 7.16 shows the equivalent ladder diagram for the set-reset function in the preceding figures with a Siemens PLC. The term *memory box* is used by them for the SET/RESET box, and the box shown is termed an SR or reset priority memory function in that reset has priority. With set priority (RS memory box), the arrangement is as shown in Figure 7.17.
The programming instructions (F indicates an internal relay) for reset priority are:

```
A    I0.0
S    F0.0
A    I0.1
R    F0.0
A    F0.0
=    Q2.0
```

Toshiba uses the term *flip-flop*, and Figure 7.18 shows the ladder diagram.

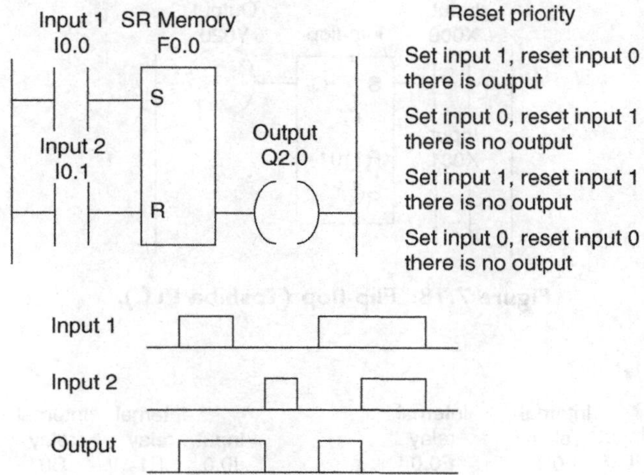

Figure 7.16: SET and RESET, with reset priority (Siemens PLC).

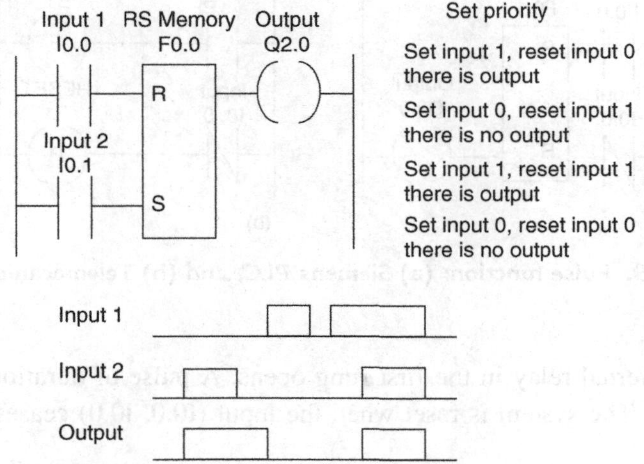

Figure 7.17: SET and RESET, with set priority (Siemens PLC).

Figure 7.19 shows how the set-reset function can be used to build the pulse (one-shot) function described in the previous section. Figure 7.19a shows it for a Siemens PLC (F indicates internal relay) and Figure 7.19b for a Telemecanique PLC (B indicates internal relay). In Figures 7.19a and 7.19b, an input (I0.0, I0,0) causes the internal relay (B0, F0.0) in the first rung to be activated. This results in the second rung, in the set/reset internal relay being set. This setting action results in the internal relay (F0.1, B1) in the first rung opening, and so, despite there being an input in the first rung, the internal relay (BO, F0.0) opens. However, because the rungs are scanned in sequence from top to bottom, a full cycle must

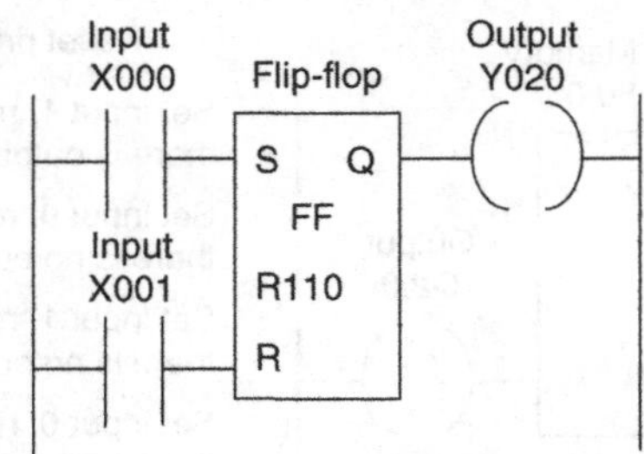

Figure 7.18: Flip-flop (Toshiba PLC).

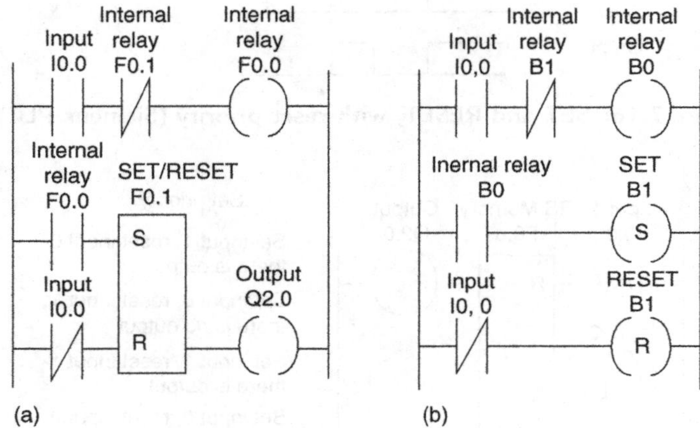

Figure 7.19: Pulse function: (a) Siemens PLC, and (b) Telemecanique PLC.

elapse before the internal relay in the first rung opens. A pulse of duration one cycle has thus been produced. The system is reset when the input (I0.0, I0,0) ceases.

7.5.1 Program Examples

An example of the basic elements of a simple program for use with a fire alarm system is shown in Figure 7.20. Fire sensors provide inputs to a SET/RESET function block so that if one of the sensors is activated, the alarm is set and remains set until it is cleared by being reset. When set it sets off the alarm.

Another example showing the basic elements of a program is shown in Figure 7.21. This could be used with a system designed to detect when a workpiece has been loaded into the correct position for some further operation. When the start contacts are closed, the output causes the workpiece to move. This continues until a light beam is interrupted and resets, causing the output to cease. A stop button is available to stop the movement at any time.

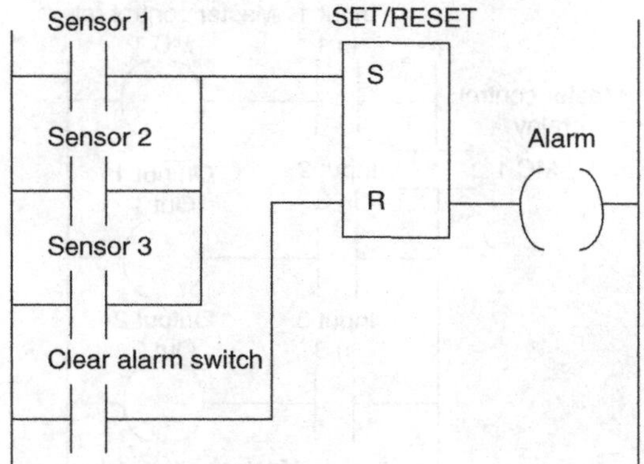

Figure 7.20: An alarm system.

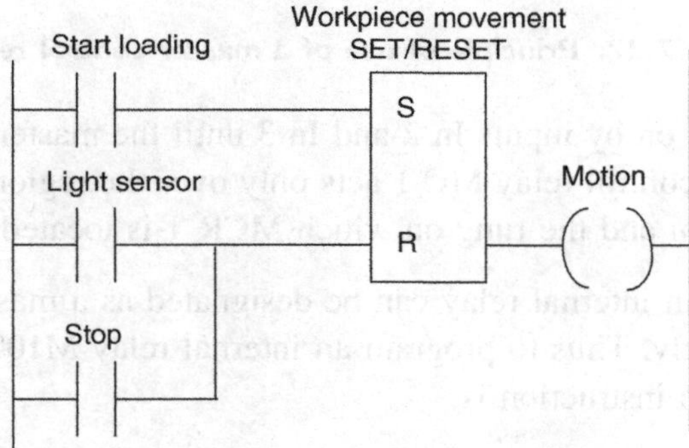

Figure 7.21: Loading system.

7.6 Master Control Relay

When large numbers of outputs have to be controlled, it is sometimes necessary for whole sections of ladder diagrams to be turned on or off when certain criteria are realized. This could be achieved by including the contacts of the same internal relay in each of the rungs so that its operation affects all of them. An alternative is to use a *master control relay*. Figure 7.22 illustrates the use of such a relay to control a section of a ladder program.

With no input to input In 1, the output internal relay MC 1 is not energized, and so its contacts are open. This means that all the rungs between where it is designated to operate and the rung on which its reset MCR or another master control relay is located are switched off. Assuming that it is designated to operate from its own rung, we can imagine it to be located in the power line in the position shown, and so rungs 2 and 3 are off. When input In 1 contacts close, the master relay MC 1 is energized. When this happens, all the rungs between it and the rung with its reset MCR 1 are switched on. Thus outputs Out 1 and

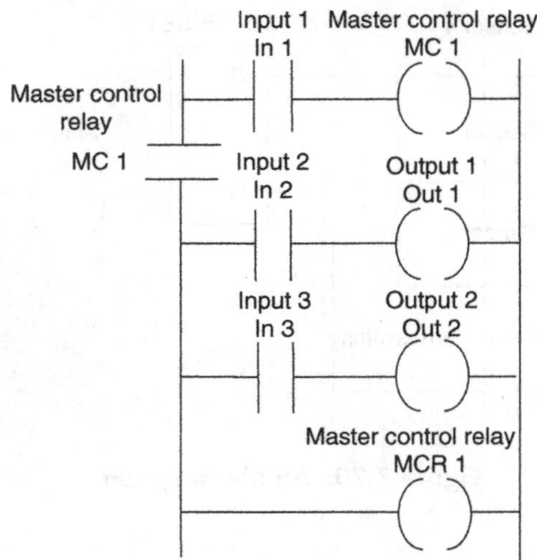

Figure 7.22: Principle of use of a master control relay.

Out 2 cannot be switched on by inputs In 2 and In 3 until the master control relay has been switched on. The master control relay MC 1 acts only over the region between the rung it is designated to operate from and the rung on which MCR 1 is located.

With a Mitsubishi PLC, an internal relay can be designated as a master control relay by programming it accordingly. Thus to program an internal relay M100 to act as a master control relay, the program instruction is:

MC M100

To program the resetting of that relay, the program instruction is:

MCR M100

Thus for the ladder diagram shown in Figure 7.23, which is Figure 7.22 with Mitsubishi addresses, the program instructions are:

```
LD    X400
OUT   M100
MC    M100
LD    X401
OUT   Y430
LD    X402
OUT   Y431
MC    M100
```

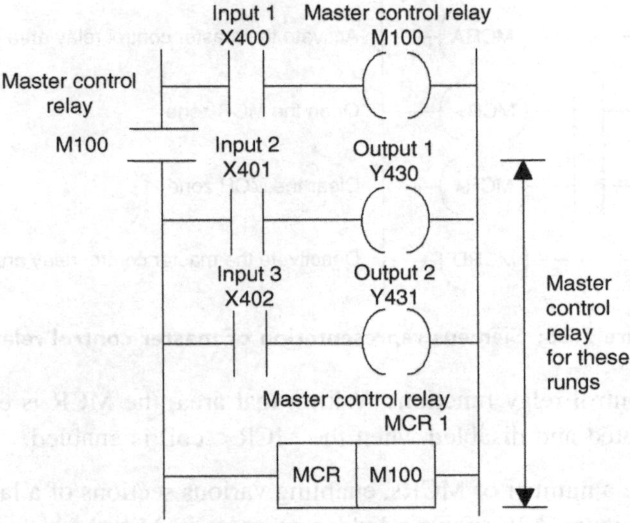

Figure 7.23: MCR with Mitsubishi PLC.

Figure 7.24 shows the format used by Allen-Bradley. To end the control of one *master control relay* (MCR), a second master control relay (MCR) is used with no contacts or logic preceding it. It is said to be programmed unconditionally.

The representation used for MCRs in Siemens ladder programs is shown in Figure 7.25. An area in which an MCR is to operate is defined by the activate master control area and

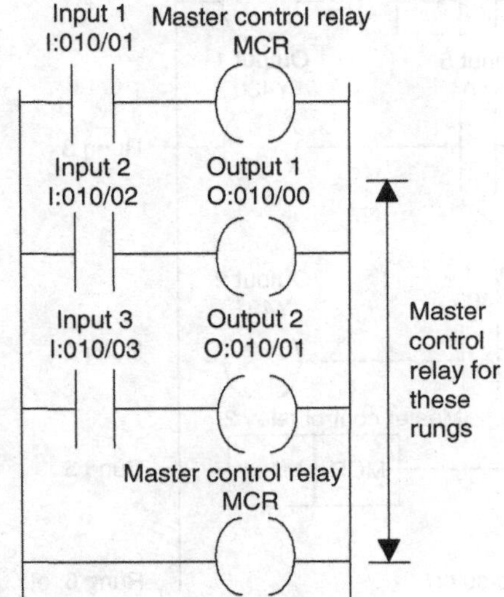

Figure 7.24: An MCR with Allen-Bradley PLC.

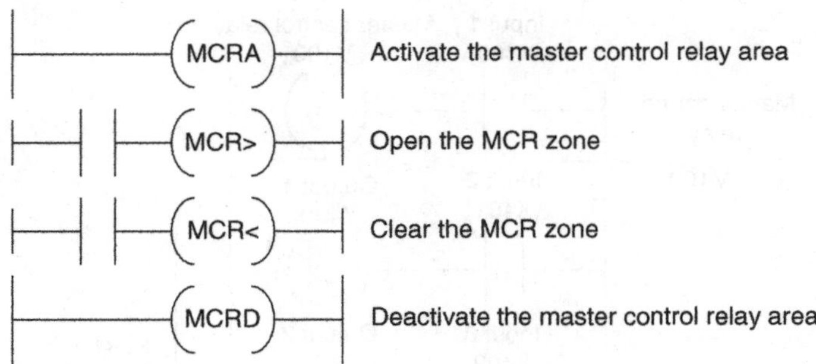

Figure 7.25: Siemens representation of master control relays.

deactivate master control relay functions. Within that area, the MCR is enabled when the MCR> coil is activated and disabled when the MCR< coil is enabled.

A program might use a number of MCRs, enabling various sections of a ladder program to be switched in or out. Figure 7.26 shows a ladder program in Mitsubishi format involving two MCRs. With M100 switched on but M101 off, the sequence is: rungs 1, 3, 4, 6, and so on.

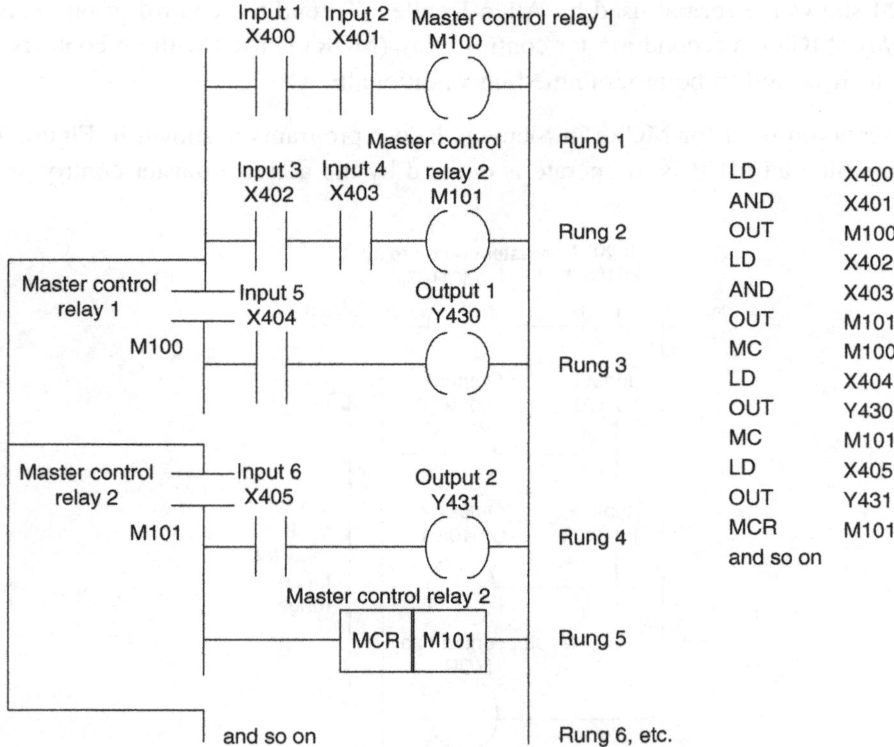

Figure 7.26: Example showing more than one master control relay.

The end of the M100 controlled section is indicated by the occurrence of the other MCR, M101. With M101 switched on but M100 off, the sequence is: rungs 2, 4, 5, 6, and so on. The end of this section is indicated by the presence of the reset. This reset has to be used since the rung is not followed immediately by another MCR. Such an arrangement could be used to switch on one set of ladder rungs if one type of input occurs and another set of ladder rungs if a different input occurs.

7.6.1 Examples of Programs

The following looks at a program that illustrates the uses of MCRs. The program is being developed for use with a pneumatic valve system involving the movement of pistons in cylinders to give a particular sequence of piston actions. First, however, we show how latching might be used with such systems to maintain actions.

Consider a pneumatic system with single-solenoid controlled valves and involving two cylinders A and B with limit switches a–, a+, b–, b+ detecting the limits of the piston rod movements (Figure 7.27), with the requirement to give the sequence A+, B+, A–, B–. Figure 7.28 shows the ladder diagram that can be used.

The solenoid A+ is energized when the start switch and limit switch b– are closed. This provides latching to keep A+ energized as long as the normally closed contacts for limit switch b+ are not activated. When limit switch a+ is activated, solenoid B+ is energized. This provides latching that keeps B+ energized as long as the normally closed contacts for limit switch a– are not activated. When cylinder B extends, the limit switch b+ opens its normally closed contacts and unlatches the solenoid A+. Solenoid A thus retracts. When it has retracted and opened the normally closed contacts a–, solenoid B+ becomes unlatched and cylinder B retracts.

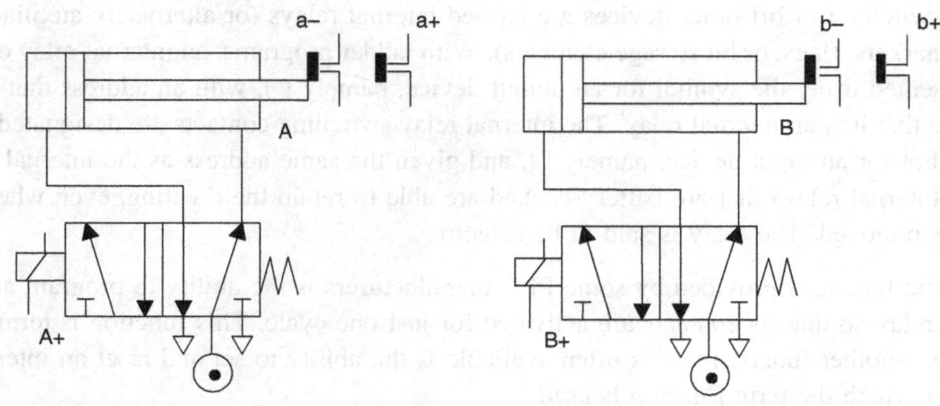

Figure 7.27: A valve system.

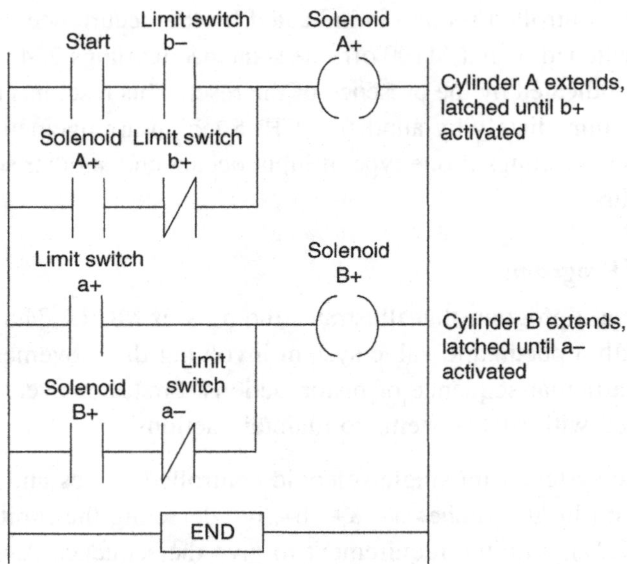

Figure 7.28: A ladder program.

Now consider the ladder program that could be used with the pair of single-solenoid-controlled cylinders in Figure 7.27 to give, when and only when the start switch is momentarily triggered, the sequence A+, B+, A–, then a 10 s time delay, B–, and stop at that point until the start switch is triggered again. Figure 7.29 shows how such a program can be devised using a MCR. The MCR is activated by the start switch and remains on until switched off by the rung containing just MCR. (See Chapter 9 for a discussion of timers.)

Summary

In PLCs, elements that are used to hold data, that is, bits, and behave like relays and so are able to switch on or off other devices are termed internal relays (or alternately auxiliary relays, markers, flags, or bit storage elements). With ladder programs, an internal relay output is represented using the symbol for an output device, namely ( ), with an address that indicates that it is an internal relay. The internal relay switching contacts are designated with the symbol for an input device, namely | |, and given the same address as the internal relay output. Internal relays that are battery-backed are able to retain their setting, even when the power is removed. The relay is said to be *retentive*.

One of the functions provided by some PLC manufacturers is the ability to program an internal relay so that its contacts are activated for just one cycle. This function is termed *one-shot*. Another function that is often available is the ability to set and reset an internal relay, for which the term *flip-flop* is used.

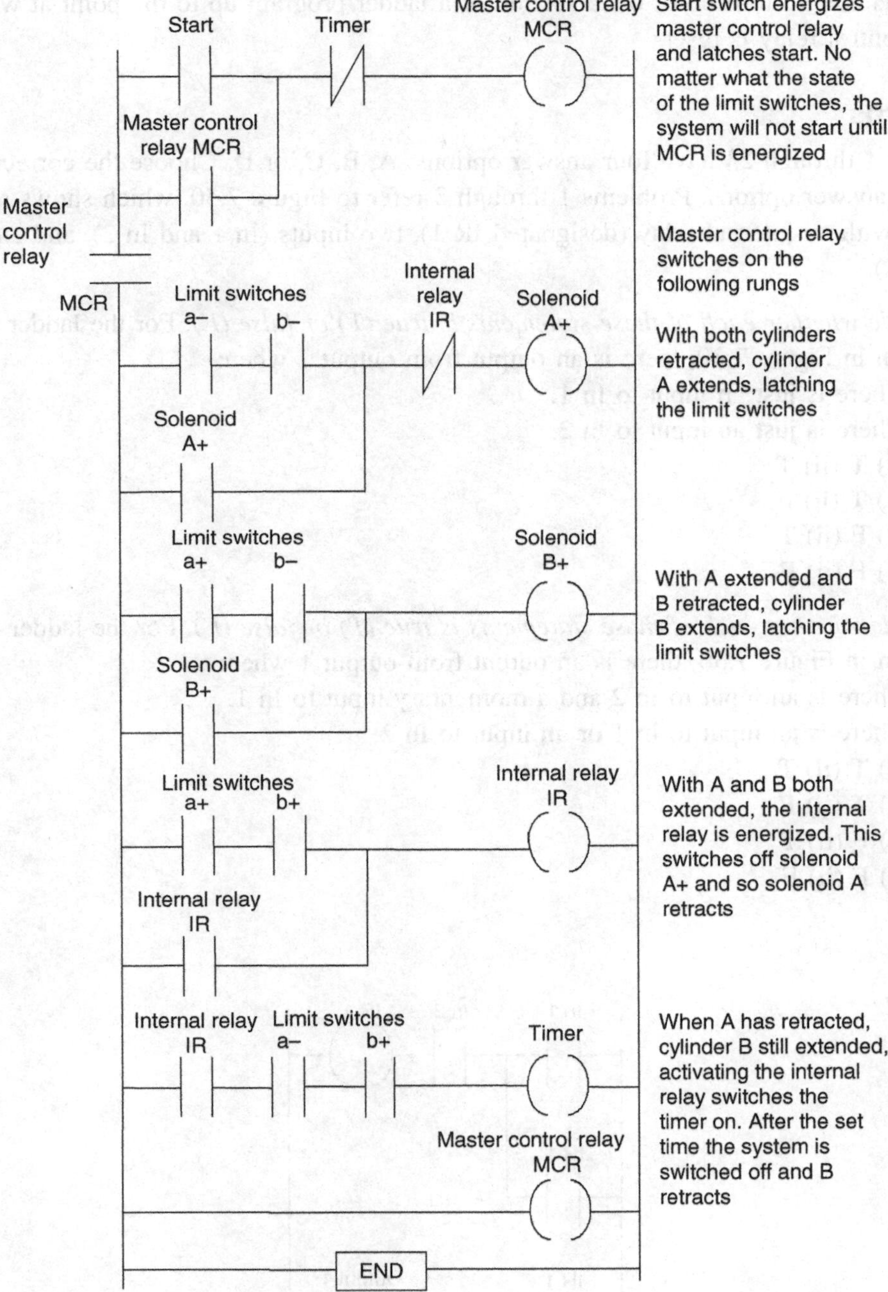

Figure 7.29: A ladder program.

A MCR is able to turn on or off a section of a ladder program up to the point at which the master control relay is reset.

Problems

Problems 1 through 23 have four answer options: A, B, C, or D. Choose the correct answer from the answer options. Problems 1 through 3 refer to Figure 7.30, which shows a ladder diagram with an internal relay (designated IR 1), two inputs (In 1 and In 2), and an output (Output 1).

1. *Decide whether each of these statements is true (T) or false (F).* For the ladder diagram shown in Figure 7.30, there is an output from output 1 when:
   (i) There is just an input to In 1.
   (ii) There is just an input to In 2.
   A. (i) T (ii) T
   B. (i) T (ii) F
   C. (i) F (ii) T
   D. (i) F (ii) F

2. *Decide whether each of these statements is true (T) or false (F).* For the ladder diagram shown in Figure 7.30, there is an output from output 1 when:
   (i) There is an input to In 2 and a momentary input to In 1.
   (ii) There is an input to In 1 or an input to In 2.
   A. (i) T (ii) T
   B. (i) T (ii) F
   C. (i) F (ii) T
   D. (i) F (ii) F

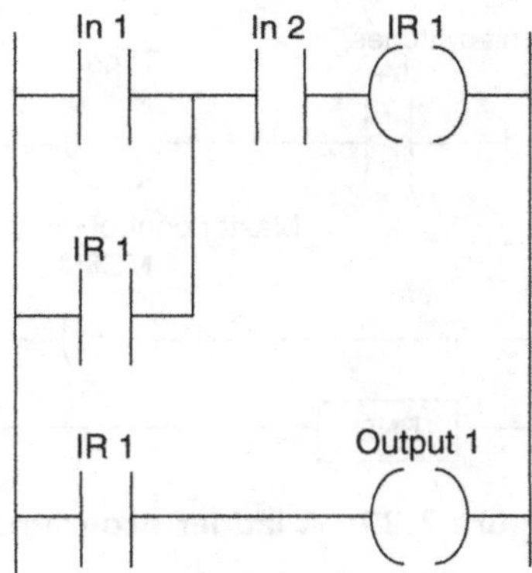

Figure 7.30: Diagram for Problems 1, 2, and 3.

3. *Decide whether each of these statements is true (T) or false (F).* For the ladder diagram shown in Figure 7.30, the internal relay:
   (i) Switches on when there is just an input to In 1.
   (ii) Switches on when there is an input to In 1 and to In 2.
   A. (i) T (ii) T
   B. (i) T (ii) F
   C. (i) F (ii) T
   D. (i) F (ii) F

Problems 4 through 6 refer to Figure 7.31, which shows a ladder diagram involving internal relays IR 1 and IR 2, inputs In 1, In 2, In 3, and In 4, and output Output 1.

4. *Decide whether each of these statements is true (T) or false (F).* For the ladder diagram shown in Figure 7.31, the internal relay IR 1 is energized when:
   (i) There is an input to In 1.
   (ii) There is an input to In 3.
   A. (i) T (ii) T
   B. (i) T (ii) F
   C. (i) F (ii) T
   D. (i) F (ii) F

5. *Decide whether each of these statements is true (T) or false (F).* For the ladder diagram shown in Figure 7.31, the internal relay IR 2 is energized when:
   (i) Internal relay IR 1 is energized.
   (ii) Input 4 is energized.
   A. (i) T (ii) T
   B. (i) T (ii) F

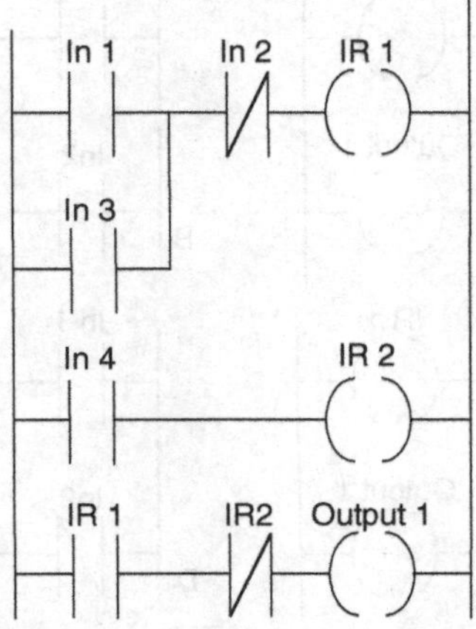

Figure 7.31: Diagram for Problems 4, 5, and 6.

C. (i) F (ii) T
D. (i) F (ii) F

6. *Decide whether each of these statements is true (T) or false (F).* For the ladder diagram shown in Figure 7.31, there is an output from Output 1 when:
   (i) There are inputs to only In 1, In 2, and In 4.
   (ii) There are inputs to only In 3 and In 4.
   A. (i) T (ii) T
   B. (i) T (ii) F
   C. (i) F (ii) T
   D. (i) F (ii) F

7. Which one of the programs in Figure 7.32 can obtain an output from Out 1 when just input In 1 occurs?

8. Which one of the programs in Figure 7.33 will give an output from Out 1 in the same program scan as there is an input to In 1?

Problems 9 and 10 refer to Figure 7.34, which shows a ladder diagram involving a battery-backed relay IR 1, two inputs (In 1 and In 2), and an output (Output 1).

9. *Decide whether each of these statements is true (T) or false (F).* For the ladder diagram shown in Figure 7.34, there is an output from Output 1 when:
   (i) There is a short duration input to In 1.
   (ii) There is no input to In 2.
   A. (i) T (ii) T
   B. (i) T (ii) F

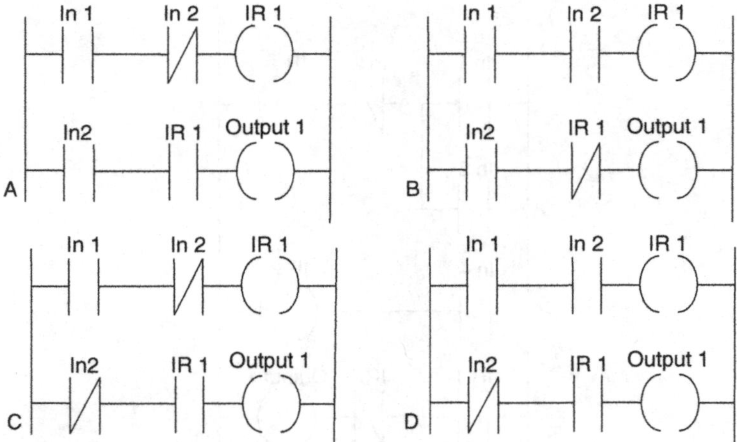

Figure 7.32: Diagram for Problem 7.

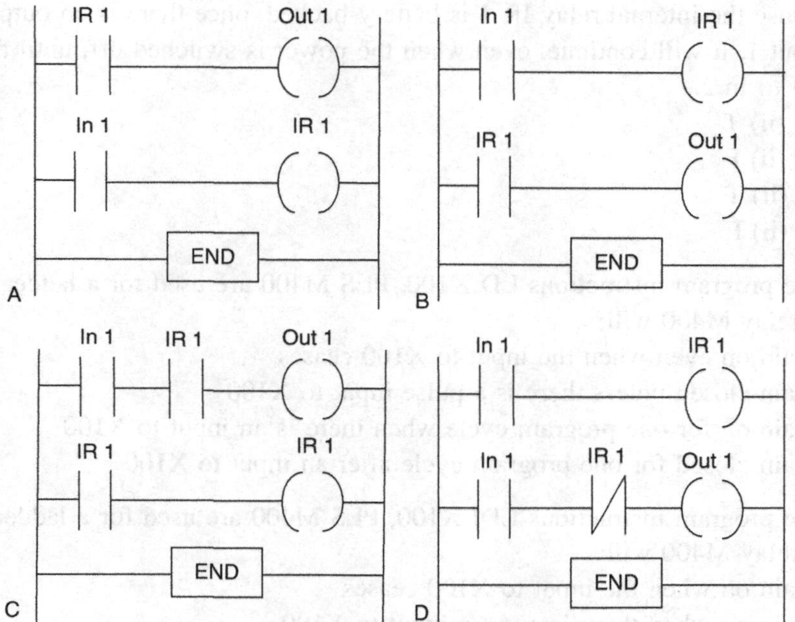

Figure 7.33: Diagram for Problem 8.

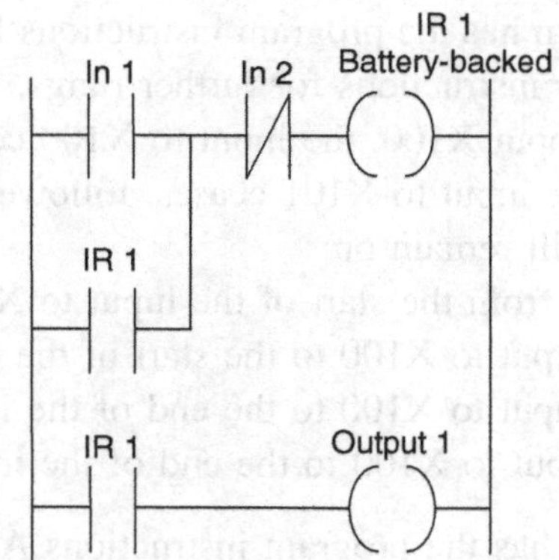

Figure 7.34: Diagram for Problems 9 and 10.

C. (i) F (ii) T
D. (i) F (ii) F

10. *Decide whether each of these statements is true (T) or false (F).* For the ladder diagram shown in Figure 7.34:
    (i) The input In 1 is latched by the internal relay so that the internal relay IR 1 remains energized, even when the input In 1 ceases.

(ii) Because the internal relay IR 1 is battery-backed, once there is an output from Output 1, it will continue, even when the power is switched off, until there is an input to In 2.

A. (i) T (ii) T
B. (i) T (ii) F
C. (i) F (ii) T
D. (i) F (ii) F

11. When the program instructions LD X100, PLS M400 are used for a ladder rung, the internal relay M400 will:
    A. Remain on even when the input to X100 ceases
    B. Remain closed unless there is a pulse input to X100
    C. Remain on for one program cycle when there is an input to X100
    D. Remain closed for one program cycle after an input to X100

12. When the program instructions LDI X100, PLS M400 are used for a ladder rung, the internal relay M400 will:
    A. Remain on when the input to X100 ceases
    B. Remain on when there is a pulse input to X100
    C. Remain on for one program cycle when there is an input to X100
    D. Remain on for one program cycle after the input to X100 ceases

13. A Mitsubishi ladder program has the program instructions LD X100, S M200, LD X101, R M200, followed by other instructions for further rungs. There is the following sequence: an input to the input X100, the input to X100 ceases, some time elapses, an input to the input X101, the input to X101 ceases, followed by inputs to later rungs. The internal relay M200 will remain on:
    A. For one program cycle from the start of the input to X100
    B. From the start of the input to X100 to the start of the input to X101
    C. From the start of the input to X100 to the end of the input to X101
    D. From the end of the input to X100 to the end of the input to X101

14. A Siemens ladder program has the program instructions A I0.0, S F0.0, A I0.1, R F0.0, A F0.0, = Q2.0, followed by other instructions for further rungs. There is the sequence: an input to input I0.0, the input to I0.0 ceases, some time elapses, an input to input I0.1, the input to I0.1 ceases, followed by inputs to later rungs. The internal relay F0.0 will remain on:
    A. For one program cycle from the start of the input to I0.0
    B. From the start of the input to I0.0 to the start of the input to I0.1
    C. From the start of the input to I0.0 to the end of the input to I0.1
    D. From the end of the input to I0.0 to the end of the input to I0.1

15. A Telemecanique ladder program has the program instructions L I0,0, S O0,0, L I0,1, R O0,0, followed by other instructions for further rungs. There is the following sequence: an input to input I0,0, the input to I0,0 ceases, some time elapses, an input to input I0,1, the input to I0,1 ceases, followed by inputs to later rungs. The internal relay O0,0 will remain on:
    A. For one program cycle from the start of the input to I0,0
    B. From the start of the input to I0,0 to the start of the input to I0,1
    C. From the start of the input to I0,0 to the end of the input to I0,1
    D. From the end of the input to I0,0 to the end of the input to I0,1
16. An output is required from output Y430 that lasts for one cycle after an input to X100 starts. This can be given by a ladder program with the instructions:
    A. LD X100, Y430
    B. LD X100, M100, LD M100, Y430
    C. LD X100, PLS M100, LD M100, Y430
    D. LD X400, PLS M100, LDI M100, Y430

Problems 17 and 18 refer to Figure 7.35, which shows two versions of the same ladder diagram according to two different PLC manufacturers. In Figure 7.35a, which uses Siemens notation, I is used for inputs, F for internal relays, and Q for the output. In Figure 7.35b, which uses Telemecanique notation, I is used for inputs and B for internal relays.

17. For the ladder diagram shown in Figure 7.35a, when there is an input to I0.0, the output Q2.0:
    A. Comes on and remains on for one cycle.
    B. Comes on and remains on.
    C. Goes off and remains off for one cycle.
    D. Goes off and remains off.

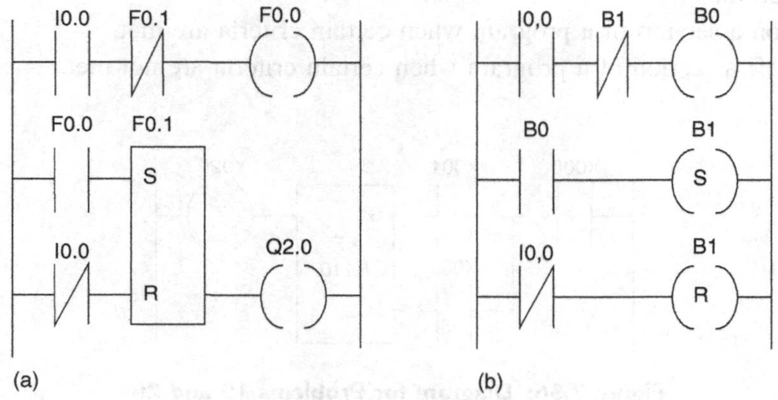

Figure 7.35: Diagram for Problems 17 and 18.

18. For the ladder diagram shown in Figure 7.35b, when there is an input to I0,0, the internal relay B1:
    A. Comes on and remains on for one cycle.
    B. Comes on and remains on.
    C. Goes off and remains off for one cycle.
    D. Goes off and remains off.

Problems 19 and 20 refer to Figure 7.36, which shows a Toshiba ladder program with inputs X000, X001, and X002, an output Y020, and a flip-flop R110.

19. *Decide whether each of these statements is true (T) or false (F).* For there to be an output from Y020, there must be an input to:
    (i) X000.
    (ii) X001.
    A. (i) T (ii) T
    B. (i) T (ii) F
    C. (i) F (ii) T
    D. (i) F (ii) F

20. *Decide whether each of these statements is true (T) or false (F).* With an input to X000, then:
    (i) An input to X001 causes the output to come on.
    (ii) An input to X002 causes the output to come on.
    A. (i) T (ii) T
    B. (i) T (ii) F
    C. (i) F (ii) T
    D. (i) F (ii) F

21. *Decide whether each of these statements is true (T) or false (F).* A master control relay can be used to:
    (i) Turn on a section of a program when certain criteria are met.
    (ii) Turn off a section of a program when certain criteria are not met.

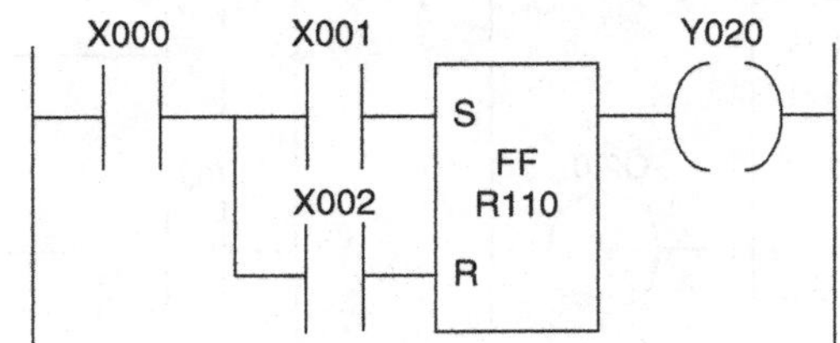

Figure 7.36: Diagram for Problems 19 and 20.

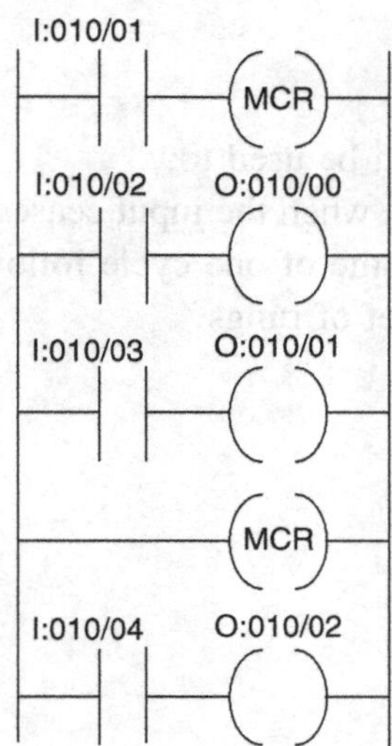

Figure 7.37: Diagram for Problems 22 and 23.

A. (i) T (ii) T
B. (i) T (ii) F
C. (i) F (ii) T
D. (i) F (ii) F

Problems 22 and 23 refer to Figure 7.37, which shows a ladder program in Allen-Bradley format.

22. *Decide whether each of these statements is true (T) or false (F).* When there is an input to I:010/01:
    (i) An input to I:010/02 gives an output from O:010/00.
    (ii) An input to I:010/03 gives an output from O:010/01.
    A. (i) T (ii) T
    B. (i) T (ii) F
    C. (i) F (ii) T
    D. (i) F (ii) F

23. *Decide whether each of these statements is true (T) or false (F).* When there is no input to I:010/01:
    (i) An input to I:010/02 gives no output from O:010/00.
    (ii) An input to I:010/04 gives no output from O:010/02.
    A. (i) T (ii) T
    B. (i) T (ii) F

C. (i) F (ii) T
D. (i) F (ii) F

24. Devise ladder programs that can be used to:
    (a) Maintain an output on, even when the input ceases and when there is a power failure
    (b) Switch on an output for a time of one cycle following a brief input
    (c) Switch on the power to a set of rungs

CHAPTER 8

Jump and Call

This chapter considers the *jump instruction*, which enables part of a program to be jumped over, and the way in which subroutines in ladder programs can be called up. Subroutines enable commonly occurring operations in a program to be repeatedly called up and used over again.

8.1 Jump

A function often provided with PLCs is the *conditional jump*. We can describe this as:

```
IF (some condition occurs) THEN
  perform some instructions
ELSE
  perform some other instructions
```

Such a facility enables programs to be designed such that if certain conditions are met, certain events occur, and if they are not met, other events occur. Thus, for example, we might need to design a system so that if the temperature is above 60°C a fan is switched on, and if below that temperature no action occurs.

Thus, if the appropriate conditions are met, this function enables part of a ladder program to be jumped over. Figure 8.1 illustrates this concept in a general manner. When there is an input to Input 1, its contacts close and there is an output to the jump relay. This then results in the program jumping to the rung in which the jump end occurs and skipping the intermediate program rungs. Thus, in this case, when there is an input to Input 1, the program jumps to rung 4 and then proceeds with rungs 5, 6, and so on. When there is no input to Input 1, the jump relay is not energized and the program then proceeds to rungs 2, 3, and so on.

Figure 8.2a shows the preceding ladder program in the form used by Mitsubishi. The jump instruction is denoted by conditional jump (CJP) and the place to which the jump occurs is denoted by end of jump (EJP). The condition that the jump will occur is that there is an input to X400. When that happens, the rungs involving inputs X401 and X403 are ignored and the program jumps to continue with the rungs following the end-jump instruction with the same number as the start-jump instruction—in this case, EJP 700.

W. Bolton: Programmable Logic Controllers, Sixth Edition. http://dx.doi.org/10.1016/B978-0-12-802929-9.00008-X

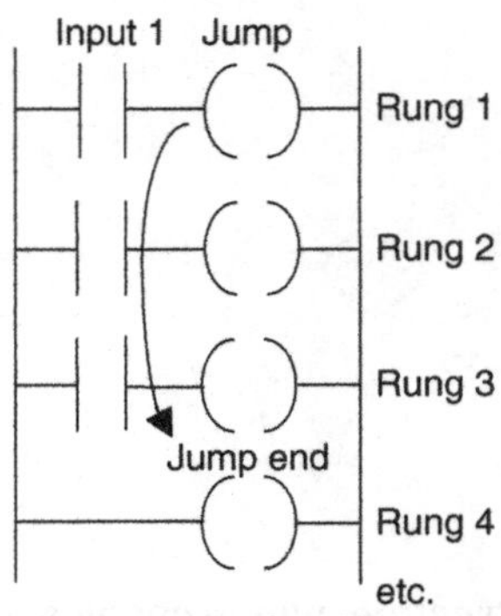

Figure 8.1: Jump.

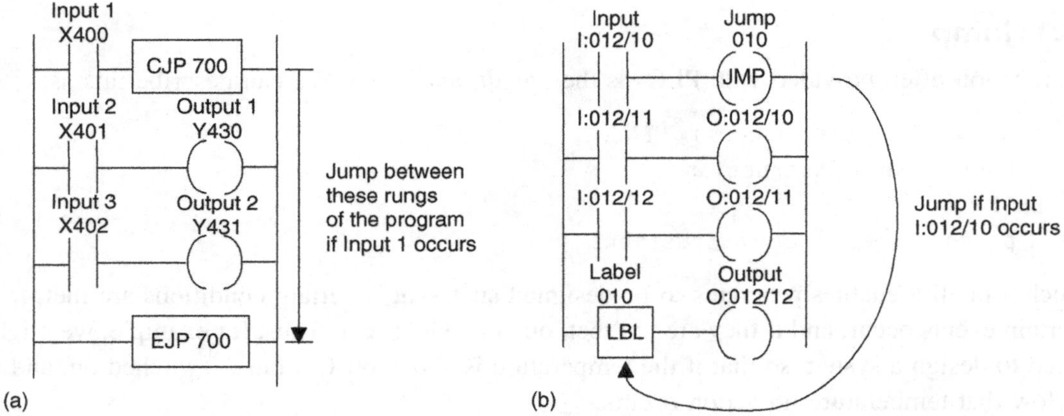

Figure 8.2: Jump: (a) a Mitsubishi program, and (b) an Allen-Bradley program.

With the Allen-Bradley PLC-5 format, the jump takes place from the jump instruction (JMP) to the label instruction (LBL). The JMP instruction is given a three-digit number from 000 to 255 and the LBL instruction the same number. Figure 8.2b shows a ladder program in this format.

With Siemens' programs, conditional jumps are represented as shown in Figure 8.3, there being a jump instruction JMP that is executed if the input is a 1 and another jump instruction JMPN that is executed if the input is 0. The end of both instructions is the label DEST.

8.1.1 Jumps Within Jumps

Jumps within jumps are possible. For example, we might have the situation shown in Figure 8.4. If the condition for jump instruction 1 is realized, the program jumps to rung 8. If the condition is not met, the program continues to rung 3. If the condition for

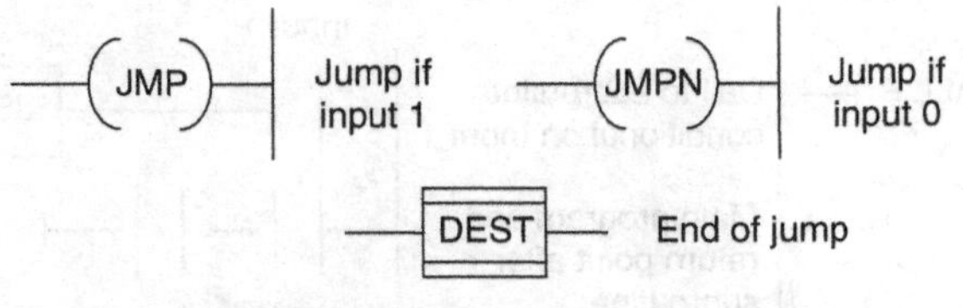

JMP Jump if input 1

JMPN Jump if input 0

DEST End of jump

Figure 8.3: Siemens' jump instructions.

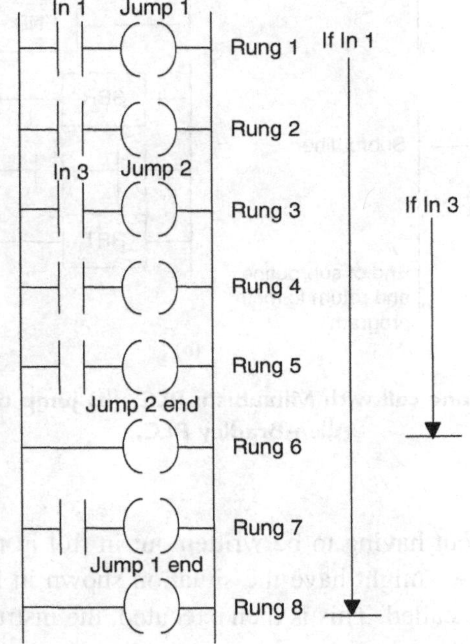

Figure 8.4: Jumps within jumps.

jump instruction 2 is realized, the program jumps to rung 6. If the condition is not met, the program continues through the rungs.

Thus if we have an input to In 1, the rung sequence is rung 1, 8, and so on. If we have no input to In 1 but we have an input to In 3, the rung sequence is 1, 2, 6, 7, 8, and so on. If we have no input to In 1 and no input to In 3, the rung sequence is 1, 2, 3, 4, 5, 6, 7, 8, and so on. The jump instruction enables different groups of program rungs to be selected, depending on the conditions occurring.

8.2 Subroutines

Subroutines are small programs to perform specific tasks that can be called for use in larger programs. The advantage of using subroutines is that they can be called repetitively to

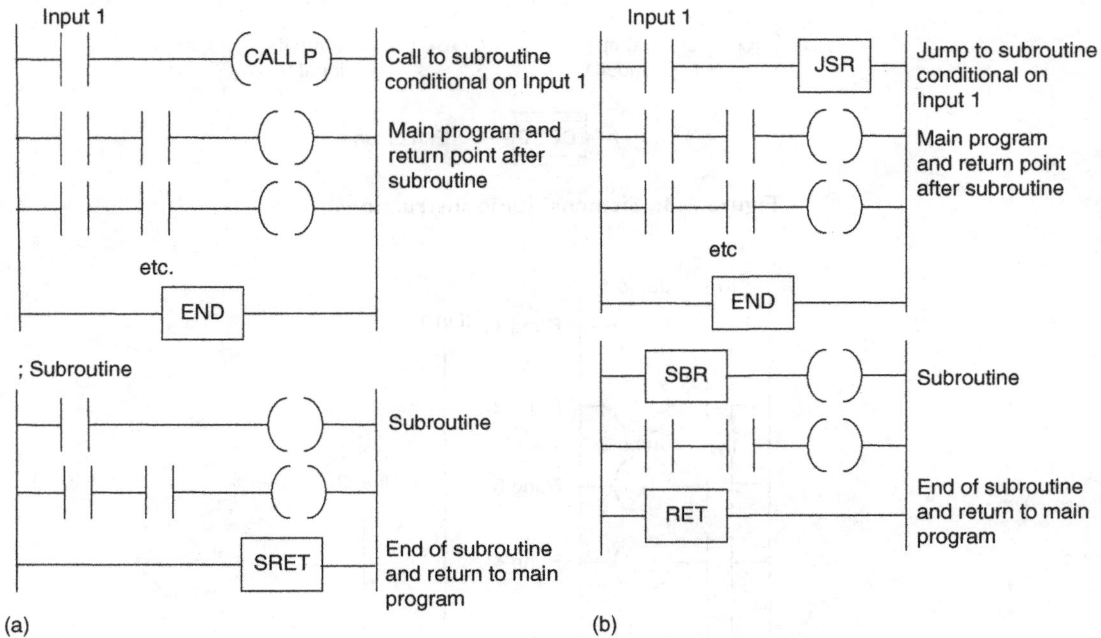

Figure 8.5: (a) Subroutine call with Mitsubishi PLC, (b) jump to subroutine call with Allen-Bradley PLC.

perform specific tasks without having to be written out in full in the larger program. Thus with a Mitsubishi program we might have the situation shown in Figure 8.5a. When input 1 occurs, the subroutine P is called. This is then executed, the instruction SRET indicating its end and the point at which the program returns to the main program.

With Allen-Bradley, subroutines are called by using a jump-to-subroutine (JSR) instruction, the start of the subroutine being indicated by SBR and its end and point of return to the main program by RET (Figure 8.5b). With other PLC manufacturers a similar format can be adopted; they might use CALL to call up a subroutine block and RET to indicate the return instruction to the main program.

8.2.1 Function Boxes

A function box approach can be used with programs and is particularly useful where there is a library of subroutine functions to be called. A *function box* is defined as being part of a program that is packaged so that it can be used a number of times in different parts of the same program or different programs. Using such boxes enables programs to be constructed from smaller, more manageable blocks. Each function box has input and output for connection to the main program, is able to store values, and contains a piece of program code

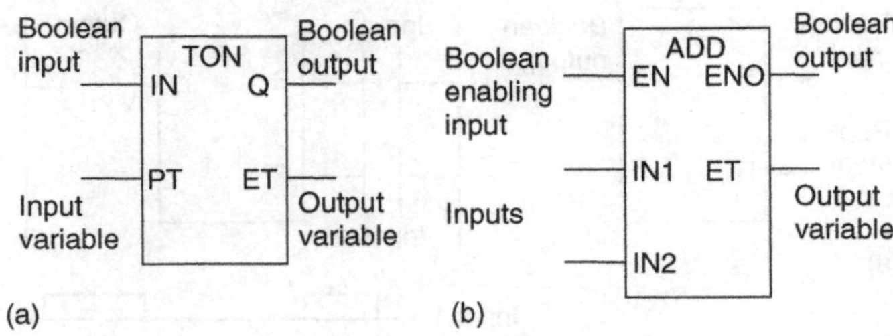

Figure 8.6: Function boxes.

that runs every time the box is used, processing the input to give the output. PLC manufacturers supply a number of function boxes that can be used within programs.

Figure 8.6a shows the form of a standard function box, such as an on-delay timer (see Chapter 9). When input IN goes to 1, output Q follows and remains 1 for the time duration set by input PT.

It is possible to control when a function box (Figure 8.6b), such as a function box to add two inputs, operates by using a special input called EN (enable). When EN is set to 1, the function is executed. If EN is set to 0, the function remains dormant and does not assign a value to its output. Such function boxes have an ENO output that is set to 1 when the function execution is successfully completed.

If the EN (enable) block input is connected directly to the left power rail, the call is without conditions and is always executed. If there is a logic operation preceding EN, the block call is executed only if the logic condition is fulfilled. In Figure 8.7 this is a closure of contacts of Input 1. Several blocks can be connected in series by connecting the ENO (enable output) of one to the EN input of the next.

As an illustration of a standard function block, consider the SR box (a bistable that is a latch described in Section 3.8; see Figure 8.8a).

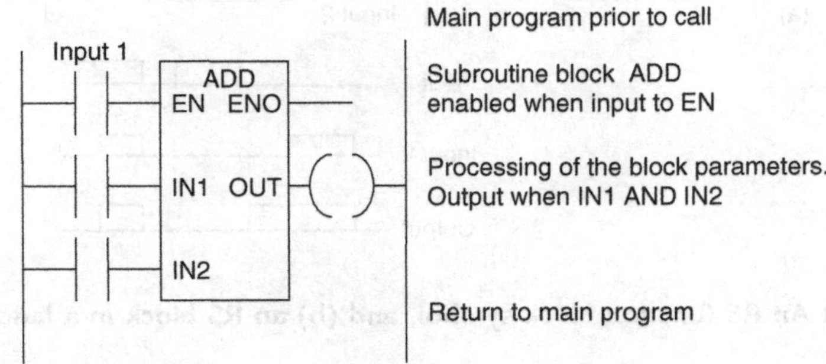

Figure 8.7: Call to subroutine function block with Siemens PLC.

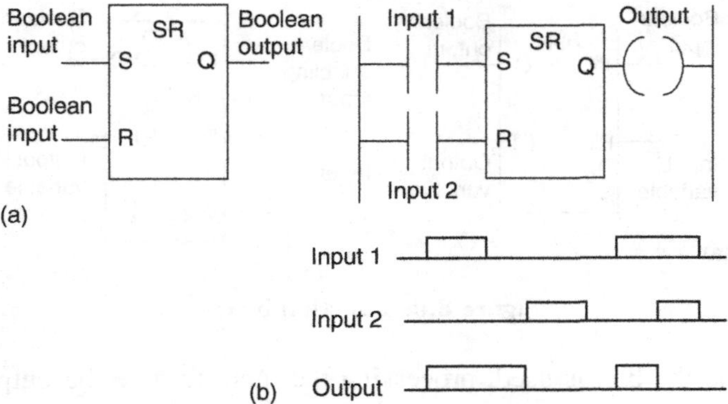

Figure 8.8: (a) An SR function block symbol, and (b) an SR block in a ladder program.

If the set input S has an input of 1 and the rest input R a 0, there is an output of 1 at Q and the block "remembers" this state until it is reset. If the set and reset signals are both 1, the output is 1. The "memory" is reset if there is a 1 input at reset R and a 0 at the set S input. Figure 8.8b shows such a block in a ladder program.

As a further illustration, Figure 8.9 shows the RS function block (a bistable latch). There is an output of 1 when the set input is 1; this then goes to 0 when reset is 1. If the set and reset inputs are both 1, the output is 0.

Other commonly used function boxes are discussed in the following chapters.

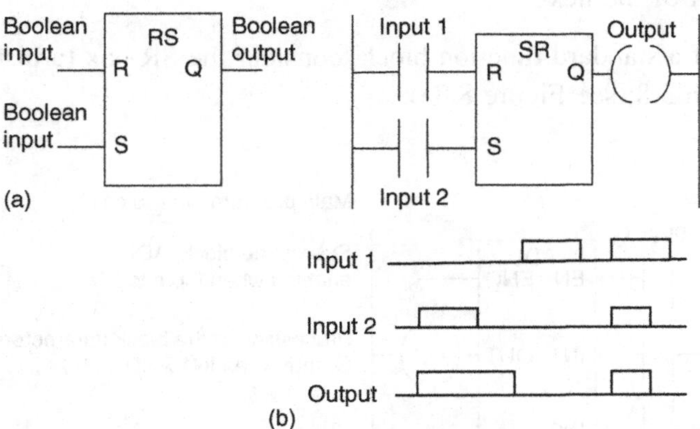

Figure 8.9: (a) An RS function block symbol, and (b) an RS block in a ladder program.

Summary

A function often provided with PLCs is the conditional jump. We can describe this as follows: IF (some condition occurs) THEN perform some instructions, ELSE perform some other instructions.

Subroutines are small programs to perform specific tasks that can be called for use in larger programs. The advantage of using subroutines is that they can be called repetitively to perform specific tasks without having to be written in full in the larger program. A function box approach can be used with programs and is particularly useful where there is a library of subroutine functions to be called. A function box is defined as being part of a program that is packaged so that it can be used a number of times in different parts of the same program or different programs. Using such boxes enables programs to be constructed from smaller, more manageable blocks. Each function box has input and output for connection to the main program, is able to store values, and contains a piece of program code that runs every time the box is used, processing the input to give the output. PLC manufacturers supply a number of function boxes that can be used within programs.

Problems

Problems 1 through 6 have four answer options: A, B, C, or D. Choose the correct answer from the answer options. Problems 1 and 2 refer to Figure 8.10, which shows a ladder diagram with inputs In 1, In 2, In 3, and In 4; outputs Out 1, Out 2, Out 3, and Out 4; and a jump instruction.

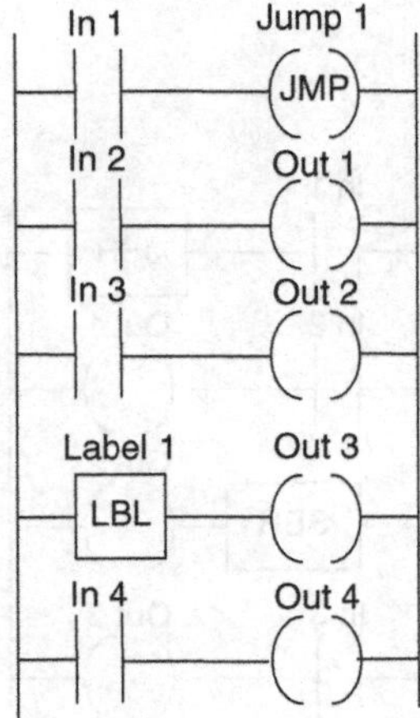

Figure 8.10: Diagram for Problems 1 and 2.

1. For the ladder diagram shown in Figure 8.10, for output Out 1 to occur:
   A. Only input In 1 must occur
   B. Both inputs In 1 and In 2 must occur
   C. Input In 1 must not occur and input 2 must occur
   D. Both inputs In 1 and In 2 must not occur

2. *Decide whether each of these statements is true (T) or false (F).* For the ladder diagram shown in Figure 8.10, following input In 1:
   (i) Output Out 1 occurs.
   (ii) Output Out 3 occurs.
   A. (i) T (ii) T
   B. (i) T (ii) F
   C. (i) F (ii) T
   D. (i) F (ii) F

Problems 3 and 4 refer to Figure 8.11, which shows a ladder diagram with inputs (In 1, In 2, and In 3), outputs (Out 1, Out 2, and Out 3), and a jump-to-subroutine instruction.

3. *Decide whether each of these statements is true (T) or false (F).* For the ladder diagram shown in Figure 8.11:
   (i) After input In 1 occurs output Out 2 occurs.
   (ii) After output Out 3 occurs the program waits for input In 2 before proceeding
   A. (i) T (ii) T
   B. (i) T (ii) F
   C. (i) F (ii) T
   D. (i) F (ii) F

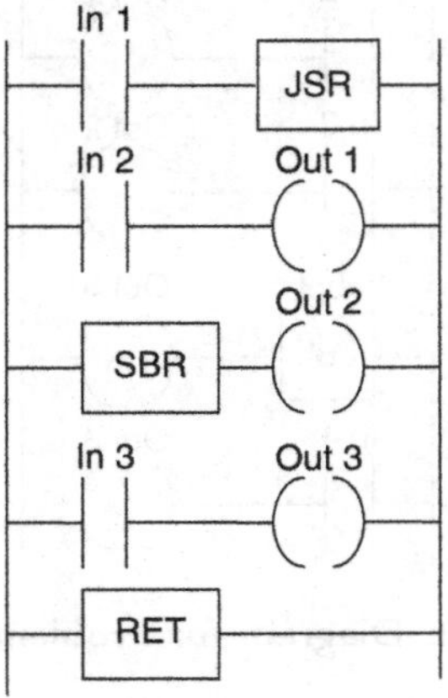

Figure 8.11: Diagram for Problems 3 and 4.

4. *Decide whether each of these statements is true (T) or false (F).* For the ladder diagram shown in Figure 8.11:
   (i) When input In 2 occurs, outputs Out 1 and Out 2 occur.
   (ii) When input In 3 occurs, output Out 3 occurs.
   A. (i) T (ii) T
   B. (i) T (ii) F
   C. (i) F (ii) T
   D. (i) F (ii) F

5. *Decide whether each of these statements is true (T) or false (F).* For the program shown in Figure 8.12, there is an output:
   (i) When input 1 is 1 and input 2 is 0.
   (ii) When input 1 is 1 and input 2 is 1.
   A. (i) T (ii) T
   B. (i) T (ii) F
   C. (i) F (ii) T
   D. (i) F (ii) F

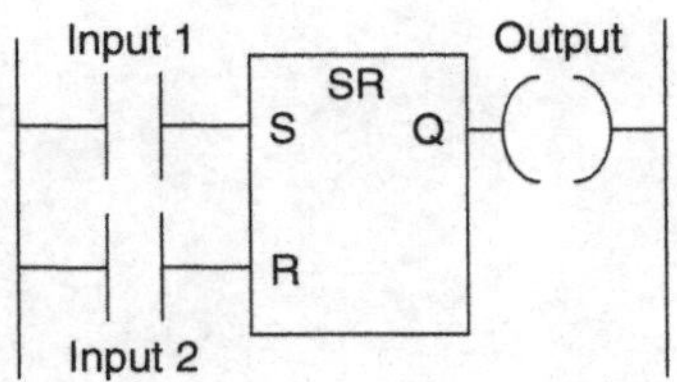

Figure 8.12: Diagram for Problem 5.

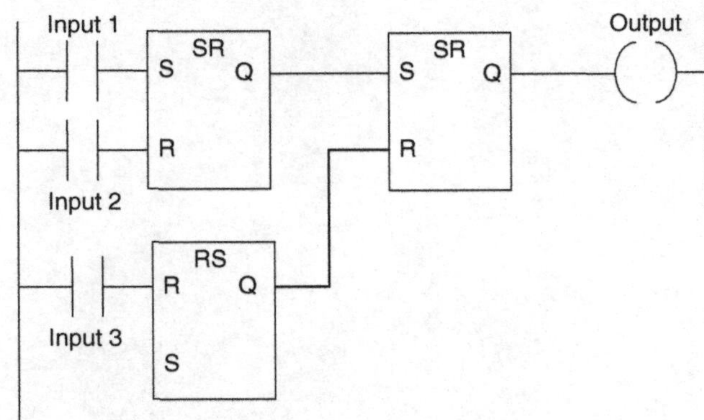

Figure 8.13: Diagram for Problem 6.

6. *Decide whether each of these statements is true (T) or false (F).* For the program shown in Figure 8.13, there is an output:
   (i) When input 1 is 1, input 2 is 0 and input 3 is 1.
   (ii) When input 1 is 0, input 2 is 1 and input 3 is 0.
   A. (i) T (ii) T
   B. (i) T (ii) F
   C. (i) F (ii) T
   D. (i) F (ii) F

7. A production plant program requires the following operations to be repeated a number of times: filling a vat, heating the liquid in the vat, and then, when the liquid is at the required temperature, emptying it. Explain how this procedure could be programmed using subroutines.

Lookup Tasks

8. For a particular PLC, determine what function boxes are available.

9. For a particular PLC, determine the programming method to be used to call up a subroutine.

CHAPTER 9

Timers

In many control tasks there is a need to control time. For example, a motor or a pump might need to be controlled to operate for a particular interval of time or perhaps be switched on after some time interval. PLCs thus have timers as built-in devices. Timers count seconds or fractions of seconds using the internal CPU clock. This chapter shows how such timer function blocks can be programmed to carry out control tasks.

9.1 Types of Timers

PLC manufacturers differ on how timers should be programmed and hence how they can be considered. A common approach is to consider timers to behave like relays with coils that when energized, result in the closure or opening of contacts after some preset time. The timer is thus treated as an output for a rung, with control being exercised over pairs of contacts elsewhere (Figure 9.1a). This is the predominant approach used in this book. Some treat a timer as a delay block that when inserted in a rung, delays signals in that rung from reaching the output (Figure 9.1b).

There are a number of different forms of timers that can be found with PLCs: *on-delay*, *off-delay*, and *pulse*. With small PLCs there is likely to be just one form, the on-delay timers. Figure 9.2 shows the IEC symbols. TON is used to denote on-delay, TOF off-delay, and TP pulse timers. On-delay is also represented by T−0 and off-delay by 0−T.

On-delay timers (TON) come on after a particular time delay (Figure 9.3a). Thus as the input goes from 0 to 1, the elapsed time starts to increase, and when it reaches the time specified by the input PT, the output goes to 1. An *off-delay timer* (TOF) is on for a fixed period of time before turning off (Figure 9.3b). The timer starts when the input signal changes from 1 to 0. Another type of timer is the *pulse timer* (TP). This timer gives an output of 1 for a fixed period of time (Figure 9.3c), starting when the input goes from 0 to 1 and switching back to 0 when the set time PT has elapsed.

The time duration for which a timer has been set is termed the *preset* and is set in multiples of the time base used. Some time bases are typically 10 ms, 100 ms, 1 s, 10 s, and 100 s. Thus a preset value of 5 with a time base of 100 ms is a time of 500 ms. For convenience, where timers are involved in this text, a time base of 1 s has been used.

W. Bolton: Programmable Logic Controllers, Sixth Edition. http://dx.doi.org/10.1016/B978-0-12-802929-9.00009-1

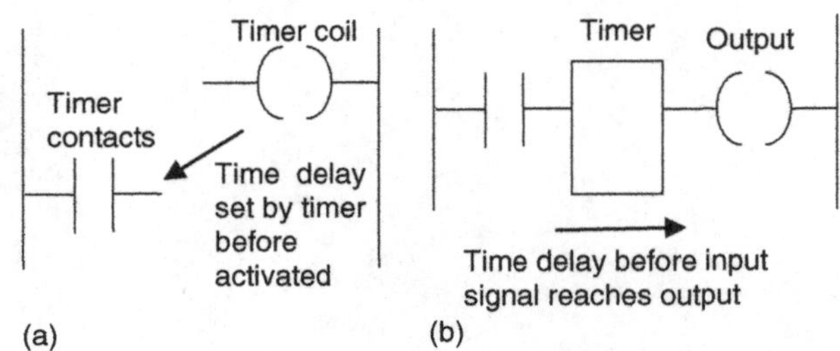

Figure 9.1: Treatment of timers.

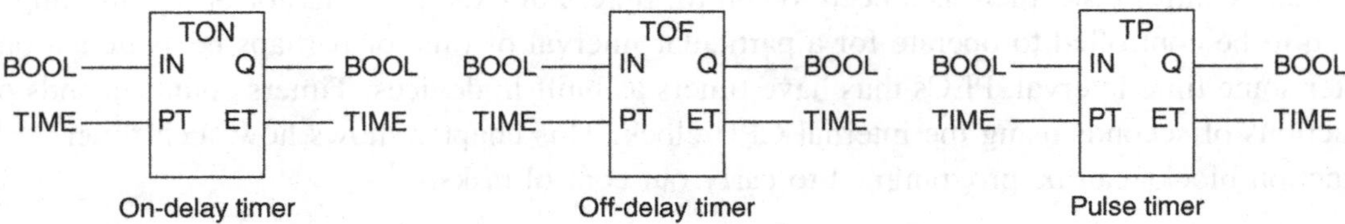

Figure 9.2: IEC 61131-1 standards: IN is the Boolean input. Q is the Boolean output. ET is the elapsed time output. PT is the input used to specify the time delay or pulse duration required.

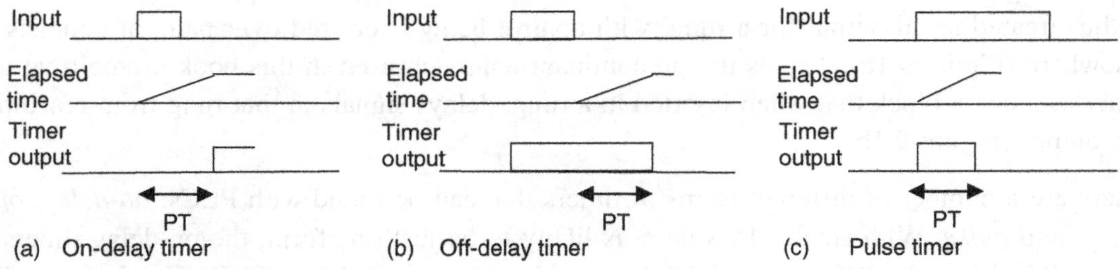

Figure 9.3: Timers: (a) on-delay, (b) off-delay, and (c) pulse.

9.2 On-Delay Timers

All PLCs generally have on-delay timers; small PLCs possibly have only this type of timer. Figure 9.4a shows a ladder rung diagram involving a on-delay timer. Figure 9.4a is typical of Mitsubishi. The timer is like a relay with a coil that is energized when input In 1 occurs (rung 1). It then closes, after some preset time delay, its contacts on rung 2. Thus the output occurs some preset time after input In 1 occurs. Figure 9.4b, an example of a possible Siemens setup, shows the timer to be a delay item in a rung, rather than a relay. When the signal at the timer's start input changes from 0 to 1, the timer starts and runs for the programmed duration, giving its output then to the output coil. The time value (TV) output can be used to ascertain the amount of time remaining at any instant. A signal input of 1 at the reset input resets the timer whether it is running or not. Techniques for the entry of

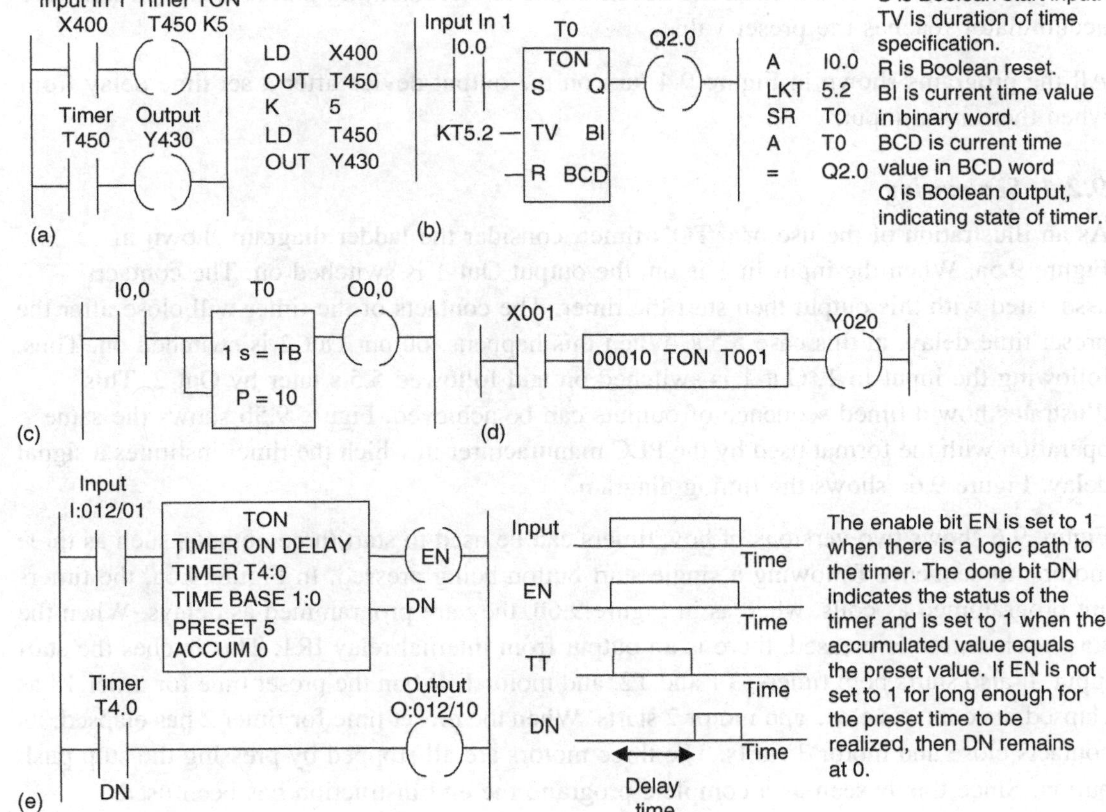

Figure 9.4: Timers: (a) Mitsubishi, (b) Siemens, (c) Telemecanique, (d) Toshiba, (e) and Allen-Bradley.

preset time values vary. Often it requires the entry of a constant K command followed by the time interval in multiples of the time base used. Figures 9.4c, 9.4d, and 9.4e show ladder diagrams for Telemecanique, Toshiba, and Allen-Bradley, respectively. The Allen-Bradley timer symbol shows the type of timer concerned, the timer address, and the time base that indicates the increments by which the timer moves to the preset value, such as 0.001 s, 0.01 s, 0.1 s or 1 s. The preset value (PRE) is the number of time increments that the timer must accumulate to reach the required time delay, and the accumulator (ACC) indicates the number of increments that the timer has accumulated while the timer is active and is reset to zero when the timer is reset (useful if a program needs to record how long a particular operation took). The Allen-Bradley timers have three Boolean bits for ladder logic control: a timer enable bit (EN), which goes on when the timer accumulator is incrementing, a timer done bit (DN), which goes on after the set time delay, and a timer timing bit (TT) that is on when the accumulator is incrementing and remains on until the accumulator reaches the preset value.

All the programs shown in Figure 9.4 turn on the output device after a set time delay from when there is an input.

9.2.1 Sequencing

As an illustration of the use of a TON timer, consider the ladder diagram shown in Figure 9.5a. When the input In 1 is on, the output Out 1 is switched on. The contacts associated with this output then start the timer. The contacts of the timer will close after the preset time delay, in this case 5.5 s. When this happens, output Out 2 is switched on. Thus, following the input In 1, Out 1 is switched on and followed 5.5 s later by Out 2. This illustrates how a timed sequence of outputs can be achieved. Figure 9.5b shows the same operation with the format used by the PLC manufacturer in which the timer institutes a signal delay. Figure 9.6c shows the timing diagram.

Figure 9.6 shows two versions of how timers can be used to start three outputs, such as three motors, in sequence following a single start button being pressed. In Figure 9.6a, the timers are programmed as coils, whereas in Figure 9.6b, they are programmed as delays. When the start push button is pressed, there is an output from internal relay IR1. This latches the start input. It also starts both timers, T1 and T2, and motor 1. When the preset time for timer 1 has elapsed, its contacts close and motor 2 starts. When the preset time for timer 2 has elapsed, its contacts close and motor 3 starts. The three motors are all stopped by pressing the stop push button. Since this is seen as a complete program, the end instruction has been used.

9.2.2 Cascaded Timers

Timers can be linked together (the term *cascaded* is used) to give longer delay times than are possible with just one timer. Figure 9.7a shows the ladder diagram for such an arrangement. Thus we might have timer 1 with a delay time of 999 s. This timer is started when there is an input to In 1. When the 999 s is up, the contacts for timer 1 close. This then starts timer 2.

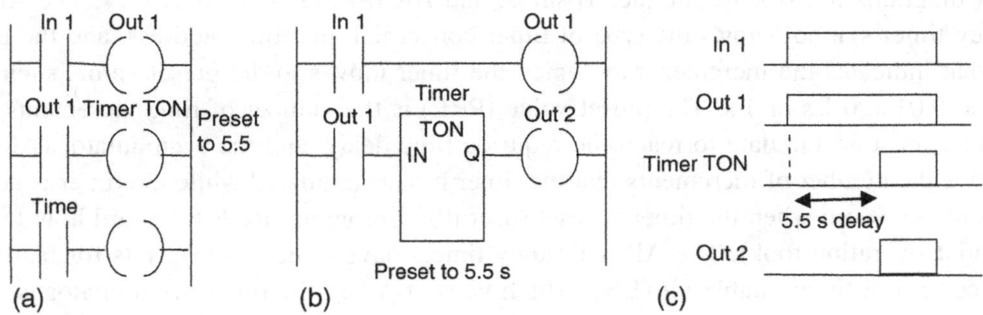

Figure 9.5: Sequenced outputs.

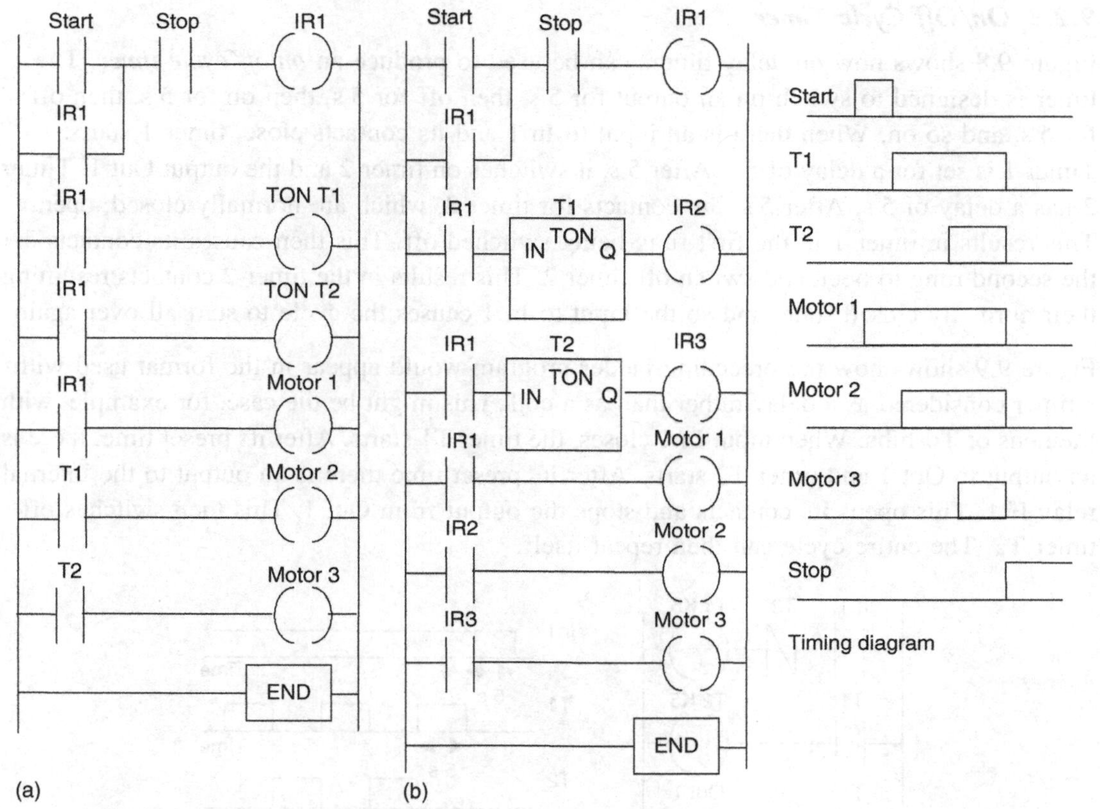

Figure 9.6: Motor sequence.

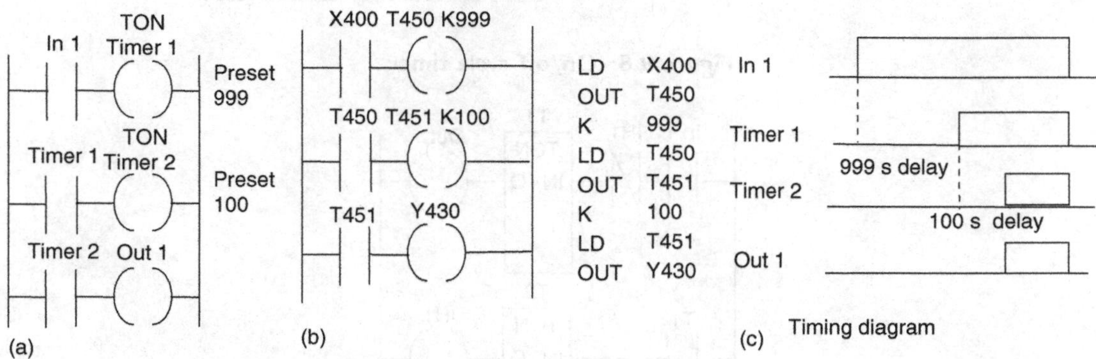

Figure 9.7: Cascaded TON timers.

This has a delay of 100 s. When this time is up, the timer 2 contacts close and there is an output from Out 1. Thus the output occurs 1099 s after the input to In 1 started. Figure 9.7b shows the Mitsubishi version of this ladder diagram with TON timers and the program instructions for that ladder.

9.2.3 On/Off Cycle Timer

Figure 9.8 shows how on-delay timers can be used to produce an *on/off cycle timer*. The timer is designed to switch on an output for 5 s, then off for 5 s, then on for 5 s, then off for 5 s, and so on. When there is an input to In 1 and its contacts close, timer 1 starts. Timer 1 is set for a delay of 5 s. After 5 s, it switches on timer 2 and the output Out 1. Timer 2 has a delay of 5 s. After 5 s, the contacts for timer 2, which are normally closed, open. This results in timer 1 in the first rung being switched off. This then causes its contacts in the second rung to open and switch off timer 2. This results in the timer 2 contacts resuming their normally closed state, and so the input to In 1 causes the cycle to start all over again.

Figure 9.9 shows how the preceding ladder program would appear in the format used with a timer considered as a delay rather than as a coil. This might be the case, for example, with Siemens or Toshiba. When input In 1 closes, the timer T1 starts. After its preset time, there is an output to Out 1 and timer T2 starts. After its preset time there is an output to the internal relay IR1. This opens its contacts and stops the output from Out 1. This then switches off timer T2. The entire cycle can then repeat itself.

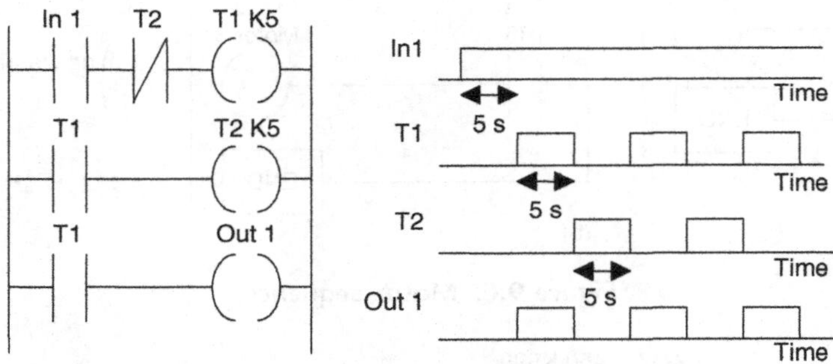

Figure 9.8: On/off cycle timer.

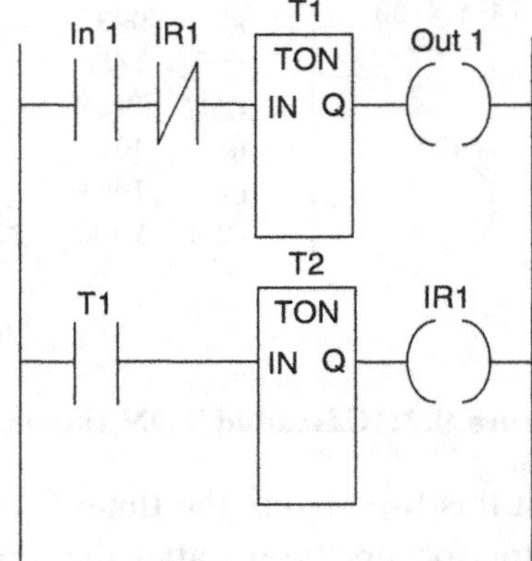

Figure 9.9: On/off cycle timer using TON timers.

9.3 Off-Delay Timers

Figure 9.10 shows how a on-delay timer can be used to produce an *off-delay timer*. With such an arrangement, when there is a momentary input to In 1, both the output Out 1 and the timer are switched on. Because the input is latched by the Out 1 contacts, the output remains on. After the preset timer delay, the timer contacts, which are normally closed, open and switch off the output. Thus the output starts as on and remains on until the time delay has elapsed.

Some PLCs have built-in off-delay timers as well as on-delay timers and thus there is no need to use an on-delay timer to produce an off-delay timer. Figure 9.11a illustrates this concept for an Allen-Bradley PLC and Figure 9.11b for a Siemens PLC, illustrating the ladder diagram and the instruction list. Note that with Siemens, the timer is considered a delay item in a rung rather than a relay. In the rectangle symbol used for the timer, the 0 precedes the T and indicates that it is an off-delay timer.

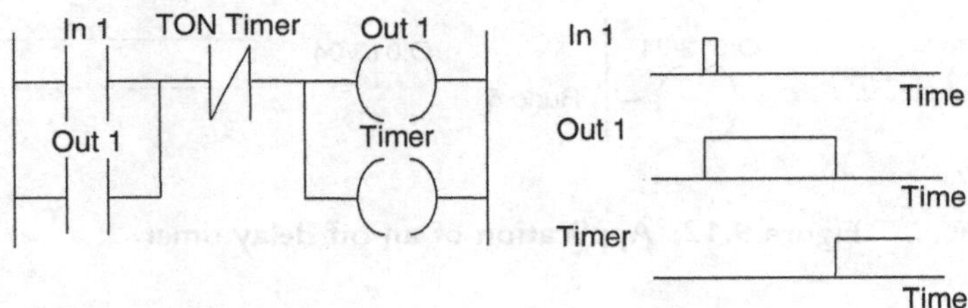

Figure 9.10: Off-delay timer using TON.

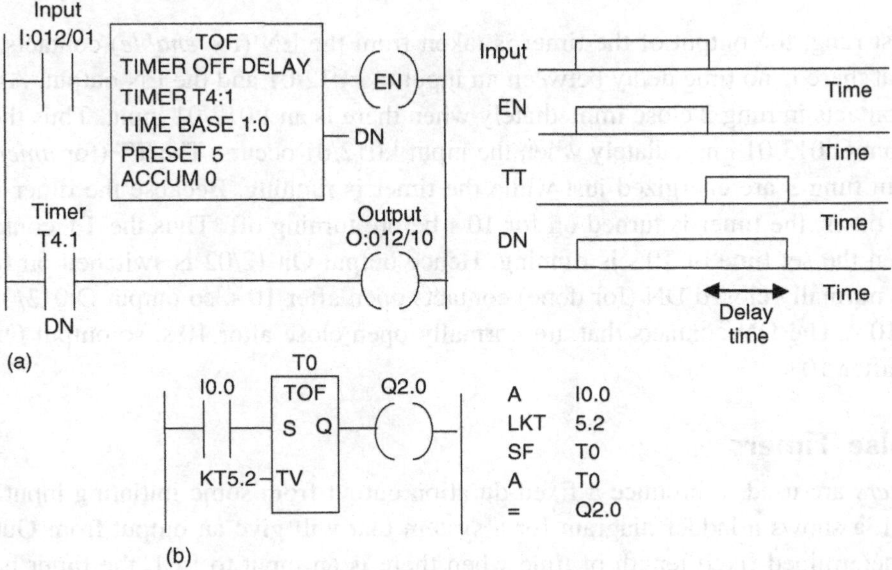

Figure 9.11: Off-delay timers: (a) Allen-Bradley, and (b) Siemens.

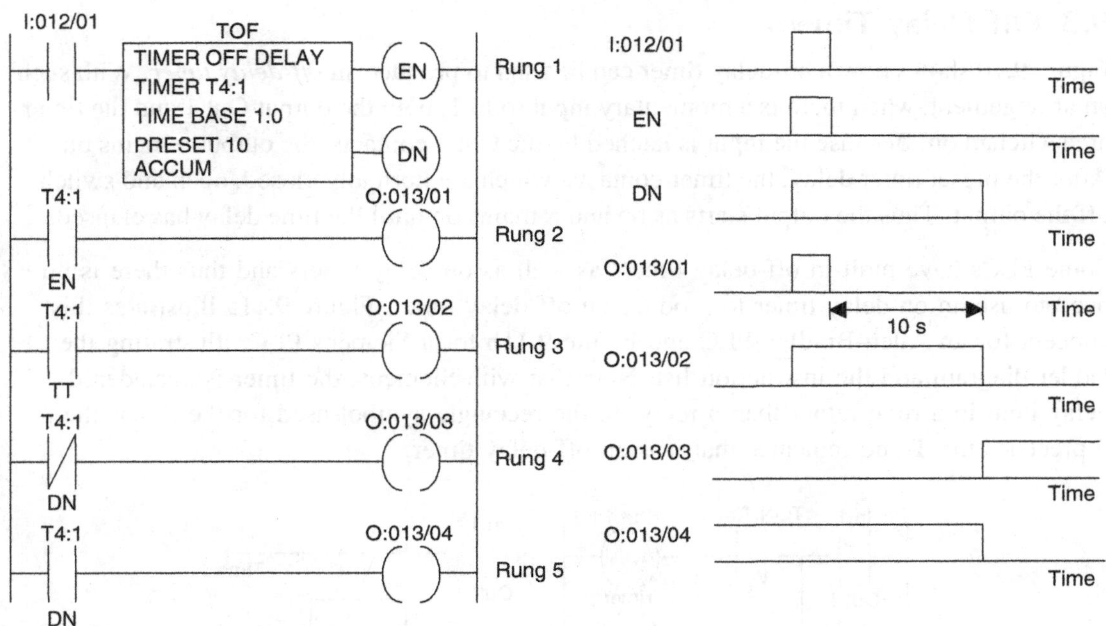

Figure 9.12: Application of an off-delay timer.

As an illustration of the use of an off-delay timer, consider the Allen-Bradley program shown in Figure 9.12. TOF is used to indicate that it is an off-delay rather than on-delay (TON) timer. The time base is set to 1:0, which is 1 s. The preset is 10, so the timer is preset to 10 s.

In the first rung, the output of the timer is taken from the EN (for *enable*) contacts. This means that there is no time delay between an input to I:012/01 and the EN output. As a result the EN contacts in rung 2 close immediately when there is an I:012/01 input. Thus there is an output from O:013/01 immediately when the input I:012/01 occurs. The TT (for *timer timing*) contacts in rung 3 are energized just while the timer is running. Because the timer is an off-delay timer, the timer is turned on for 10 s before turning off. Thus the TT contacts will close when the set time of 10 s is running. Hence output O:012/02 is switched on for this 10 s. The normally closed DN (for done) contacts open after 10 s, so output O:013/03 comes on after 10 s. The DN contacts that are normally open close after 10 s, so output O:013/04 goes off after 10 s.

9.4 Pulse Timers

Pulse timers are used to produce a fixed-duration output from some initiating input. Figure 9.13a shows a ladder diagram for a system that will give an output from Out 1 for a predetermined fixed length of time when there is an input to In 1, the timer being one involving a coil. There are two outputs for the input In 1. When there is an input to In 1,

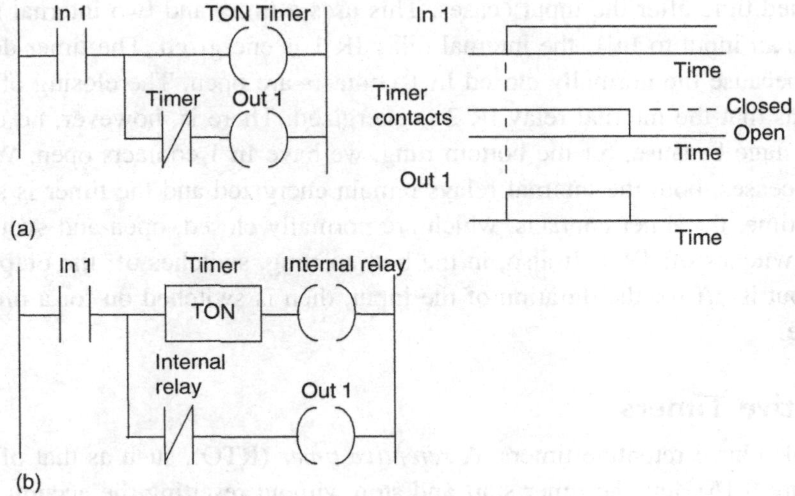

Figure 9.13: Pulse-on timer.

there is an output from Out 1 and the timer starts. When the predetermined time has elapsed, the timer contacts open. This switches off the output. Thus the output remains on for only the time specified by the timer.

Figure 9.13b shows an equivalent ladder diagram to Figure 9.13a but employs a timer that produces a delay in the time taken for the signal to reach the output.

In Figure 9.13, the pulse timer has an output switched on by an input for a predetermined time, then switched off. Figure 9.14 shows another pulse timer that switches on an output for

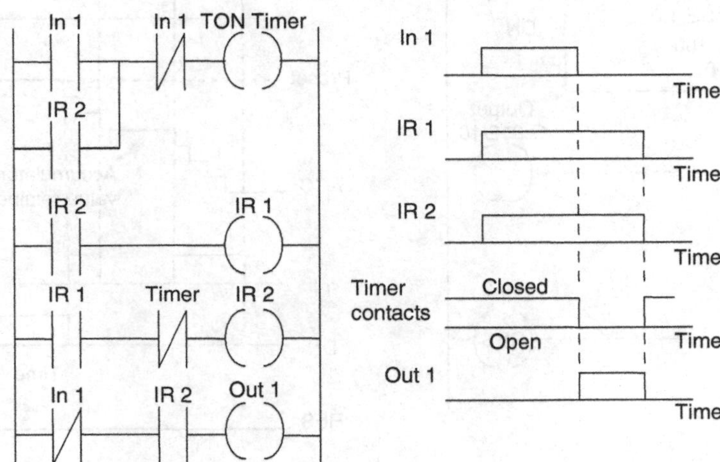

Figure 9.14: Pulse timer on, when output ceases.

a predetermined time after the input ceases. This uses a timer and two internal relays. When there is an input to In 1, the internal relay IR 1 is energized. The timer does not start at this point because the normally closed In 1 contacts are open. The closing of the IR 1 contacts means that the internal relay IR 2 is energized. There is, however, no output from Out 1 at this stage because, for the bottom rung, we have In 1 contacts open. When the input to In 1 ceases, both the internal relays remain energized and the timer is started. After the set time, the timer contacts, which are normally closed, open and switch off IR 2. This in turn switches off IR 1. It also, in the bottom rung, switches off the output Out 1. Thus the output is off for the duration of the input, then is switched on for a predetermined length of time.

9.5 Retentive Timers

Some PLCs also have retentive timers. A *retentive timer* (RTO), such as that of Allen-Bradley (Figure 9.15), lets the timer start and stop without resetting the accumulated value. The retentive timer operates the same as a TON timer except that the accumulator is not reset when the timer enable goes to 0 and continues to increment whenever the enable bit goes to 1. When the accumulator bit reaches the preset value, the timer timing bit goes to 0 and the done bit to 1 and the accumulator bit remains in that state until a reset instruction is received.

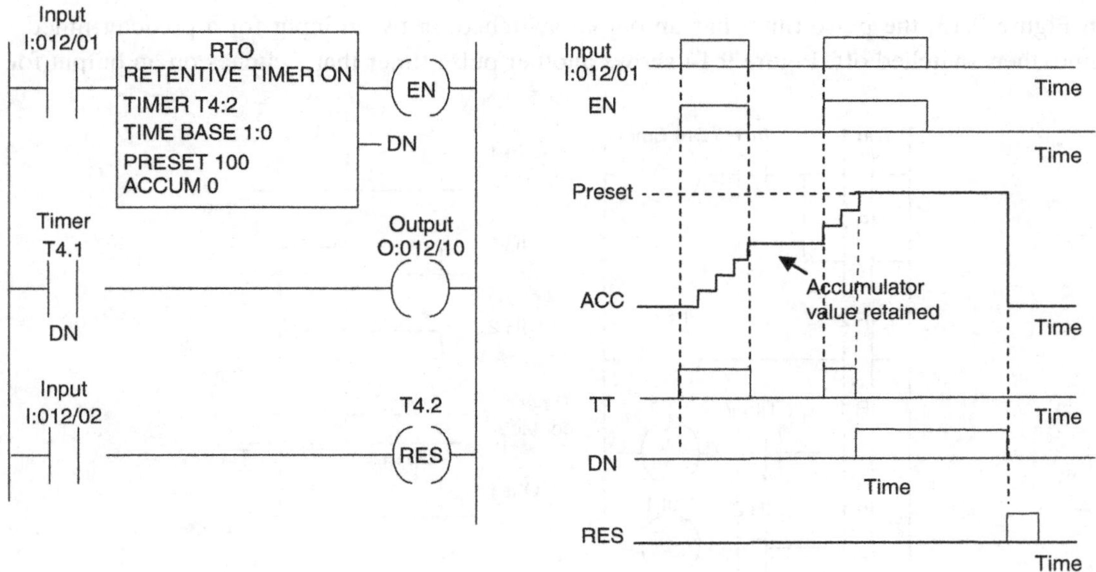

Figure 9.15: Allan-Bradley retentive timer.

9.6 Programming Examples

Consider a program (Figure 9.16) that could be used to flash a light on and off as long as there is some output occurring. Thus we might have both timer 0 and timer 1 set to 1 s. When the output occurs, timer 0 starts and switches on after 1 s. This closes the timer 0 contacts and starts timer 1. This switches on after 1 s and, in doing so, switches off timer 0. In so doing, it switches off itself. The lamp is on only when timer 0 is on, and so we have a program to flash the lamp on and off as long as there is an output.

As an illustration of programming involving timers, consider the sequencing of traffic lights to give the sequence red only, red plus amber, green, and amber, then repeat. A simple system might just have the sequence triggered by time, with each of the possible states occurring in sequence for a fixed amount of time. Figure 9.17 shows the sequential function chart and a possible ladder program to give the sequence.

As another illustration of the use of a timer, consider Figure 9.18, which is a program to initially switch on motor 1 and its signal lamp 1, then after a time set by the timer to switch off motor 1 and signal lamp 1 and switch on motor 2 and signal lamp 2. When there is an input, the timer starts. Until the timer has timed out, the closed timer contacts remain closed and so motor 1 and lamp 1 are on; the open timer contacts remain open and so motor 2 and lamp 2 are off. When the timer has timed out, the closed timer contacts open and switch off motor 1 and lamp 1; the open timer contacts close and switch on motor 2 and lamp 2.

Figure 9.19 shows an example of where a TON timer and a TOF timer are both used to control a motor. The first rung starts with a start switch that is latched by the TON timer so that the input to it remains on, even when the start switch is no longer activated. After the delay time of the TON timer the motor is started. The motor is stopped by input to the Stop input, but the motor continues running for a time delay set by the TOF timer before finally stopping.

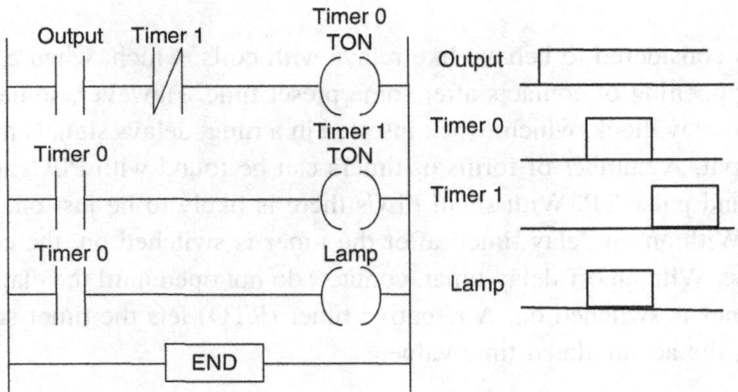

Figure 9.16: Flashing light.

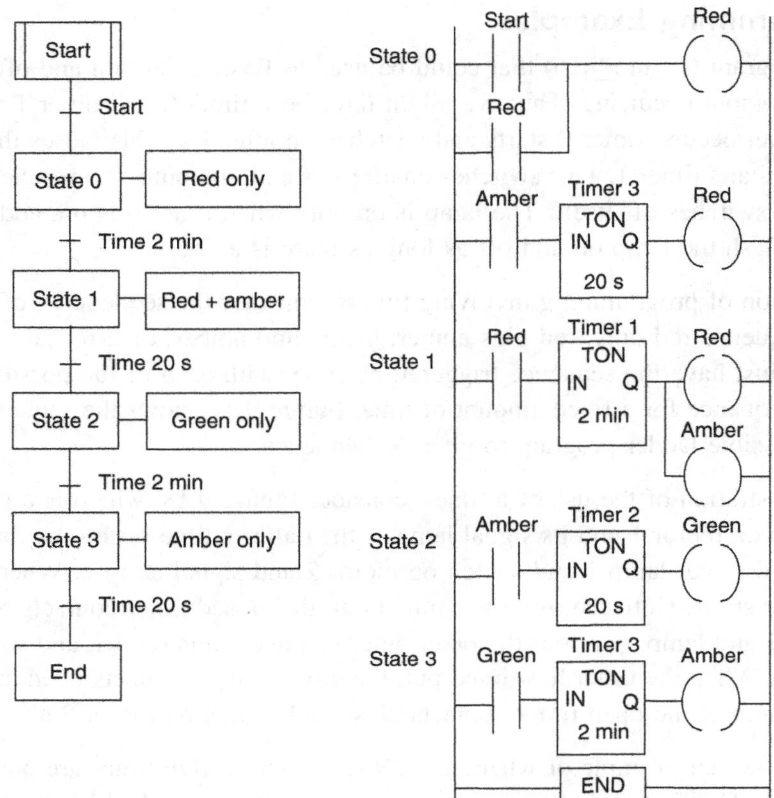

Figure 9.17: Traffic light sequence.

Summary

Timers are often considered to behave like relays with coils, which, when energized, result in the closing or opening of contacts after some preset time. However, some manufacturers treat a timer as a delay block, which, when inserted in a rung, delays signals in that rung from reaching the output. A number of forms of timers can be found with PLCs: on-delay TON, off-delay TOF, and pulse TP. With small PLCs there is likely to be just one form: the on-delay timer. With an on-delay timer, after the timer is switched on, the contacts close after a preset time. With an off-delay timer, contacts do not open until the elapse of the preset time after the timer is switched on. A retentive timer (RTO) lets the timer start and stop without resetting the accumulated time value.

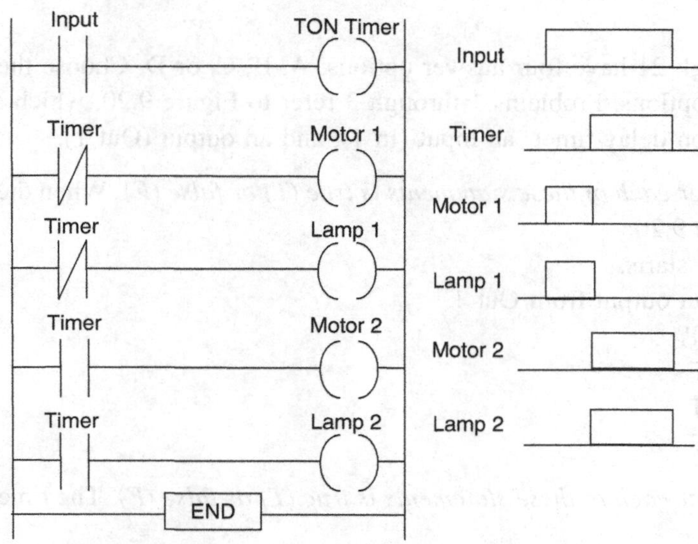

Figure 9.18: Program to switch on in sequence motor 1 and then switch it off and switch motor 2 on.

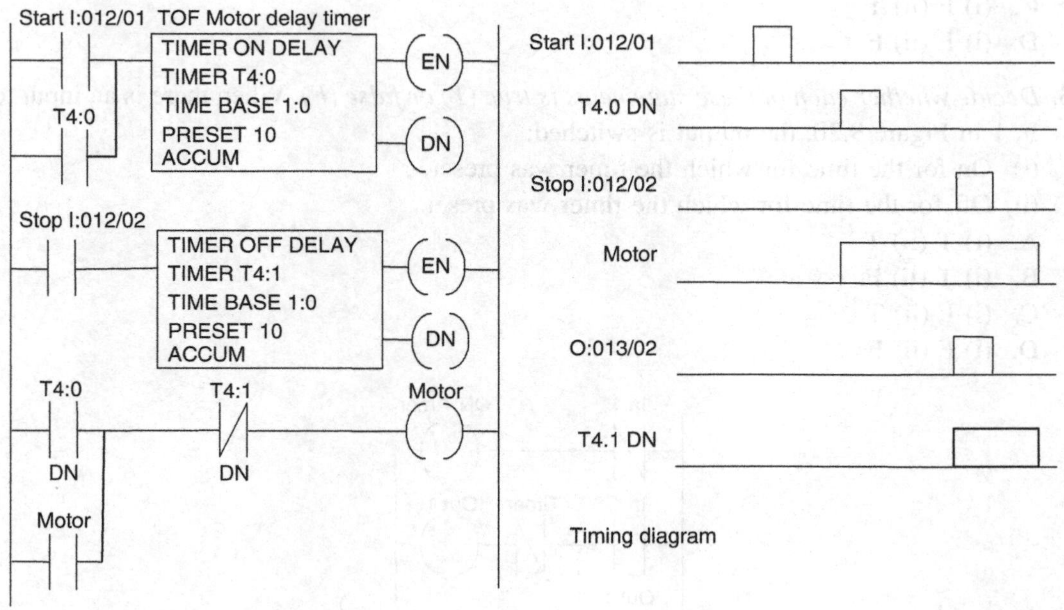

Figure 9.19: Motor delayed start and delayed switch off.

Problems

Problems 1 through 21 have four answer options: A, B, C, or D. Choose the correct answer from the answer options. Problems 1 through 3 refer to Figure 9.20, which shows a ladder diagram with an on-delay timer, an input (In 1), and an output (Out 1).

1. *Decide whether each of these statements is true (T) or false (F).* When there is an input to In 1 in Figure 9.20:
   (i) The timer starts.
   (ii) There is an output from Out 1.
   A. (i) T (ii) T
   B. (i) T (ii) F
   C. (i) F (ii) T
   D. (i) F (ii) F

2. *Decide whether each of these statements is true (T) or false (F).* The timer in Figure 9.20 starts when:
   (i) There is an output.
   (ii) The input ceases.
   A. (i) T (ii) T
   B. (i) T (ii) F
   C. (i) F (ii) T
   D. (i) F (ii) F

3. *Decide whether each of these statements is true (T) or false (F).* When there is an input to In 1 in Figure 9.20, the output is switched:
   (i) On for the time for which the timer was preset.
   (ii) Off for the time for which the timer was preset.
   A. (i) T (ii) T
   B. (i) T (ii) F
   C. (i) F (ii) T
   D. (i) F (ii) F

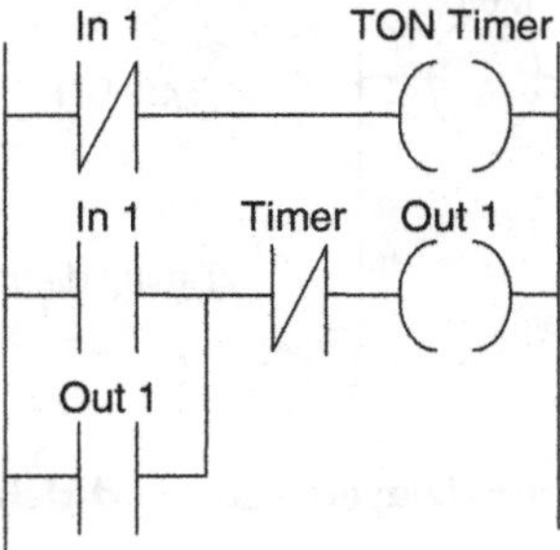

Figure 9.20: Diagram for Problems 1 through 3.

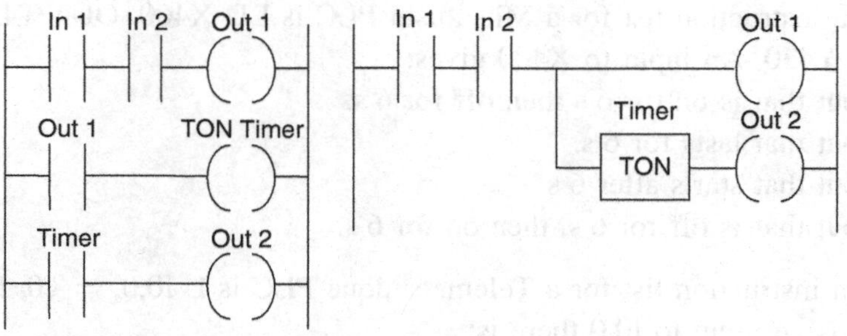

Figure 9.21: Diagram for Problems 4 through 6.

Problems 4 through 6 refer to Figure 9.21, which shows two alternative versions of a ladder diagram with two inputs (In 1 and In 2), two outputs (Out 1 and Out 2), and an on-delay timer.

4. *Decide whether each of these statements is true (T) or false (F).* When there is just an input to In 1 in Figure 9.21:
   (i) The timer starts.
   (ii) There is an output from Out 2.
   A. (i) T (ii) T
   B. (i) T (ii) F
   C. (i) F (ii) T
   D. (i) F (ii) F

5. *Decide whether each of these statements is true (T) or false (F).* When there is just an input to In 2 in Figure 9.21:
   (i) The timer starts.
   (ii) There is an output from Out 2.
   A. (i) T (ii) T
   B. (i) T (ii) F
   C. (i) F (ii) T
   D. (i) F (ii) F

6. *Decide whether each of these statements is true (T) or false (F).* When there is an input to In 1 and no input to In 2 in Figure 9.21, there is an output from Out 2 that:
   (i) Starts immediately.
   (ii) Ceases after the timer preset time.
   A. (i) T (ii) T
   B. (i) T (ii) F
   C. (i) F (ii) T
   D. (i) F (ii) F

7. The program instruction list for a Mitsubishi PLC is LD X400, OUT T450, K 6, LD T450, OUT Y430. An input to X400 gives:
   A. An output that is on for 6 s then off for 6 s.
   B. An output that lasts for 6 s.
   C. An output that starts after 6 s.
   D. An output that is off for 6 s, then on for 6 s.

8. The program instruction list for a Telemecanique PLC is L I0,0, = T0, L T0, = O0,0. When there is an input to I0,0 there is:
   A. An output that is on for 6 s then off for 6 s.
   B. An output that lasts for 6 s.
   C. An output that starts after 6 s.
   D. An output that is off for 6 s, then on for 6 s.

Problems 9 and 10 refer to the program instruction list for a Mitsubishi PLC: LD X400, OR Y430, ANI T450, OUT Y430, LD X401, OR M100, AND Y430, OUT T450, K 10, OUT M100.

9. *Decide whether each of these statements is true (T) or false (F).* When there is a momentary input to X400:
   (i) The output from Y430 starts.
   (ii) The output from Y430 ceases after 10 s.
   A. (i) T (ii) T
   B. (i) T (ii) F
   C. (i) F (ii) T
   D. (i) F (ii) F

10. *Decide whether each of these statements is true (T) or false (F).* The output from Y430:
    (i) Starts when there is a momentary input to X401.
    (ii) Ceases 10 s after the input to X401.
    A. (i) T (ii) T
    B. (i) T (ii) F
    C. (i) F (ii) T
    D. (i) F (ii) F

Problems 11 and 12 refer to Figure 9.22, which shows a system with an input (In 1), an on-delay timer, and an output (Out 1). The timer is set for a time of 6 s. The graph shows how the signal to the input varies with time.

11. *Decide whether each of these statements is true (T) or false (F).* The output from Out 1 in Figure 9.22:
    (i) Starts when the input starts.
    (ii) Ceases 6 s after the start of the input.

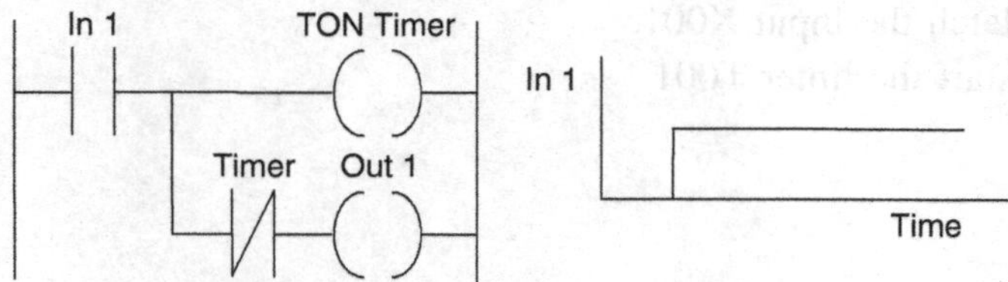

Figure 9.22: Diagram for Problems 11 and 12.

A. (i) T (ii) T
B. (i) T (ii) F
C. (i) F (ii) T
D. (i) F (ii) F

12. *Decide whether each of these statements is true (T) or false (F).* The timer contacts in Figure 9.22:
 (i) Remain closed for 6 s after the start of the input.
 (ii) Open 6 s after the input starts.
 A. (i) T (ii) T
 B. (i) T (ii) F
 C. (i) F (ii) T
 D. (i) F (ii) F

Problems 13 through 15 refer to Figure 9.23, which shows a ladder program for a Toshiba PLC involving internal relays, denoted by the letter R, and a TON timer with a preset of 20 s.

13. *Decide whether each of these statements is true (T) or false (F).* The internal relay R000 in Figure 9.23:

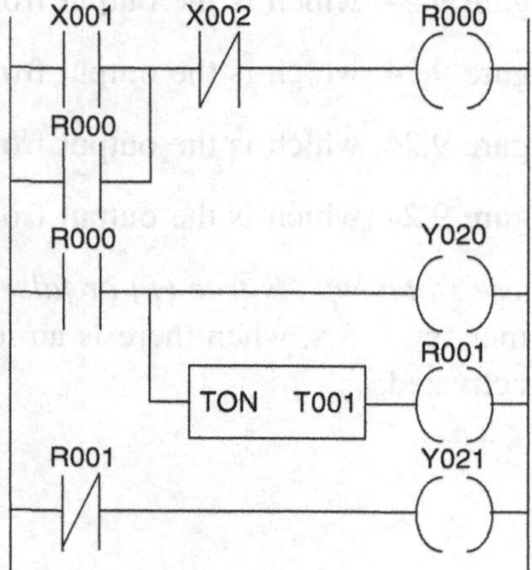

Figure 9.23: Diagram for Problems 13 through 15.

(i) Is used to latch the input X001.
(ii) Is used to start the timer T001.
A. (i) T (ii) T
B. (i) T (ii) F
C. (i) F (ii) T
D. (i) F (ii) F

14. *Decide whether each of these statements is true (T) or false (F).* With no input to X002 in Figure 9.23, the output Y020 is:
(i) Energized when there is an input to X001.
(ii) Ceases when there is no input to X001.
A. (i) T (ii) T
B. (i) T (ii) F
C. (i) F (ii) T
D. (i) F (ii) F

15. *Decide whether each of these statements is true (T) or false (F).* With no input to X002 in Figure 9.23:
(i) The output Y021 is switched on 20 s after the input X001.
(ii) The output Y020 is switched off 20 s after the input X001.
A. (i) T (ii) T
B. (i) T (ii) F
C. (i) F (ii) T
D. (i) F (ii) F

Problems 16 through 19 refer to Figure 9.24, which shows an Allen-Bradley program, and Figure 9.25, which shows a number of time charts for a particular signal input to I:012/01.

16. For the input shown in Figure 9.24, which is the output from O:013/01 in Figure 9.25?

17. For the input shown in Figure 9.24, which is the output from O:013/02 in Figure 9.25?

18. For the input shown in Figure 9.24, which is the output from O:013/03 in Figure 9.25?

19. For the input shown in Figure 9.24, which is the output from O:013/04 in Figure 9.25?

20. *Decide whether each of these statements is true (T) or false (F).* For the ladder program in Figure 9.26, with the timer set to 5 s, when there is an input to In 1:
(i) Out 1 is immediately activated.
(ii) Out 2 is activated after 5 s.
A. (i) T (ii) T
B. (i) T (ii) F
C. (i) F (ii) T
D. (i) F (ii) F

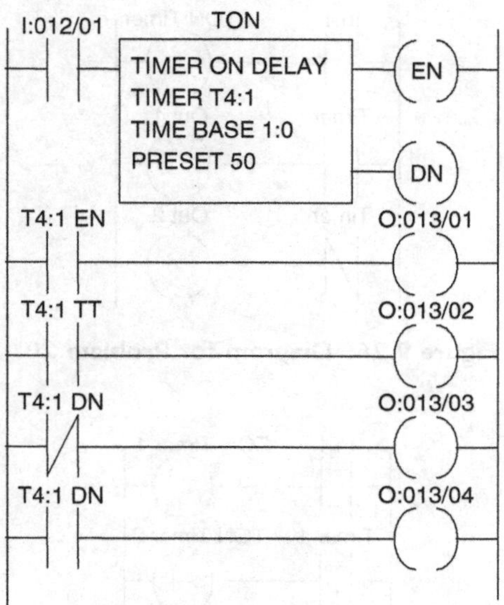

Figure 9.24: Diagram for Problems 16 through 19.

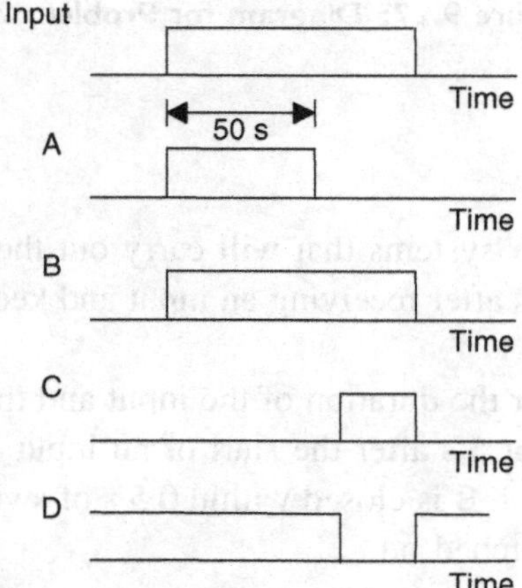

Figure 9.25: Diagram for Problems 16 through 19.

21. *Decide whether each of these statements is true (T) or false (F).* For the ladder program in Figure 9.27, with both timers set to 10 s, when there is an input to In 1:
    (i) Out 1 is activated 20 s after the input occurs.
    (ii) Out 1 is activated for a time of 20 s.
    A. (i) T (ii) T
    B. (i) T (ii) F

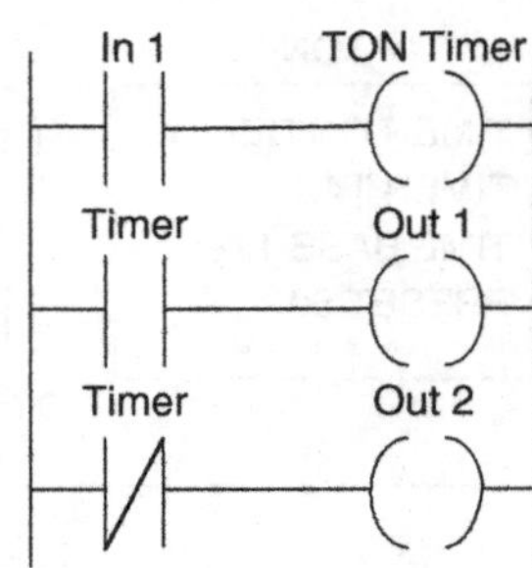

Figure 9.26: Diagram for Problem 20.

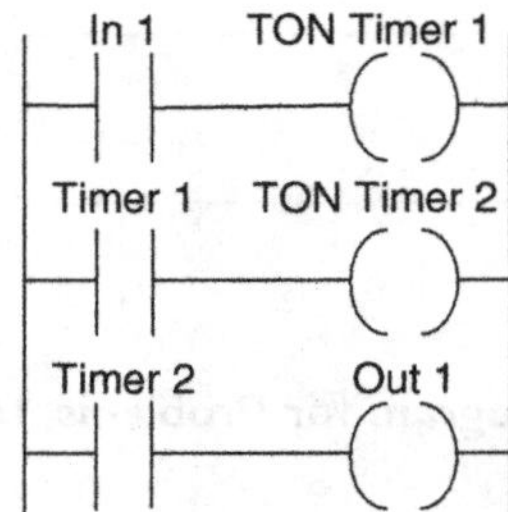

Figure 9.27: Diagram for Problem 21.

C. (i) F (ii) T
D. (i) F (ii) F

22. Devise ladder programs for systems that will carry out the following tasks:
 (a) Switch on an output 5 s after receiving an input and keep it on for the duration of that input.
 (b) Switch on an output for the duration of the input and then keep it on for a further 5 s.
 (c) Switch on an output for 5 s after the start of an input signal.
 (d) Start a machine if switch B is closed within 0.5 s of switch A being closed; otherwise the machine is not switched on.

Lookup Tasks

23. Look up the timers available for a particular range of PLCs.

CHAPTER 10

Counters

Counters are provided as built-in elements in PLCs and allow the number of occurrences of input signals to be counted. Some uses might include where items have to be counted as they pass along a conveyor belt, the number of revolutions of a shaft, or perhaps the number of people passing through a door. This chapter describes how such counters can be programmed.

10.1 Forms of Counter

A counter is set to some preset number value and, when this value of input pulses has been received, it will operate its contacts. Normally open contacts would be closed, normally closed contacts opened.

There are two basic types of counter: down-counters and up-counters. *Down-counters* count down from the preset value to zero, that is, events are subtracted from the set value. When the counter reaches the zero value, its contacts change state. Most PLCs offer down-counting. *Up-counters* count from zero up to the preset value, that is, events are added until the number reaches the preset value. When the counter reaches the set value, its contacts change state. Some PLCs offer the facility for both down- and up-counting. Figure 10.1 shows the IEC symbols for such counters.

Different PLC manufacturers deal with counters in slightly different ways. Some count down (CTD) or up (CTU) and reset and treat the counter as though it is a relay coil and so a rung output. In this way, counters can be considered to consist of two basic elements: one relay coil to count input pulses and one to reset the counter, the associated contacts of the counter being used in other rungs. Figure 10.2a illustrates this method. Mitsubishi is one of the manufacturers that uses this method. Others treat the counter as an intermediate block in a rung from which signals emanate when the count is attained. Figure 10.2b illustrates this method, used by Siemens, among others.

10.2 Programming

Figure 10.3 shows a basic counting circuit. Each time there is a transition from 0 to 1 at input In 1, the counter is reset. When there is an input to In 2 and a transition from 0 to 1, the counter starts counting. If the counter is set for, say, 10 pulses, then when 10 pulse inputs,

W. Bolton: Programmable Logic Controllers, Sixth Edition. http://dx.doi.org/10.1016/B978-0-12-802929-9.00010-8

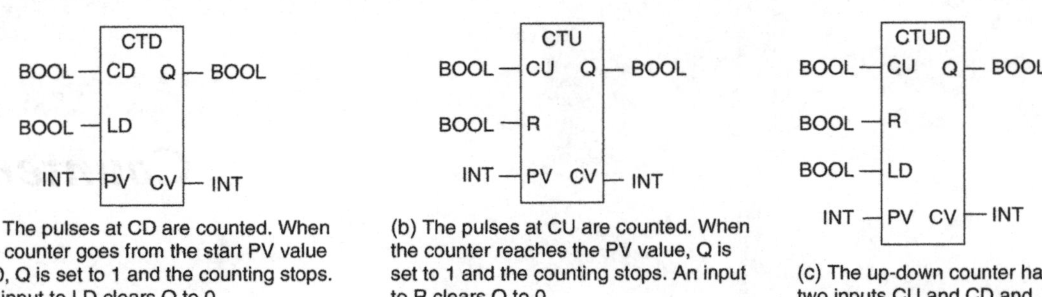

(a) The pulses at CD are counted. When the counter goes from the start PV value to 0, Q is set to 1 and the counting stops. An input to LD clears Q to 0.

(b) The pulses at CU are counted. When the counter reaches the PV value, Q is set to 1 and the counting stops. An input to R clears Q to 0.

(c) The up-down counter has two inputs CU and CD and can be used to count up on one input and down on the other

Figure 10.1: IEC symbols for counters: (a) down-counter, (b) up-counter, and (c) up-down counter.

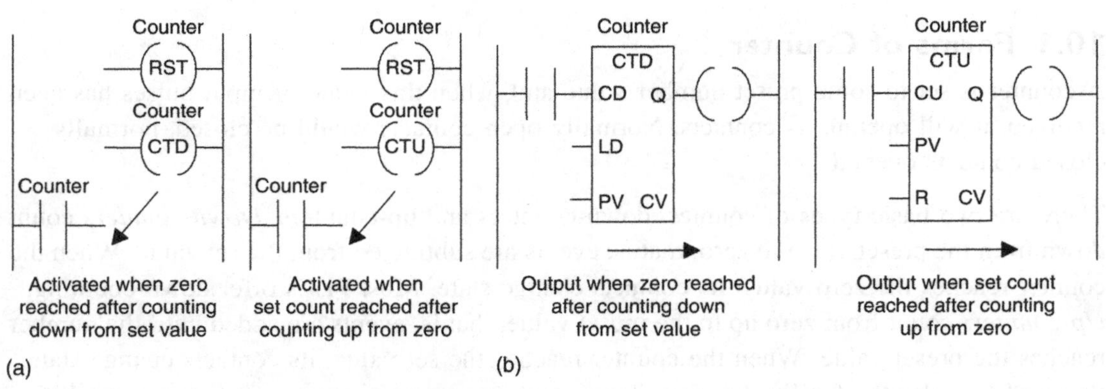

Figure 10.2: Different counter representations. (a) Counter as coils with contacts in another rung, RST is reset. (b) The IEC 61131-3 representation as an element in a rung.

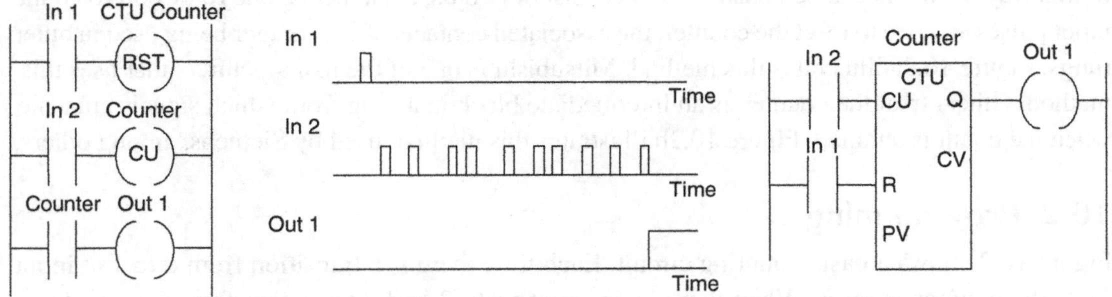

Figure 10.3: Basic counter program.

that is, 10 transitions from 0 to 1, have been received at In 2, the counter's contacts will close and there will be an output from Out 1. If at any time during the counting there is an input to In 1, the counter will be reset, start all over again, and count for 10 pulses.

Figure 10.4a shows how the preceding program and its program instruction list would appear with a Mitsubishi PLC and a CTU counter. The reset and counting elements are combined in a single box spanning the two rungs. You can consider the rectangle to enclose the two counter ( ) outputs in Figure 10.3. The count value is set by a K program instruction. Figure 10.4b shows the same program with a Siemens PLC. With this ladder program, the counter is considered a delay element in the output line (as shown in Figure 10.1b). The counter is reset by an input to I0.1 and counts the pulses into input I0.0. The CU indicates that it is a up-count counter; a CD indicates a down-count counter. The counter set value is indicated by the LKC number. Figure 10.4c illustrates the program for a Toshiba PLC.

Figures 10.5a and 10.5b show the program for Allen-Bradley with up-count and down-count counters. The following are terms associated with such counters:

- The preset value (PRE) is the count value that the counter must accumulate before the counter output is 1.
- The accumulated value (ACC) is the accumulated number of 0 to 1 transitions of the counter rung. The count-up (CU) enable bit is 1 when the input logic makes the up-counter rung 1 and 0 when the rung is 0. The count-down (CD) enable bit is 1 when the input logic makes the down-counter rung 1 and 0 when the rung is 0.
- The done (DN) bit is 1 for both counters when the ACC value is equal to or greater than the PRE value and 0 when it is less.

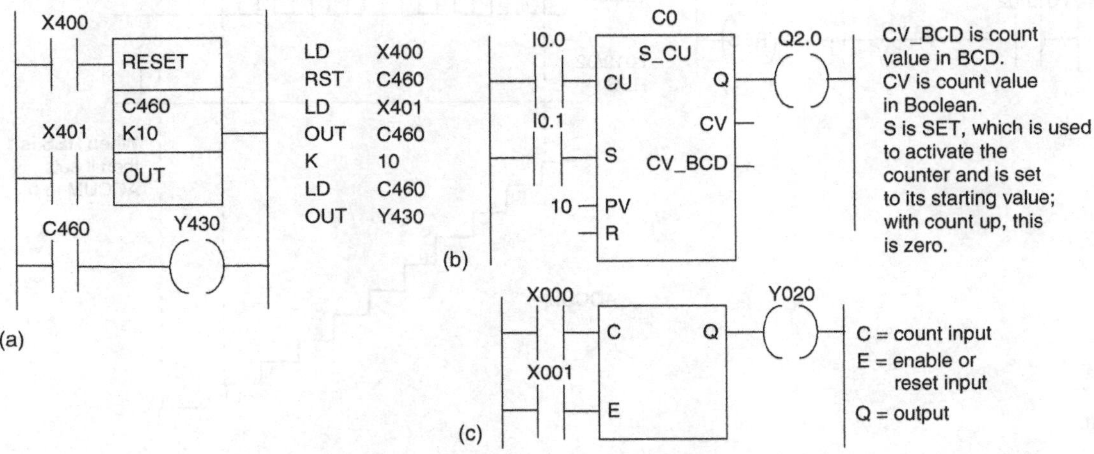

Figure 10.4: (a) A Mitsubishi program, (b) a Siemens program, and (c) a Toshiba program.

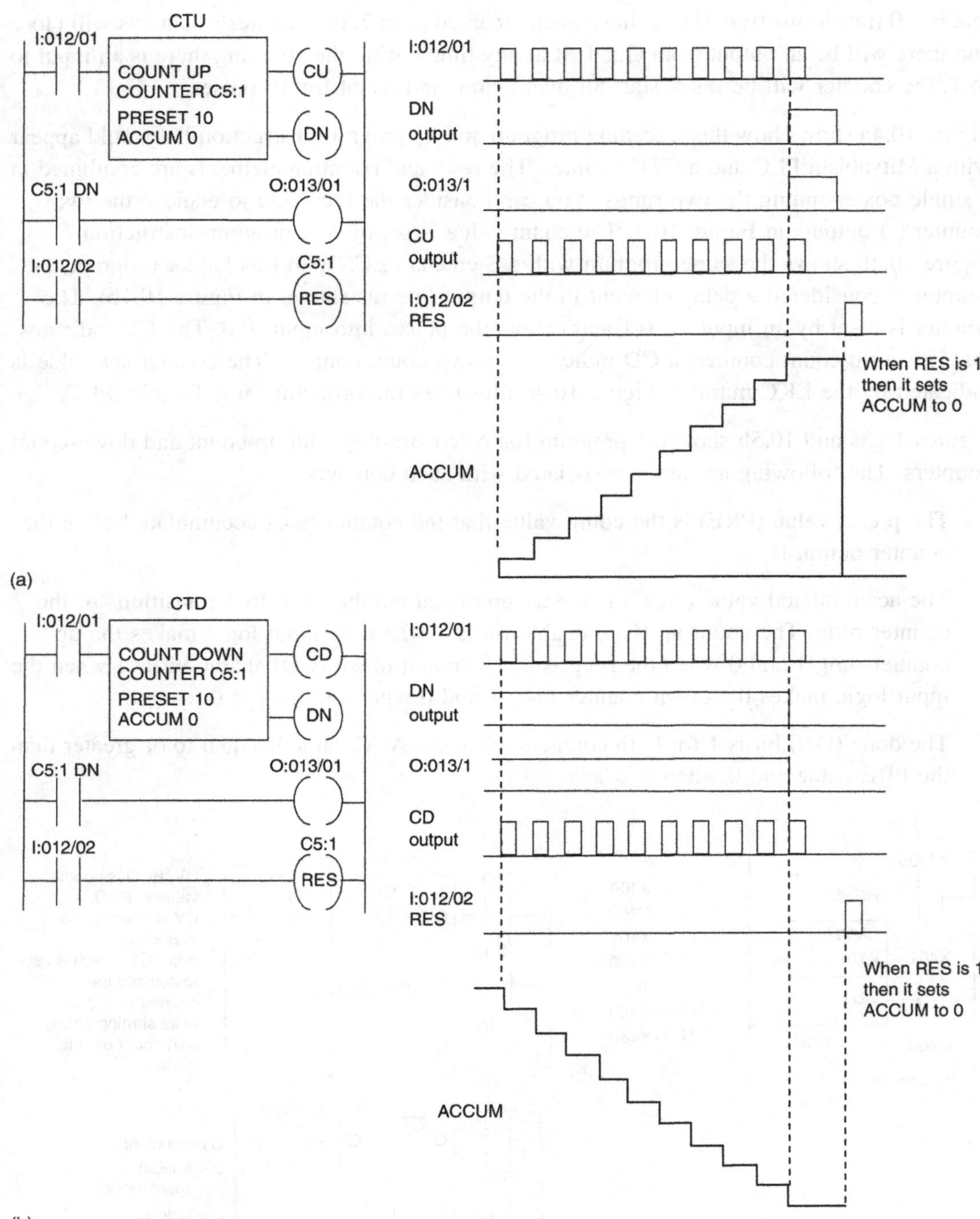

Figure 10.5: Allen-Bradley: (a) count-up, and (b) count-down.

- The count-up overflow (OV) bit is 1 when the up-counter increments above the maximum positive value.
- The count-down underflow (UN) bit is 1 when the counter decrements below the maximum negative value.
- Reset (RES) returns counter accumulator values to zero. As long as RES is 1, ACC and all output bits are held at 0. When RES is 0, the counter is able to start counting.

To ensure that the input pulses to a counter input are short duration, the ladder program shown in Figure 10.6 can be used. When there is an input to In 1, the internal relay IR 1 is activated; when the next rung is scanned a short while later, internal relay IR 2 is activated. When IR 2 is activated it switches off the input to IR 1. Thus IR 1 gives only a short duration pulse, which is then used as the input to a counter.

10.2.1 Counter Application

As an illustration of the use that can be made of a counter, consider the problem of items passing along a conveyor belt. The passage of an item past a particular point is registered by the interruption of a light beam to a photoelectric cell, and after a set number there is to be a signal sent informing that the set count has been reached and the conveyor stopped. Figure 10.7a shows the basic elements of a Siemens program that could be used. A reset signal causes the counter to reset and start counting again. The set signal is used to make the counter active. Figure 10.7b shows the basic elements of the comparable Allen-Bradley program. When the count reaches the preset value, the done bit is set to 1, and so O:013/01 occurs, the corresponding contacts are opened, and the conveyor stopped.

As a further illustration of the use of a counter, consider the problem of the control of a machine that is required to direct six tins along one path for packaging in a box and then 12 tins along another path for packaging in another box (Figure 10.8a). A deflector plate

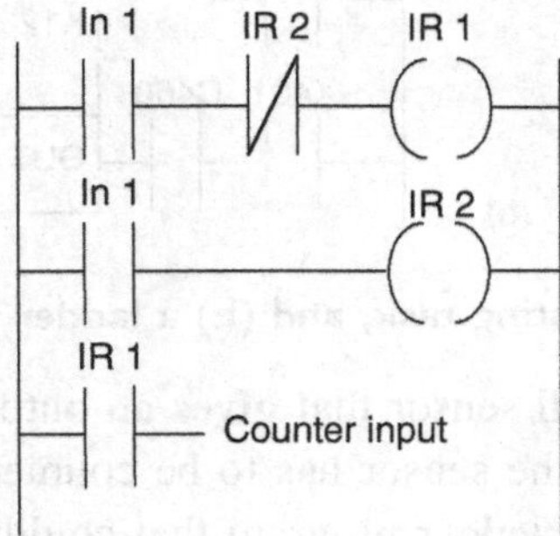

Figure 10.6: Short duration input pulses for a counter.

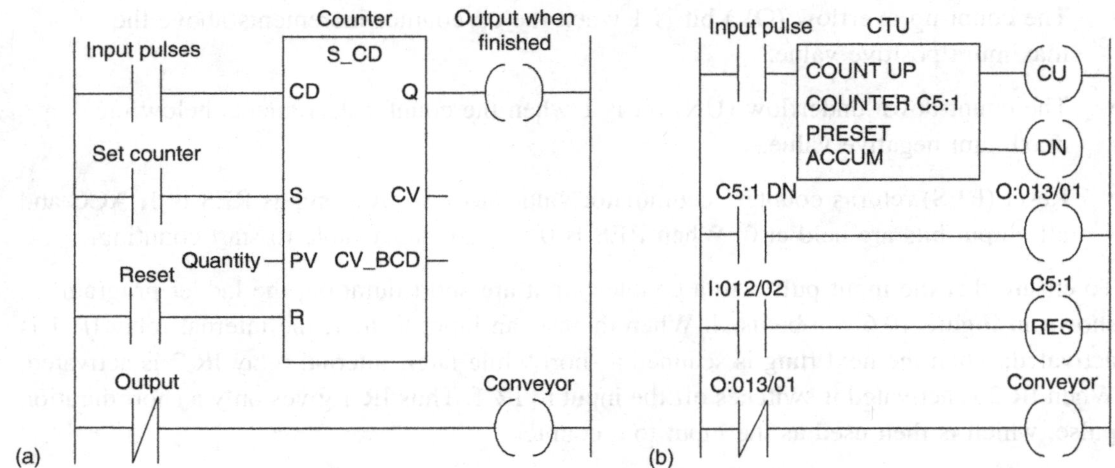

Figure 10.7: (a) Siemens and (b) Allen-Bradley counting programs.

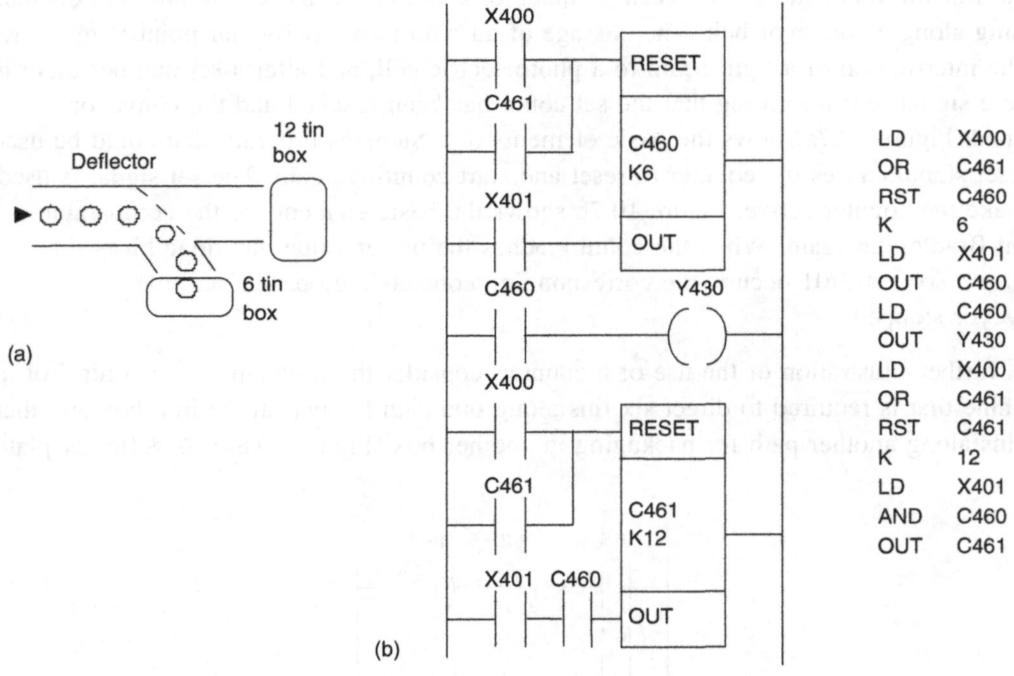

Figure 10.8: (a) A counting task, and (b) a ladder program for the task.

might be controlled by a photocell sensor that gives an output every time a tin passes it. Thus the number of pulses from the sensor has to be counted and used to control the deflector. Figure 10.8b shows the ladder program that could be used, with Mitsubishi notation.

When there is a pulse input to X400, both the counters are reset. The input to X400 could be the push-button switch used to start the conveyor moving. The input that is counted is X401. This might be an input from a photocell sensor that detects the presence of tins passing along the conveyor. C460 starts counting after X400 is momentarily closed. When C460 has counted six items, it closes its contacts and so gives an output at Y430. This might be a solenoid that is used to activate a deflector to deflect items into one box or another. Thus the deflector might be in such a position that the first six tins passing along the conveyor are deflected into the six-pack box; then the deflector plate is moved to allow tins to pass to the 12-pack box. When C460 stops counting, it closes its contacts and so allows C461 to start counting. C461 counts for 12 pulses to X401 and then closes its contacts. This results in both counters being reset, and the entire process can repeat itself.

Counters can be used to ensure that a particular part of a sequence is repeated a known number of times. This is illustrated by the following program which is designed to enable a three-cylinder, double solenoid-controlled arrangement (Figure 10.9a) to give the sequence A+, A–, A+, A–, A+, A–, B+, C+, B–, C–. The A+, A– sequence is repeated three times before B+, C+, B–, C– occur. We can use a counter to enable this repetition. Figure 10.9b shows a possible program. The counter only allows B+ to occur after it has received three inputs corresponding to three a– signals.

10.3 Up- and Down-Counting

It is possible to program up- and down-counters together. *Up-down counters* are available as single entities; see Figure 10.1 for the IEC symbol. Consider the task of counting products as they enter a conveyor line and as they leave it, or perhaps cars as they enter a multistorage parking lot and as they leave it. An output is to be triggered if the number of items/cars entering is some number greater than the number leaving, that is, the number in the parking lot has reached a "saturation" value. The output might be to illuminate a "No empty spaces" sign. Suppose we use the up-counter for items entering and the down-counter for items leaving. Figure 10.10a shows the basic form a ladder program for such an application can take. When an item enters, it gives a pulse on input In 1. This increases the count by 1. Thus each item entering increases the accumulated count by 1. When an item leaves, it gives an input to In 2. This reduces the number by 1. Thus each item leaving reduces the accumulated count by 1. When the accumulated value reaches the preset value, the output Out 1 is switched on. Figure 10.10b shows how the preceding system might appear for a Siemens PLC and the associated program instruction list. CU is the count up input and CD the count down. R is the reset. The set accumulator value is loaded via F0.0, this being an internal relay.

Figure 10.11 shows the implementation of this program with an Allen-Bradley program and an up- and down-counter.

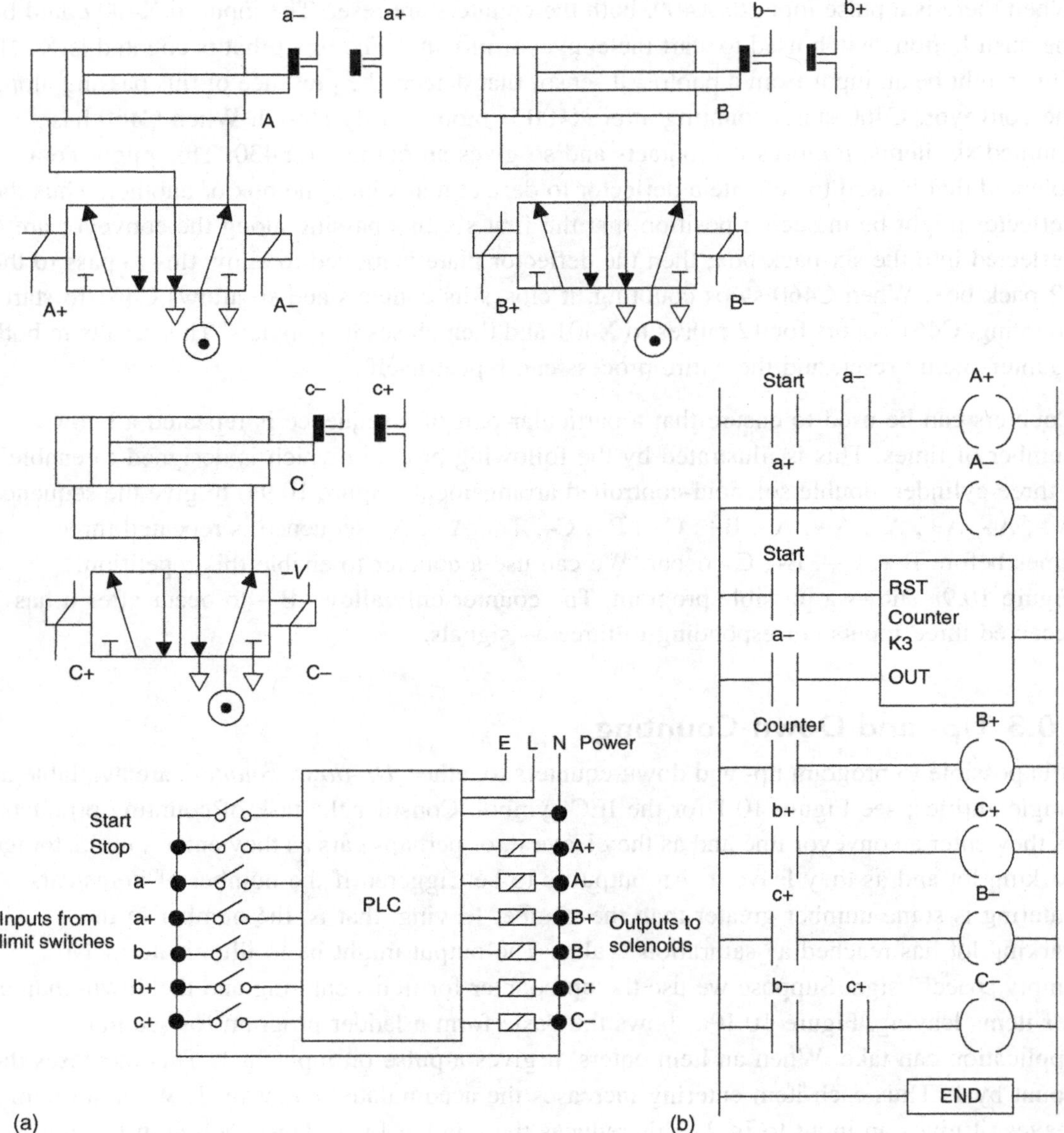

Figure 10.9: (a) A three-cylinder system, and (b) a program.

10.4 Timers with Counters

A typical timer can count up to 16 binary bits of data, this corresponding to 32,767 base time units. Thus, if we have a time base of 1 s, the maximum time that can be dealt with by a timer is just over 546 minutes, or 9.1 hours. If the time base is to be 0.1 s, the maximum time is 54.6 minutes, or just short of an hour. By combining a timer with a counter, longer

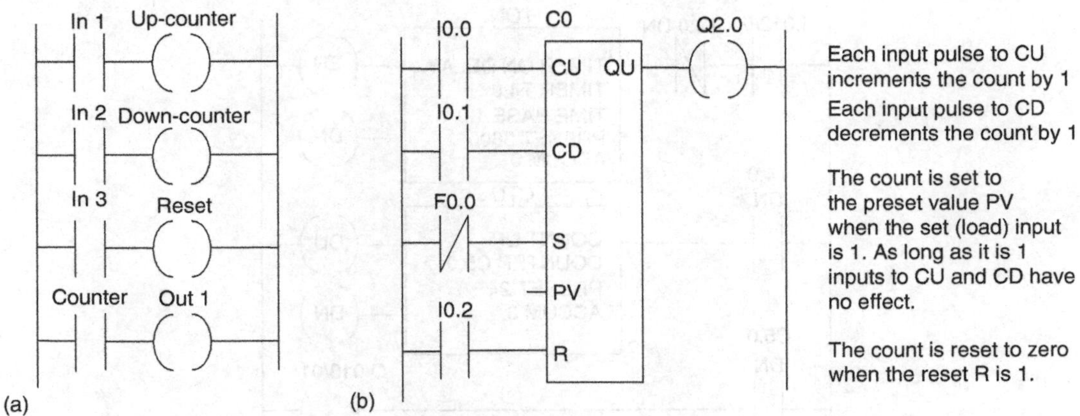

Figure 10.10: (a) Using up- and down-counters; (b) a Siemens program.

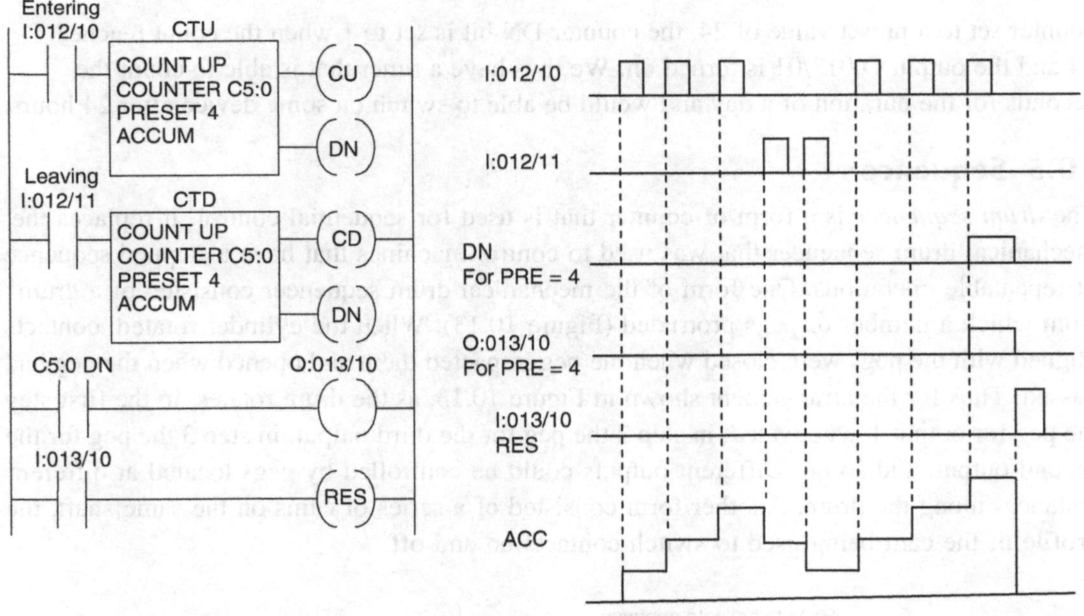

Figure 10.11: Allen-Bradley program.

times can be counted. Figure 10.12 illustrates this with an Allen-Bradley program. If the timer has a time base of 1 s and a preset value of 3600, it can count for up to 1 hour. When input I:012/01 is activated, the timer starts to time in 1-second increments. When the time reaches the preset value of 1 hour, the DN bit is set to 1 and the counter increments by 1. Setting the DN bit to 1 also resets the timer and the timer starts to time again. When it next reaches its preset time of 1 hour, the DN bit is again set to 1 and the counter increments by 1. With the

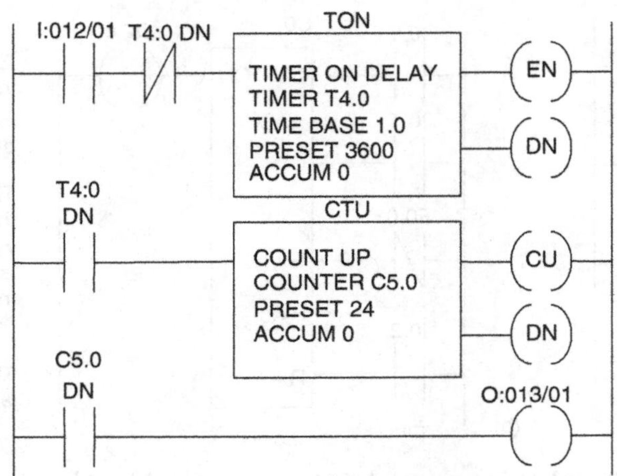

Figure 10.12: Using a counter to extend the range of a timer.

counter set to a preset value of 24, the counter DN bit is set to 1 when the count reaches 24 and the output O:013/01 is turned on. We thus have a timer that is able to count the seconds for the duration of a day and would be able to switch on some device after 24 hours.

10.5 Sequencer

The *drum sequencer* is a form of counter that is used for sequential control. It replaces the mechanical drum sequencer that was used to control machines that have a stepped sequence of repeatable operations. One form of the mechanical drum sequencer consisted of a drum from which a number of pegs protruded (Figure 10.13). When the cylinder rotated, contacts aligned with the pegs were closed when the peg impacted them and opened when the peg had passed. Thus for the arrangement shown in Figure 10.13, as the drum rotates, in the first step the peg for output 1 is activated, in step 2 the peg for the third output, in step 3 the peg for the second output, and so on. Different outputs could be controlled by pegs located at different distances along the drum. Another form consisted of a series of cams on the same shaft, the profile of the cam being used to switch contacts on and off.

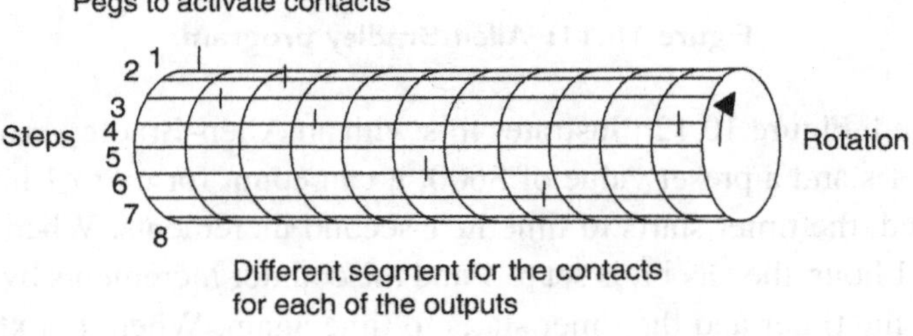

Figure 10.13: Drum sequencer.

The PLC sequencer consists of a master counter that has a range of preset counts corresponding to the various steps; so as it progresses through the count, when each preset count is reached it can be used to control outputs. Each step in the count sequence relates to a certain output or group of outputs. The outputs are internal relays, which are in turn used to control the external output devices.

Suppose we want output 1 to be switched on 5 s after the start and remain on until the time reaches 10 s, output 2 to be switched on at 10 s and remain on until 20 s, output 3 to be switched on at 15 s and remain on until 25 s, and so on. We can represent these requirements by a time sequence diagram, shown in Figure 10.14, demonstrating the required time sequence.

We can transform the timing diagram into a drum sequence requirement. Taking each drum sequence step to take 5 s gives the requirement diagram shown in Table 10.1. Thus at step 1 we require output 1 to be switched on and to remain on until step 2. At step 2 we require output 2 to be switched on and remain on until step 4. At step 3 we require output 3 to be switched on and remain on until step 5. At step 5 we require output 4 to be switched on and remain on until step 6.

With a PLC such as a Toshiba, the sequencer is set up by switching on the Step Sequence Initialize (STIZ) function block R500 (Figure 10.15). This sets up the program for step 1 and

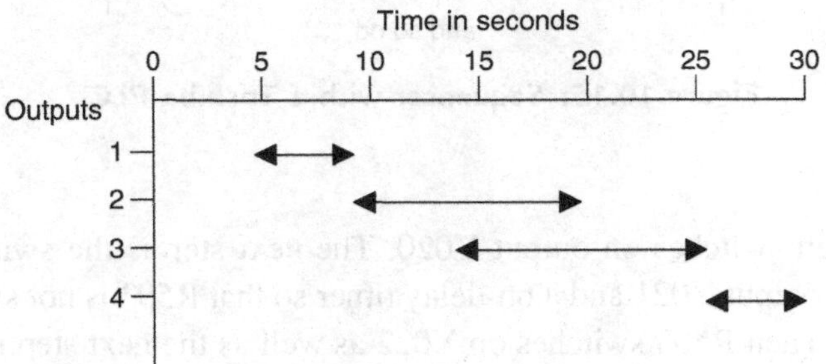

Figure 10.14: Timing diagram.

Table 10.1: Sequence Requirements

Step	Time (s)	Output 1	Output 2	Output 3	Output 4
0	0	0	0	0	0
1	5	1	0	0	0
2	10	0	1	0	0
3	15	0	1	1	0
4	20	0	0	1	0
5	25	0	0	1	1
6	30	0	0	0	0

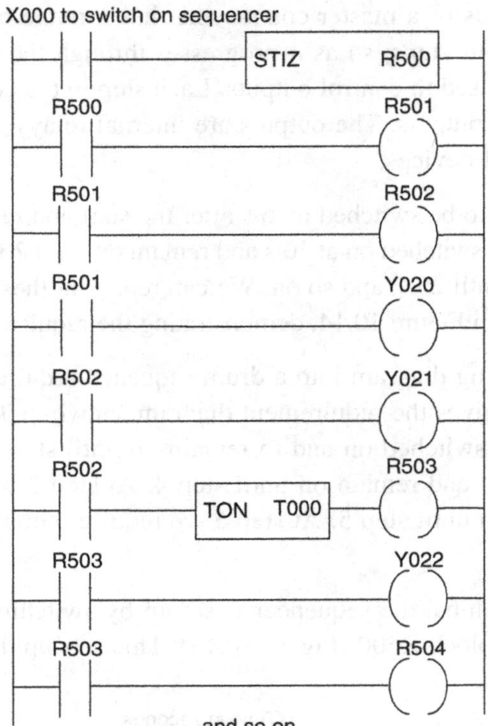

Figure 10.15: Sequencer with a Toshiba PLC.

R501. This relay then switches on output Y020. The next step is the switching on of R502. This switches on the output Y021 and a on-delay timer so that R503 is not switched on until the timer has timed out. Then R503 switches on Y022 as well as the next step in the sequence.

With the Allen-Bradley form of PLC the sequencer is programmed using a sequence of binary words in the form of the outputs required. The term *sequencer output* (SQO) is used by Allen-Bradley for the output instruction that uses a file or an array to control various output devices. Thus we might have the following binary word sequence entered into a file for each sequencer step using the programming device and so for six steps and five outputs in step 0 all of them would have the output 0, in step 1 output 1 would be 1 and the other outputs 0, in step 2 output 2 would be 1 and the other outputs 0, and so on.

Terms used with the sequencer are:

- *File* for the starting address for registers in the sequencer file, the indexed file indicator # being used for this address. Thus we might have #N7.0 the file for step 0 and so #N7.1 for step 1, #N7.2 for step 2, and so on.

- *Mask* for the bit pattern through which the sequencer instruction moves source data to the destination address. An h is placed behind the parameter to indicate the mask is an hexadecimal number or a B to indicate binary, decimal notation being entered without any indicator. Thus we might have 0001Fh which in binary is 0000 0000 0001 1111 and so allows the just the first five bits from the sequencer to be passed to the destination address all remaining bits being masked, i.e. not transmitted to the destination. As a further illustration we could have a mask word of 1110011011111110 and so, because bits 1, 9, 12, and 13 are 0, these bits in the sequencer words are not passed to the destination. Mask bit patterns are used as a means of selectively screening out data from the sequencer file for the required output.
- *Destination* is the address of the output word or file to which the instruction moves data from its sequencer file. Thus we might have O4.0 to indicate the output to which the output from step 1 goes.
- *Control* is the address that contains parameters with control information for the instruction and discrete outputs to indicate sequencer instruction results and status. The control file address is in control area R of the processor memory and the default file is 6.
- *Length* is the number of steps in in the sequencer file starting at step 1. As there is a step 0, the total number of words in the file is the length plus 1.
- *Position* is the word/step location in the sequencer file from or to which the instruction moves data. Thus if the position is 0 then after the first scan the position increments the position number by 1 and moves the sequencer file for that position number through the mask to the destination file.

As an illustration, Figure 10.16 shows a basic Allen-Bradley ladder program using such a sequencer. The timer is started by an input to I:012/1 and has a preset time of 30 s. It is reset by its DN bit. The DN bit also increments the SQO instruction to the next output word. Thus the sequencer is incremented every 30 s.

Summary

Counters are provided as built-in elements in PLCs and allow the number of occurrences of input signals to be counted. Down-counters count down from the preset value to zero, that is, events are subtracted from the set value. When the counter reaches the zero value, its contacts change state. Up-counters count from zero up to the preset value, that is, events are added until the number reaches the preset value. When the counter reaches the set value, its contacts change state. Some PLCs offer the facility for both down- and up-counting.

The PLC sequencer consists of a master counter that has a range of preset counts corresponding to the various steps; so as it progresses through the count, when each preset count is reached it can be used to control outputs.

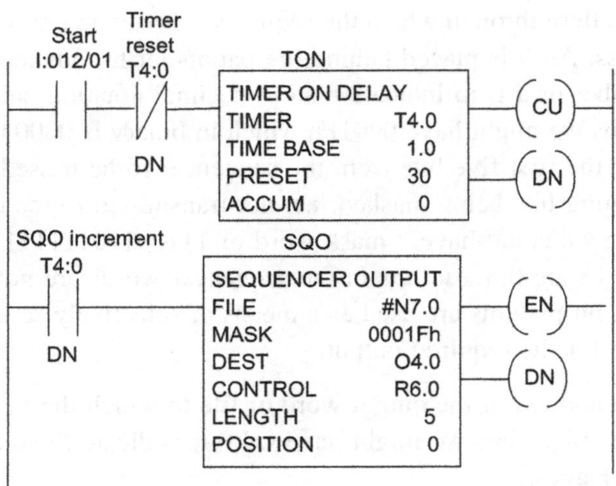

O:4.0	Destination		-	-	-	-	-	-	-	-	-	-	-	1	1	1	0	0	
			Outputs masked											↑	↑	↑	↑	↑	
000Fh	Mask		0	0	0	0	0	0	0	0	0	0	0	1	1	1	1	1	
#N7.0	File	N7.0	0	0	0	0	0	0	0	0	0	0	0	0	0	0	0	0	Positions
		N7.1	0	0	1	0	1	0	1	0	1	0	0	0	0	0	1	1	0
		N7.2	0	0	1	0	1	0	1	0	0	0	1	1	0	0	0	1	1
		N7.3	0	1	0	1	0	1	0	1	1	0	1	1	1	1	0	0	2
		N7.4	1	1	0	1	0	1	0	1	0	1	0	0	0	0	1	1	3
		N7.5	1	1	0	0	1	0	0	1	1	1	0	1	0	0	0	0	4
		N7.6	1	0	0	0	1	0	0	1	0	1	0	1	1	0	0	0	5

Figure 10.16: Allen-Bradley sequencer with the sequencer moving data from a file to an output with the positon pointer at 2 to give output 11100.

Problems

Problems 1 through 19 have four answer options: A, B, C, or D. Choose the correct answer from the answer options. Problems 1 through 3 refer to Figure 10.17, which shows a ladder diagram with a down-counter, two inputs (In 1 and In 2), and an output (Out 1).

1. *Decide whether each of these statements is true (T) or false (F).* For the ladder diagram shown in Figure 10.17, when the counter is set to 5, there is an output from Out 1 every time:
   (i) In 1 has closed 5 times.
   (ii) In 2 has closed 5 times.
   A. (i) T (ii) T

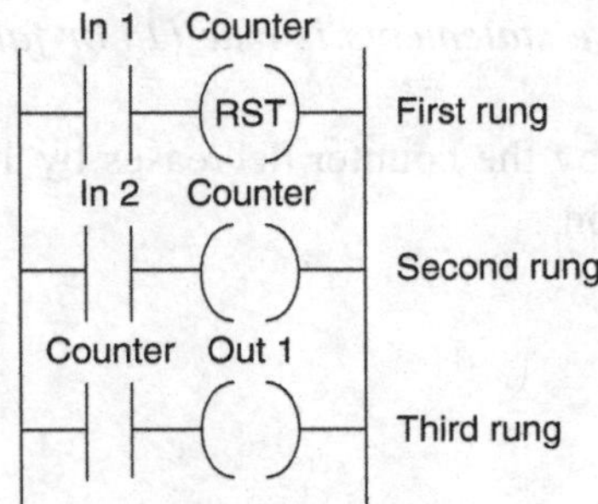

Figure 10.17: Diagram for Problems 1 through 3.

B. (i) T (ii) F
C. (i) F (ii) T
D. (i) F (ii) F

2. *Decide whether each of these statements is true (T) or false (F).* For the ladder diagram shown in Figure 10.17:
   (i) The first rung gives the condition required to reset the counter.
   (ii) The second rung gives the condition required to generate pulses to be counted.
   A. (i) T (ii) T
   B. (i) T (ii) F
   C. (i) F (ii) T
   D. (i) F (ii) F

3. *Decide whether each of these statements is true (T) or false (F).* In Figure 10.17, when there is an input to In 1:
   (i) The counter contacts in the third rung close.
   (ii) The counter is ready to start counting the pulses from In 2.
   A. (i) T (ii) T
   B. (i) T (ii) F
   C. (i) F (ii) T
   D. (i) F (ii) F

Problems 4 and 5 refer to the following program instruction list involving a down-counter:

```
LD      X400
RST     C460
LD      X401
OUT     C460
K       5
LD      460
OUT     Y430
```

4. *Decide whether each of these statements is true (T) or false (F).* Every time there is an input to X401:
   (i) The count accumulated by the counter decreases by 1.
   (ii) The output is switched on.
   A. (i) T (ii) T
   B. (i) T (ii) F
   C. (i) F (ii) T
   D. (i) F (ii) F

5. *Decide whether each of these statements is true (T) or false (F).* When there is an input to X400, the counter:
   (i) Resets to a value of 5.
   (ii) Starts counting from 0.
   A. (i) T (ii) T
   B. (i) T (ii) F
   C. (i) F (ii) T
   D. (i) F (ii) F

Problems 6 and 7 refer to the following program instruction list involving a counter C0:

```
A      I0.0
CD     C0
LKC    5
A      I0.1
R      C0
Q      2.00
```

6. *Decide whether each of these statements is true (T) or false (F).* Every time there is an input to I0.0:
   (i) The count accumulated by the counter decreases by 1.
   (ii) The output is switched on.
   A. (i) T (ii) T
   B. (i) T (ii) F
   C. (i) F (ii) T
   D. (i) F (ii) F

7. *Decide whether each of these statements is true (T) or false (F).* When there is an input to I0.1, the counter:
   (i) Resets to a value of 5.
   (ii) Starts counting from 0.
   A. (i) T (ii) T
   B. (i) T (ii) F
   C. (i) F (ii) T
   D. (i) F (ii) F

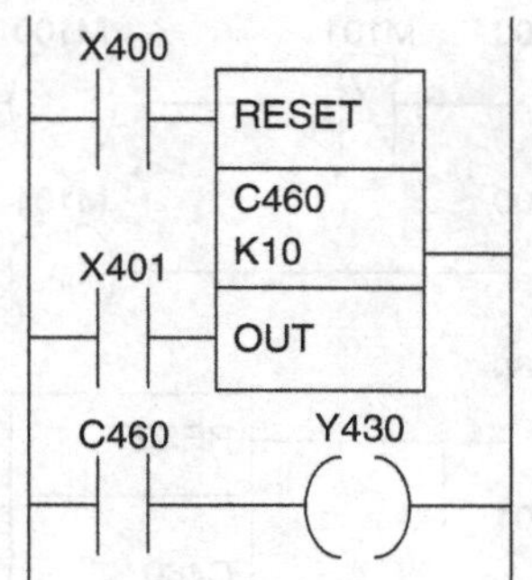

Figure 10.18: Diagram for Problems 8 and 9.

Problems 8 and 9 refer to Figure 10.18, which shows a down-counter C460 controlled by two inputs X400 and X401, with an output from Y430.

8. *Decide whether each of these statements is true (T) or false (F).* When there is an input to X400, the counter:
   (i) Resets to a value of 0.
   (ii) Starts counting.
   A. (i) T (ii) T
   B. (i) T (ii) F
   C. (i) F (ii) T
   D. (i) F (ii) F

9. *Decide whether each of these statements is true (T) or false (F).* Every time there is an input to X401, the counter:
   (i) Gives an output from Y430.
   (ii) Reduces the accumulated count by 1.
   A. (i) T (ii) T
   B. (i) T (ii) F
   C. (i) F (ii) T
   D. (i) F (ii) F

Problems 10 through 12 refer to Figure 10.19, which shows a ladder diagram involving a counter C460, inputs X400 and X401, internal relays M100 and M101, and an output Y430.

10. *Decide whether each of these statements is true (T) or false (F).* For the output Y430:
    (i) It switches on with the tenth pulse to X400.
    (ii) It switches off at the start of the eleventh pulse to X400.
    A. (i) T (ii) T
    B. (i) T (ii) F
    C. (i) F (ii) T
    D. (i) F (ii) F

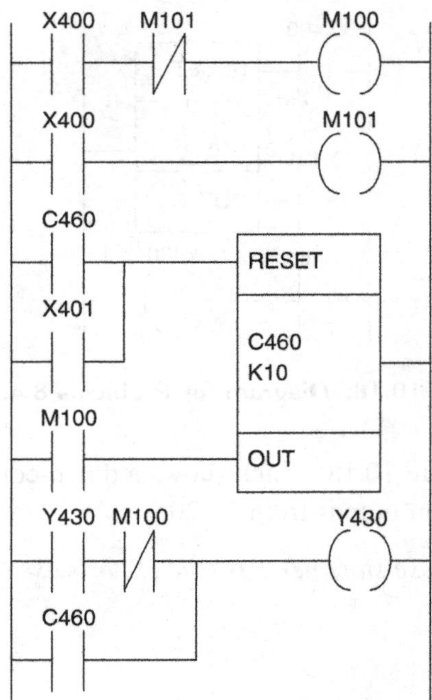

Figure 10.19: Diagram for Problems 10 through 12.

11. *Decide whether each of these statements is true (T) or false (F).* When there is an input to X400:
    (i) The internal relay M100 is energized.
    (ii) The internal relay M101 is energized.
    A. (i) T (ii) T
    B. (i) T (ii) F
    C. (i) F (ii) T
    D. (i) F (ii) F

12. *Decide whether each of these statements is true (T) or false (F).* There is an output from Y430 as long as:
    (i) The C460 contacts are closed.
    (ii) Y430 gives an output and M100 is energized.
    A. (i) T (ii) T
    B. (i) T (ii) F
    C. (i) F (ii) T
    D. (i) F (ii) F

13. *Decide whether each of these statements is true (T) or false (F).* Figure 10.20 shows a counter program in Siemens format. After 10 inputs to I0.0:
    (i) The lamp comes on.

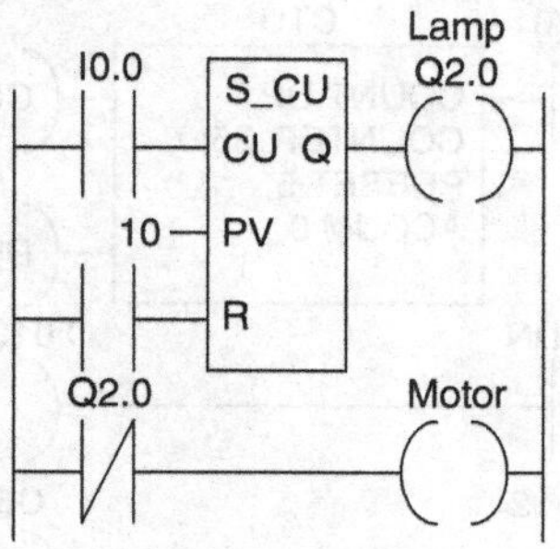

Figure 10.20: Diagram for Problem 13.

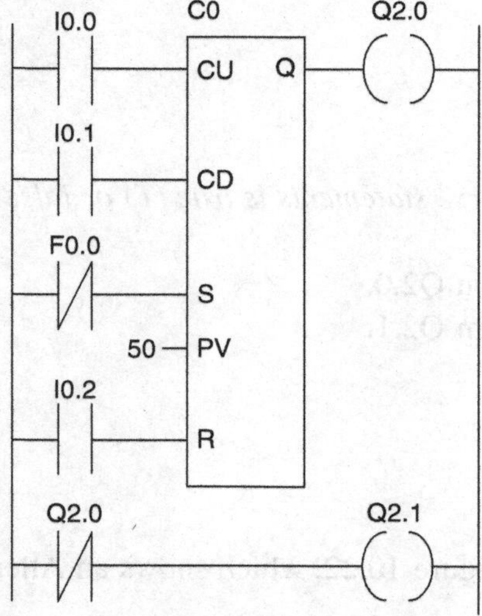

Figure 10.21: Diagram for Problems 14 and 15.

(ii) The motor starts.

A. (i) T (ii) T
B. (i) T (ii) F
C. (i) F (ii) T
D. (i) F (ii) F

Problems 14 and 15 refer to Figure 10.21, which shows a Siemens program involving an up- and down-counter.

14. *Decide whether each of these statements is true (T) or false (F).* When the count is less than 50 in Figure 10.21:

(i) There is an output from Q2.0.

(ii) There is an output from Q2.1.

A. (i) T (ii) T

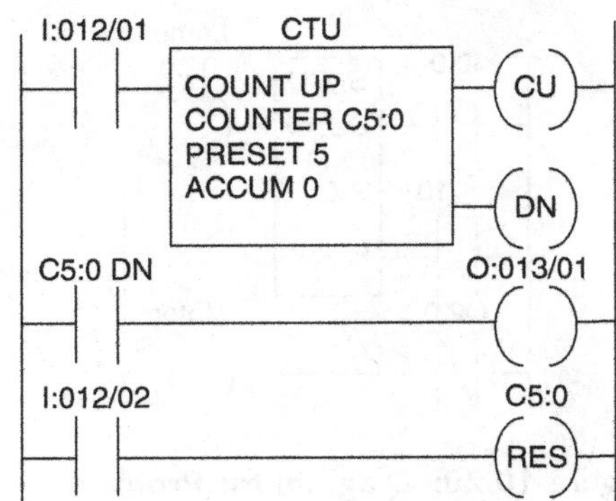

Figure 10.22: Diagram for Problems 16 and 17.

B. (i) T (ii) F
C. (i) F (ii) T
D. (i) F (ii) F

15. *Decide whether each of these statements is true (T) or false (F).* When the count reaches 50 in Figure 10.21:
(i) There is an output from Q2.0.
(ii) There is an output from Q2.1.
A. (i) T (ii) T
B. (i) T (ii) F
C. (i) F (ii) T
D. (i) F (ii) F

Problems 16 and 17 refer to Figure 10.22, which shows an Allen-Bradley program involving a count-up counter.

16. For the program shown in Figure 10.22, the counter is reset when:
A. The count reaches 5.
B. The count passes 5.
C. There is an input to I:012/01.
D. There is an input to I:012/02.

17. *Decide whether each of these statements is true (T) or false (F).* For the program shown in Figure 10.22, there is an output at O:013/01 when:
(i) There is an input to I:012/01.
(ii) There is an output from the count-up done bit DN.
A. (i) T (ii) T
B. (i) T (ii) F
C. (i) F (ii) T
D. (i) F (ii) F

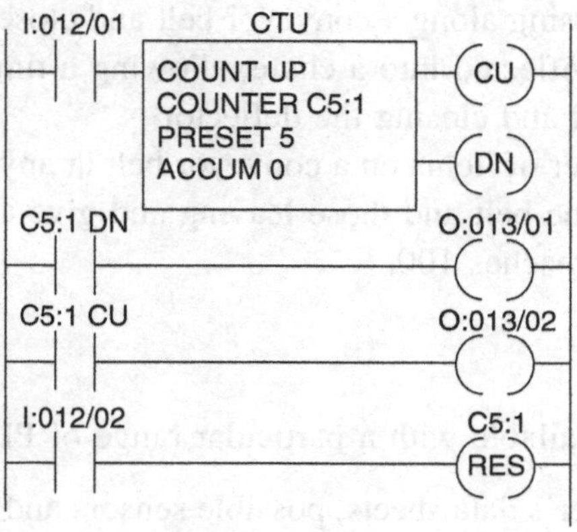

Figure 10.23: Diagram for Problems 18 and 19.

Problems 18 and 19 refer to Figure 10.23, which shows an Allen-Bradley program involving a count-up counter.

18. *Decide whether each of these statements is true (T) or false (F).* When there is a single pulse input to I:012/01 in Figure 10.23:
    (i) Output O:013/01 is switched on.
    (ii) Output O:013/02 is switched on.
    A. (i) T (ii) T
    B. (i) T (ii) F
    C. (i) F (ii) T
    D. (i) F (ii) F

19. *Decide whether each of these statements is true (T) or false (F).* When the fifth pulse input occurs to I:012/01 in Figure 10.23:
    (i) Output O:013/01 is switched on.
    (ii) Output O:013/02 is switched on.
    A. (i) T (ii) T
    B. (i) T (ii) F
    C. (i) F (ii) T
    D. (i) F (ii) F

20. Devise ladder programs for systems that will carry out the following tasks:
    (a) Give an output after a photocell sensor has given 10 pulse input signals as a result of detecting 10 objects passing in front of it.
    (b) Give an output when the number of people in a store reaches 100, there continually being people entering and leaving the store.
    (c) Show a red light when the count is less than 5 and a green light when it is equal to or greater than 5.

(d) Count 10 objects passing along a conveyor belt and close a deflecting gate when that number have been deflected into a chute, allowing a time of 5 s between the tenth object being counted and closing the deflector.

(e) Determine the number of items on a conveyor belt at any particular time by counting those moving onto the belt and those leaving and give an output signal when the number on the belt reaches 100.

Lookup Tasks

21. Look up the counters available with a particular range of PLCs.

22. Select, from manufacturer's data sheets, possible sensors and a PLC that could be used to control the counting of nontransparent objects moving along a conveyor belt.

CHAPTER 11

Shift Registers

The term *register* is used for an electronic device in which data can be stored. An internal relay (see Chapter 7) is such a device. The *shift register* is a number of internal relays grouped together that allow stored bits to be shifted from one relay to another. This chapter is about shift registers and how they can be used when a sequence of operations is required or to keep track of particular items in a production system.

11.1 Shift Registers

A register is a number of internal relays grouped together, normally 8, 16, or 32. Each internal relay is either effectively open or closed, these states being designated 0 and 1. The term *bit* is used for each such binary digit. Therefore, if we have eight internal relays in the register, we can store eight 0/1 states. Thus we might have, for internal relays:

1	2	3	4	5	6	7	8

and each relay might store an on/off signal such that the state of the register at some instant is:

1	0	1	1	0	0	1	0

that is, relay 1 is on, relay 2 is off, relay 3 is on, relay 4 is on, relay 5 is off, and so on. Such an arrangement is termed an *8-bit register*. Registers can be used for storing data that originate from input sources other than just simple, single on/off devices such as switches.

With the *shift register* it is possible to shift stored bits. Shift registers require three inputs: one to load data into the first location of the register, one as the command to shift data along by one location, and one to reset or clear the register of data. To illustrate this idea, consider the following situation where we start with an 8-bit register in the following state:

1	0	1	1	0	0	1	0

Suppose we now receive the input signal 0. This is an input signal to the first internal relay.

W. Bolton: Programmable Logic Controllers, Sixth Edition. http://dx.doi.org/10.1016/B978-0-12-802929-9.00011-X

Input 0

→ | 1 | 0 | 1 | 1 | 0 | 0 | 1 | 0 |
|---|---|---|---|---|---|---|---|

If we also receive the shift signal, the input signal enters the first location in the register, and all the bits shift along one location. The last bit overflows and is lost.

| 0 | 1 | 0 | 1 | 1 | 0 | 0 | 1 | →
|---|---|---|---|---|---|---|---|

Overflow 0

Thus a set of internal relays that were initially on, off, on, on, off, off, on, off are now off, on, off, on, on, off, off, on.

The grouping together of internal relays to form a shift register is done automatically by a PLC when the shift register function is selected. With the Mitsubishi PLC, this is done using the programming code SFT (shift) against the internal relay number that is to be the first in the register array. This then causes a block of relays, starting from that initial number, to be reserved for the shift register.

11.2 Ladder Programs

Consider a 4-bit shift register and how it can be represented in a ladder program (Figure 11.1a). The input In 3 is used to reset the shift register, that is, put all the values at 0. The input In 1 is used to input to the first internal relay in the register. The input In 2 is used to shift the states of the internal relays along by one. Each of the internal relays in the register, that is, IR 1, IR 2, IR 3, and IR 4, is connected to an output, these being Out 1, Out 2, Out 3, and Out 4.

Suppose we start by supplying a momentary input to In 3. All the internal relays are then set to 0 and so the states of the four internal relays IR 1, IR 2, IR 3, and IR 4 are 0, 0, 0, 0. When In 1 is momentarily closed, there is a 1 input into the first relay. Thus the states of the internal relays IR 1, IR 2, IR 3, and IR 4 are now 1, 0, 0, 0. The IR 1 contacts close and we thus end up with an output from Out 1. If we now supply a momentary input to In 2, the 1 is shifted from the first relay to the second. The states of the internal relays are now 0, 1, 0, 0. We now have no input from Out 1 but an output from Out 2. If we supply another momentary input to In 2, we shift the states of the relays along by one location to give 0, 0, 1, 0. Outputs Out 1 and Out 2 are now off, but Out 3 is on. If we supply another momentary input to In 2, we again shift the states of the relays along by one and have 0, 0, 0, 1. Thus now Out 1, Out 2, and Out 3 are off and Out 4 has been switched on. When another momentary input is applied to In 2, we shift the states of the relays along by one and have 0, 0, 0, 0, with the 1 overflowing and being lost. All the outputs are then off. Thus the effect of the sequence of inputs to In 2 has been to give a sequence of

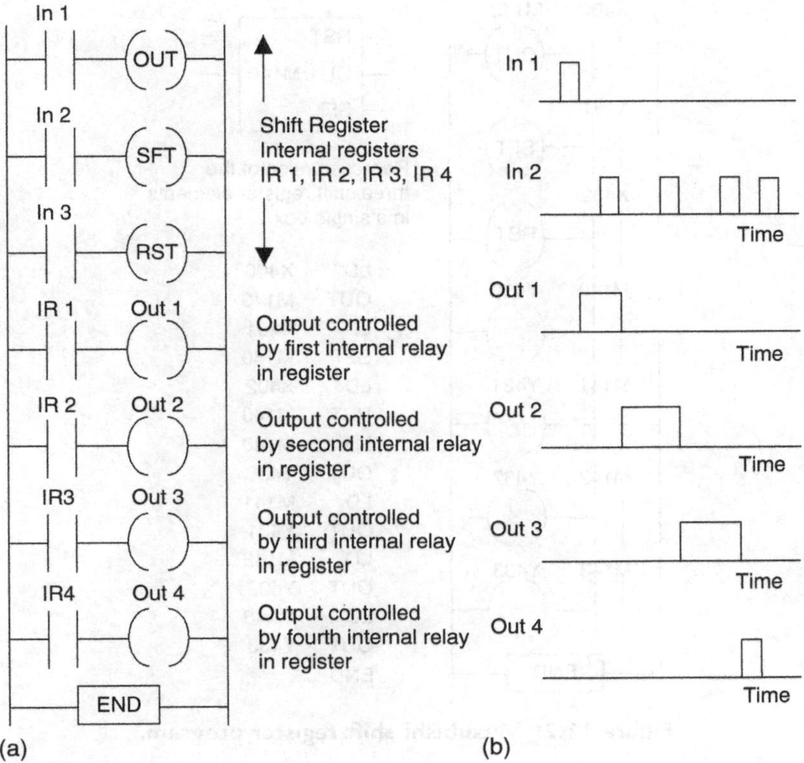

Figure 11.1: The shift register.

outputs Out 1, followed by Out 2, followed by Out 3, followed by Out 4. Figure 11.1b shows the sequence of signals.

Figure 11.2 shows the Mitsubishi version of the preceding ladder program and the associated instruction list. Instead of the three separate outputs for reset, output, and shift, the Mitsubishi shift register might appear in a program as a single function box, as shown in the figure. With the Mitsubishi shift register, the M140 is the address of the first relay in the register.

Figure 11.3 shows a shift register ladder program for a Toshiba PLC. With the Toshiba, R016 is the address of the first relay in the register. The (08) indicates that there are eight such relays. D is used for the data input, S for shift input, E for enable or reset input, and Q for output.

Figure 11.4 shows the IEC 61131-3 standard symbol for a shift register. The value to be shifted is at input IN and the number of places it is to be shifted is at input N.

Figure 11.5a shows the Siemens symbol for a shift register. If the enable input EN is 1, the shift function is executed and ENO is then 1. If EN is 0, the shift function is not executed and

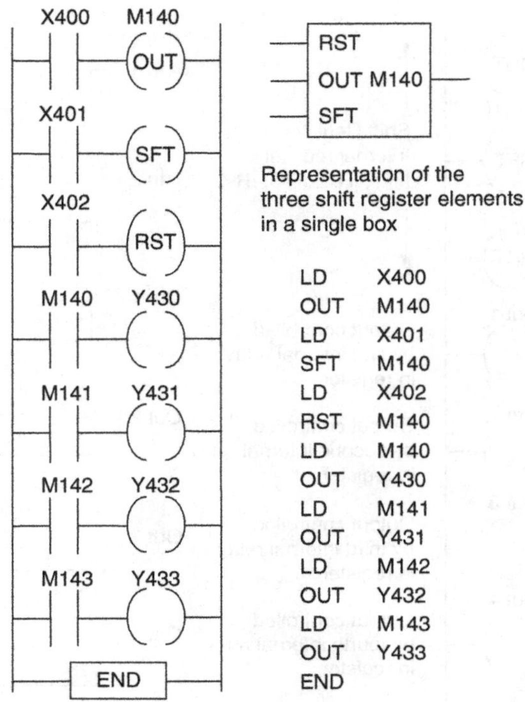

Figure 11.2: Mitsubishi shift register program.

ENO is 0. The shift function SHL_W shifts the contents of the word variable at input IN bit by bit to the left the number of positions specified by the input at N. The shifted word output is at OUT. Figure 11.5b shows the Allen-Bradley PLC 5 and SLC 500 symbols for shift registers. The FILE gives the address of the bit array that is to be shifted. CONTROL gives the address of control bits such as bit 15 (EN) as a 1 when the instruction is enabled, bit 13 (DN) as a 1 when the bits have shifted, and bit 11 (ER) as a 1 when the length is negative, and bit 10 (UL) stores the state of the bit that was shifted out of the range of bits. BIT ADDRESS is the address of the data to be shifted. LENGTH is the number of bits in the array to be shifted.

11.2.1 A Sequencing Application

Consider the requirement for a program for two double-solenoid cylinders, the arrangement as shown in Figure 11.6a, to give the sequence A+, B+, A–, B–. Figure 11.6b shows a program to achieve this sequence by the use of a shift register.

11.2.2 Keeping Track of Items

The preceding indicates how a shift register can be used for sequencing. Another application is to keep track of items. For example, a sensor might be used to detect faulty items moving along a conveyor and keep track of them so that when they reach the appropriate point, a

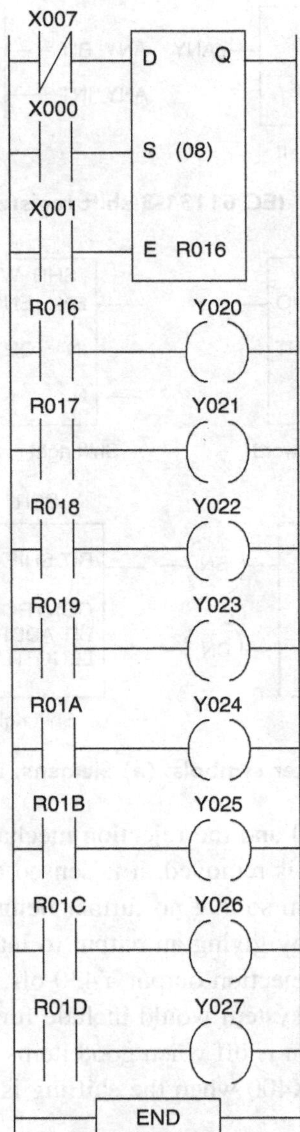

Figure 11.3: Toshiba shift register.

reject mechanism is activated to remove them from the conveyor. Figure 11.7 illustrates this arrangement and the type of ladder program that might be used.

Each time a faulty item is detected, a pulse signal occurs at input X400. This enters a 1 into the shift register at internal relay M140. When items move, whether faulty or not, there is a pulse input at X401. This shifts the 1 along the register. When the 1 reaches internal relay

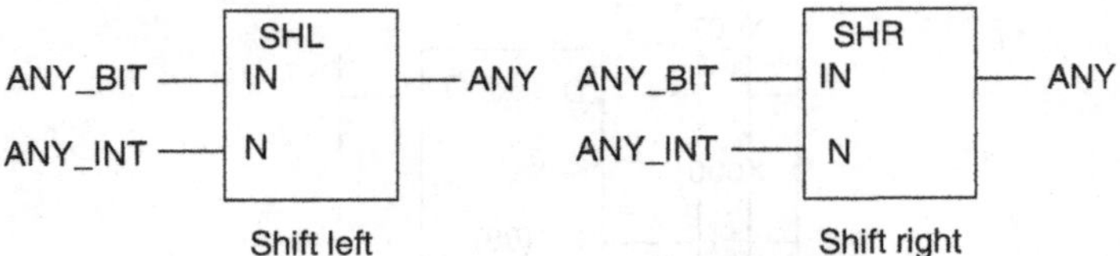

Figure 11.4: IEC 61131-3 shift register symbols.

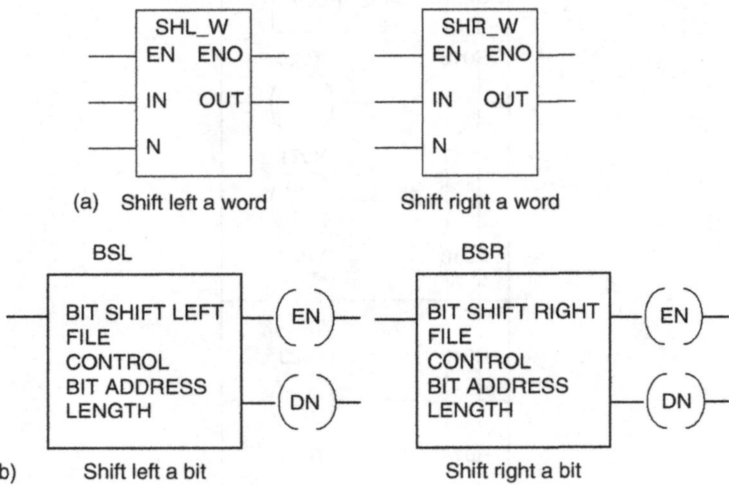

Figure 11.5: Shift register symbols: (a) Siemens, and (b) Allen-Bradley.

M144, it activates the output Y430 and the rejection mechanism removes the faulty item from the conveyor. When an item is removed, it is sensed and an input to X403 occurs. This is used to reset the mechanism so that no further items are rejected until the rejection signal reaches M144. It does this by giving an output to internal relay M100, which latches the X403 input and switches the rejection output Y430 off. This represents just the basic elements of a system. A practical system would include further internal relays to make certain that the rejection mechanism is off when good items move along the conveyor belt as well as to disable the input from X400 when the shifting is occurring.

Summary

The term *register* is used for an electronic device in which data can be stored. The shift register is a number of internal relays grouped together that allow stored bits to be shifted from one relay to another. With the shift register it is possible to shift stored bits. Shift registers require three inputs: one to load data into the first location of the register, one as the command to shift data along by one location, and one to reset or clear the register of data. The grouping together of internal relays to form a shift register is done automatically by a PLC when the shift register function is selected.

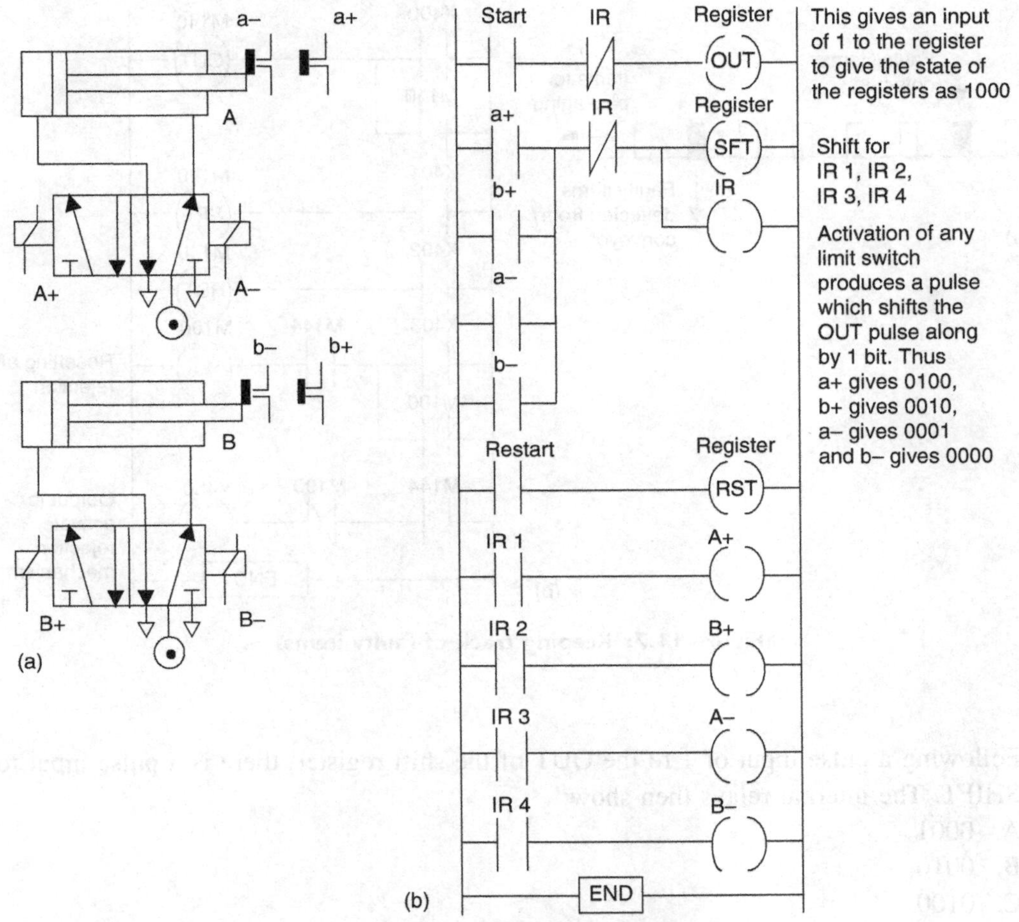

Figure 11.6: Sequencing cylinders.

Problems

Problems 1 through 9 have four answer options: A, B, C, or D. Choose the correct answer from the answer options. Problems 1 through 5 concern a 4-bit shift register involving internal relays IR 1, IR 2, IR 3, and IR 4, which has been reset to 0, 0, 0, 0.

1. When there is a pulse 1 input to the OUT of the shift register, the internal relays in the shift register show:
   A. 0001
   B. 0010
   C. 0100
   D. 1000

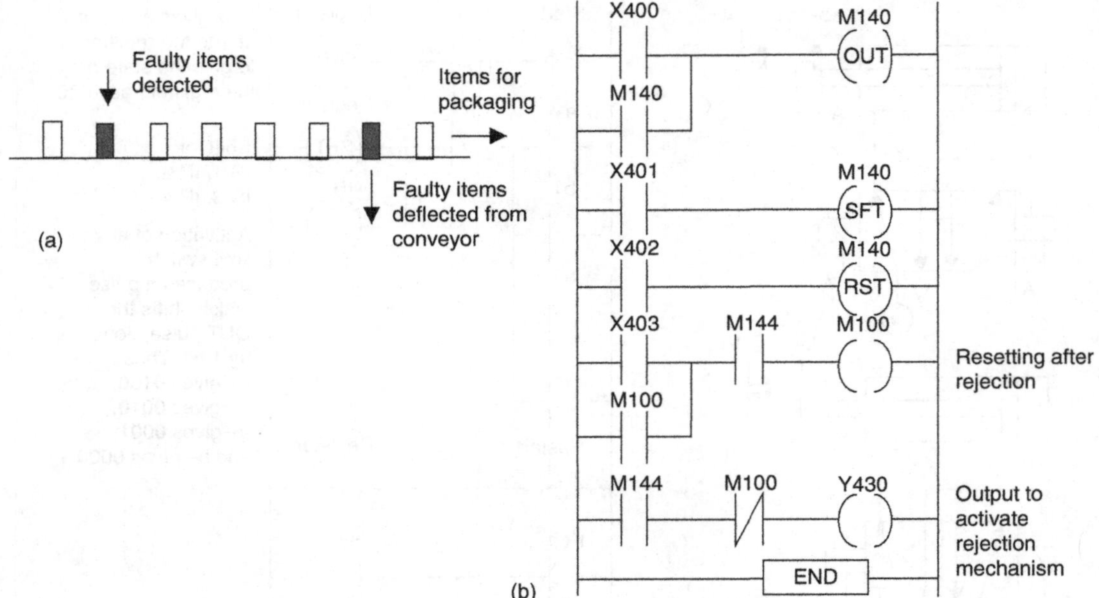

Figure 11.7: Keeping track of faulty items.

2. Following a pulse input of 1 to the OUT of the shift register, there is a pulse input to SHIFT. The internal relays then show:
   A. 0001
   B. 0010
   C. 0100
   D. 1000

3. With a continuous input of 1 to the OUT of the shift register, there is a pulse input to SHIFT. The internal relays then show:
   A. 0011
   B. 0110
   C. 1100
   D. 0010

4. With a continuous input of 1 to the OUT of the shift register, there are two pulse inputs to SHIFT. The internal relays then show:
   A. 0001
   B. 0010
   C. 1100
   D. 1110

5. With a pulse input of 1 to the OUT of the shift register, there is a pulse input to SHIFT, followed by a pulse input to RESET. The internal relays then show:
   A. 0000
   B. 0010
   C. 0100
   D. 1000

Problems 6 through 9 concern Figure 11.8, which shows a 4-bit shift register with internal relays IR 1, IR 2, IR 3, and IR 4, with three inputs (In 1, In 2, and In 3) and four outputs (Out 1, Out 2, Out 3, and Out 4).

6. *Decide whether each of these statements is true (T) or false (F).* When there is a pulse input to In 1:
   (i) The output Out 1 is energized.
   (ii) The contacts of internal relay IR 1 close.

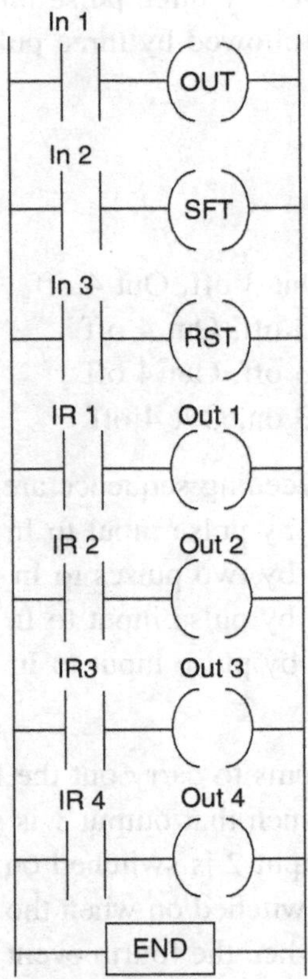

Figure 11.8: Diagram for Problems 6 through 9.

A. (i) T (ii) T
B. (i) T (ii) F
C. (i) F (ii) T
D. (i) F (ii) F

7. *Decide whether each of these statements is true (T) or false (F).* When there is a pulse input to In 1 followed by a pulse input to SFT:
   (i) Output Out 1 is energized.
   (ii) Output Out 2 is energized.
   A. (i) T (ii) T
   B. (i) T (ii) F
   C. (i) F (ii) T
   D. (i) F (ii) F

8. *Decide whether each of these statements is true (T) or false (F).* To obtain outputs Out 1, Out 2, Out 3, and Out 4 switching on in sequence and remaining on, we can have for inputs:
   (i) A pulse input to In 1 followed by three pulse inputs to SFT.
   (ii) A continuous input to In 1 followed by three pulse inputs to SFT.
   A. (i) T (ii) T
   B. (i) T (ii) F
   C. (i) F (ii) T
   D. (i) F (ii) F

9. Initially: Out 1 off, Out 2 off, Out 3 off, Out 4 off
   Next: Out 1 on, Out 2 off, Out 3 off, Out 4 off
   Next: Out 1 off, Out 2 on, Out 3 off, Out 4 off
   Next: Out 1 on, Out 2 off, Out 3 on, Out 4 off

The inputs required to obtain the preceding sequence are:
   A. Pulse input to In 1 followed by pulse input to In 2.
   B. Pulse input to In 1 followed by two pulses to In 2.
   C. Pulse input to In 1 followed by pulse input to In 2, then by pulse input to In 1.
   D. Pulse input to In 1 followed by pulse input to In 2, then by pulse inputs to In 1 and In 2.

10. Devise ladder programs for systems to carry out the following tasks:
    (a) A sequence of four outputs such that output 1 is switched on when the first event is detected and remains on, output 2 is switched on when the second event is detected and remains on, output 3 is switched on when the third event is detected and remains on, output 4 is switched on when the fourth event is detected and remains on, and all outputs are switched off when one particular input signal occurs.

(b) Control of a paint sprayer in a booth through which items pass on an overhead conveyor so that the paint is switched on when a part is in front of the paint gun and off when there is no part. The items are suspended from the overhead conveyor by hooks; not every hook has an item suspended from it.

Lookup Tasks

11. Find out the details of shift registers available with PLCs from a particular range from a specific manufacturer.

CHAPTER 12

Data Handling

Timers, counters, and individual internal relays are all concerned with the handling of individual bits, that is, single on/off signals. Shift registers involve a number of bits with a group of internal relays being linked (see Chapter 11). The block of data in the register is manipulated. This chapter is about PLC operations involving blocks of data representing a value; such blocks are called *words*. A block of data is needed if we are to represent numbers rather than just a single on/off input. Data handling consists of operations involving moving or transferring numeric information stored in one memory word location to another word in a different location, comparing data values, and carrying out simple arithmetic operations. For example, there might be the need to compare a numeric value with a set value and initiate action if the actual value is less than the set value. This chapter is an introductory discussion of such operations.

12.1 Registers and Bits

A register is where data can be stored (see Section 8.1 for an initial discussion of registers). In a PLC there are a number of such registers. Each data register can store a *binary word* of usually 8 or 16 bits. The number of bits determines the size of the number that can be stored. The *binary system* uses only two symbols, 0 and 1 (see Chapter 3). Thus we might have the 4-bit number 1111. This is the denary number, that is, the familiar number system based on 10s, of $2^0 + 2^1 + 2^2 + 2^3 = 1 + 2 + 4 + 8 = 15$. Thus a 4-bit register can store a positive number between 0 and $2^0 + 2^1 + 2^2 + 2^3$ or $2^4 - 1 = 15$. An 8-bit register can store a positive number between 0 and $2^0 + 2^1 + 2^2 + 2^3 + 2^4 + 2^5 + 2^6 + 2^7$ or $2^8 - 1$, that is, 255. A 16-bit register can store a positive number between 0 and $2^{16} - 1$, that is, 65,535.

Thus a 16-bit word can be used for positive numbers in the range 0 to 65,535. If negative numbers are required, the most significant bit is used to represent the sign, a 1 representing a negative number and a 0 a positive number; the format used for negative numbers is *two's complement*. Two's complement is a way of writing negative numbers so that when we add, say, the signed equivalent of +5 and –5, we obtain 0. Thus in this format, 1011 represents the negative number −5 and 0101 the positive number +5; 1011 + 0101 = (1)0000 with the (1) for the 4-bit number being lost. See Chapter 3 for further discussion.

W. Bolton: Programmable Logic Controllers, Sixth Edition. http://dx.doi.org/10.1016/B978-0-12-802929-9.00012-1

The *binary coded decimal* (BCD) format is often used with PLCs when they are connected to devices such as digital displays. With the natural binary number there is no simple link between the separate symbols of a denary number and the equivalent binary number. You have to work out the arithmetic to decipher one number from the other. With the BCD system, each denary digit is represented, in turn, by a 4-bit binary number (four is the smallest number of binary bits that gives a denary number greater than 10, that is, $2^n > 10$). To illustrate this idea, consider the denary number 123. The 3 is represented by the 4-bit binary number 0011, the 2 by the 4-bit number 0010, and the 1 by 0001. Thus the BCD number of 123 is 0001 0010 0011. BCD is a convenient system for use with external devices that are arranged in denary format, such as decade switches (thumbwheel switches) and digital displays. Then four binary bits can be used for each denary digit. PLCs therefore often have inputs or outputs that can be programmed to convert BCD from external input devices to the binary format needed inside the PLC and from the binary format used internally in the PLC to BCD for external output devices (see Section 12.3).

The thumbwheel switch is widely used as a means of inputting BCD data manually into a PLC. It has four contacts that can be opened or closed to give the four binary bits to represent a denary number (Figure 12.1). The contacts are opened or closed by rotating a wheel using one's thumb. By using a number of such switches, data can be input in BCD format.

12.2 Data Handling

The following are examples of data-handling instructions to be found with PLCs.

12.2.1 Data Movement

For moving data from one location or register to another, Figure 12.2 illustrates a common practice of using one rung of a ladder program for each move operation, showing the form used by three manufacturers: Mitsubishi, Allen-Bradley, and Siemens. For the rung shown,

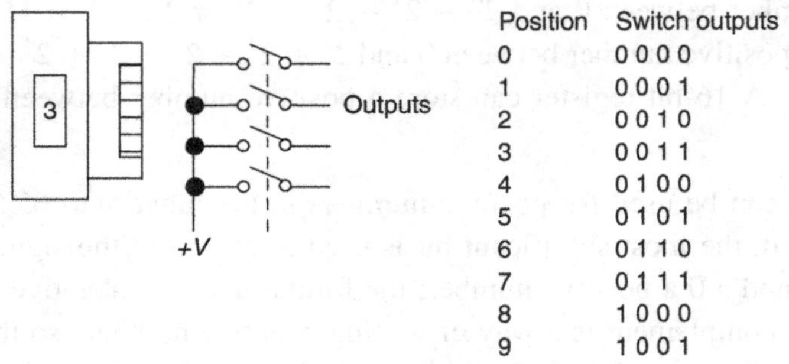

Figure 12.1: Thumbwheel switch.

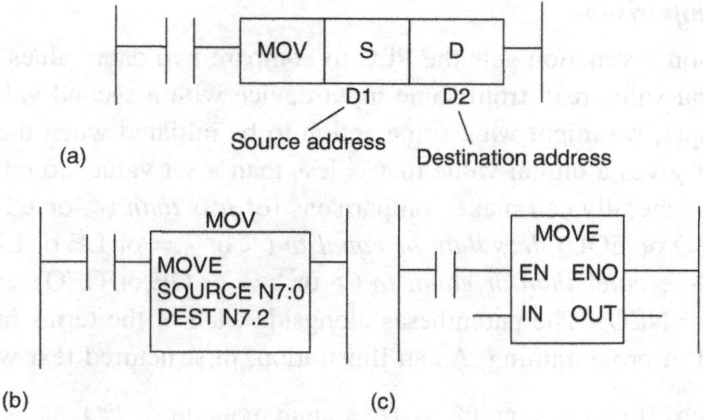

Figure 12.2: Data movement: (a) Mitsubishi, (b) Allen-Bradley, and (c) Siemens.

when there is an input to | | in the rung, the move occurs from the designated source address to the designated destination address. For data handling with these PLCs, the typical ladder program data-handling instruction contains the data-handling instruction, the source (S) address from where the data is to be obtained, and the destination (D) address to where it is to be moved. The approach that is used by some manufacturers, such as Siemens, is to regard data movement as two separate instructions, loading data from the source into an accumulator and then transferring the data from the accumulator to the destination. Figure 12.2c shows the Siemens symbol for the MOVE function. The data is moved from the IN input to the OUT output when EN is enabled.

Data transfers might be to move a preset value to a timer or counter, or a time or counter value to some register for storage, or data from an input to a register or a register to output. Figure 12.3 shows the rung, in the Allen-Bradley format, that might be used to transfer a number held at address N7:0 to the preset of timer T4:6 when the input conditions for that rung are met. A data transfer from the accumulated value in a counter to a register would have a source address of the form C5:18.ACC and a destination address of the form N7:0. A data transfer from an input to a register might have a source address of the form I:012 and a destination address of the form N7:0. A data transfer from a register to an output might have a source address of the form N7:0 and a destination address of the form O:030.

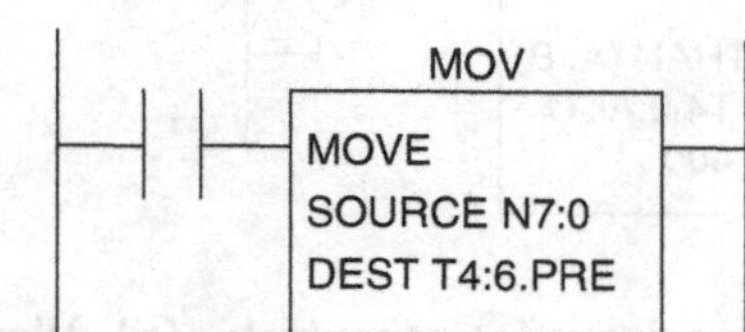

Figure 12.3: Moving number to timer preset.

12.2.2 Data Comparison

The data comparison instruction gets the PLC to compare two data values. Thus it might be to compare a digital value read from some input device with a second value contained in a register. For example, we might want some action to be initiated when the input from a temperature sensor gives a digital value that is less than a set value stored in a data register in the PLC. PLCs generally can make comparisons for *less than* (< or LT or LES), *equal to* (= or = = or EQ or EQU), *less than or equal to* (≤ or <= or LE or LEQ), *greater than* (> or GT or GRT), *greater than or equal to* (≥ or >= or GE or GEQ), and *not equal to* (≠ or <> or NE or NEQ). The parentheses alongside each of the terms indicates common abbreviations used in programming. As an illustration, in structured text we might have:

```
(*Check that boiler pressure P2 is less than pressure P1*)
Output := P2 < P1;
```

With ladder programs, for data comparison the typical instruction will contain the data-transfer instruction to compare data, the source (S) address from which the data is to be obtained for the comparison, and the destination (D) address of the data against which it is to be compared. The instructions commonly used for the comparison are the terms indicated in the preceding parentheses. Figure 12.4 shows the type of formats used by three manufacturers using the greater-than form of comparison. Similar forms apply to the other forms of comparison. In Figure 12.4a the format is that used by Mitsubishi, S indicating the source of the data value for the comparison and D the destination or value against which the comparison is to be made. Thus if the source value is greater than the destination value, the output is 1. In Figure 12.4b the Allen-Bradley format has been used. Here the source of the data being compared is given as the accumulated value in timer 4.0 and the data against

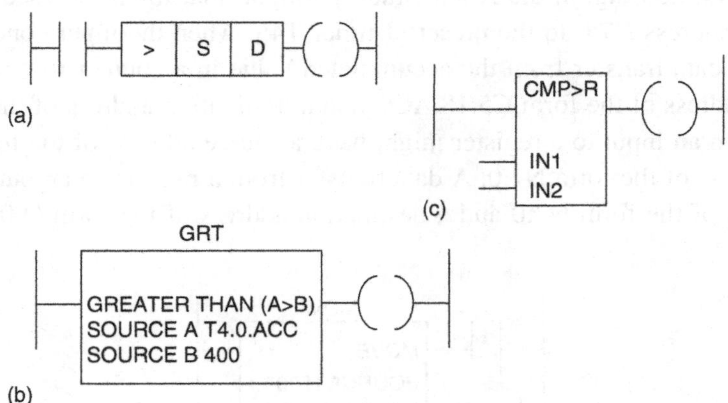

Figure 12.4: Greater than comparison: (a) Mitsubishi, (b) Allen-Bradley, and (c) Siemens.

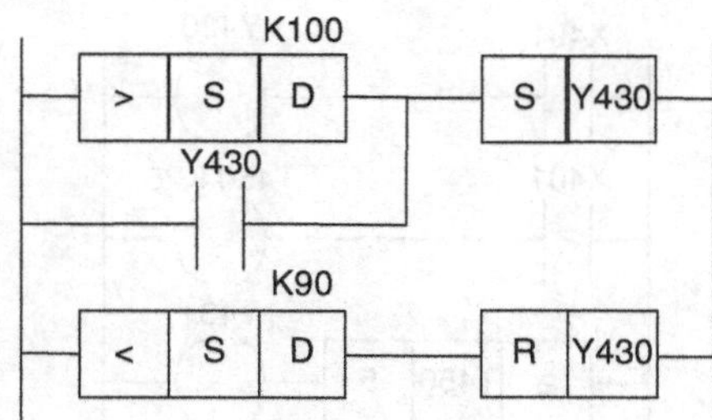

Figure 12.5: Alarm program.

which it is being compared is the number 400. Figure 12.4c shows the Siemens format. The values to be compared are at inputs IN1 and IN2 and the result of the comparison is at the output: 1 if the comparison is successful, otherwise 0. The R is used to indicate real numbers, that is, floating point numbers, I being used for integers, that is, fixed-point numbers involving 16 bits, and D for fixed-point numbers involving 32 bits. Both the inputs need to be of the same data type, such as REAL.

As an illustration of the use of such a comparison, consider the task of sounding an alarm if a sensor indicates that a temperature has risen above some value, say, 100°C. The alarm is to remain sounding until the temperature falls below 90°C. Figure 12.5 shows the ladder diagram that might be used. When the temperature rises to become equal to or greater than 100°C, the greater-than comparison element gives a 1 output and so sets an internal relay. There is then an output. This output latches the greater-than comparison element, so the output remains on, even when the temperature falls below 100°C. The output is not switched off until the less-than 90°C element gives an output and resets the internal relay.

Another example of the use of comparison is when, say, four outputs need to be started in sequence, that is, output 1 starts when the initial switch is closed, followed sometime later by output 2, sometime later by output 3, and sometime later by output 4. Though this could be done using three timers, another possibility is to use one timer with greater-than or equal elements. Figure 12.6 shows a possible ladder diagram. When the X401 contacts close, the output Y430 starts. The timer is also started. When the timer-accumulated value reaches 5 s, the greater-than or equal-to element switches on Y431. When the timer-accumulated value reaches 15 s, the greater-than or equal-to element switches on Y432. When the timer reaches 25 s, its contacts switch on Y433.

12.2.3 Data Selection

There are a number of selection function blocks available with PLCs. Figure 12.7 shows the standard IEC symbols.

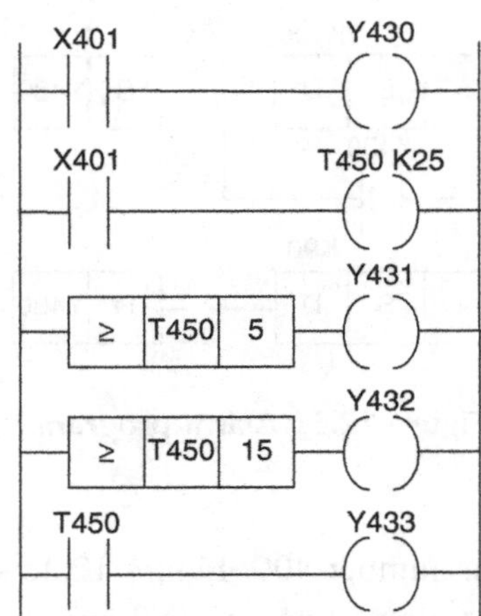

Figure 12.6: Sequential switching on.

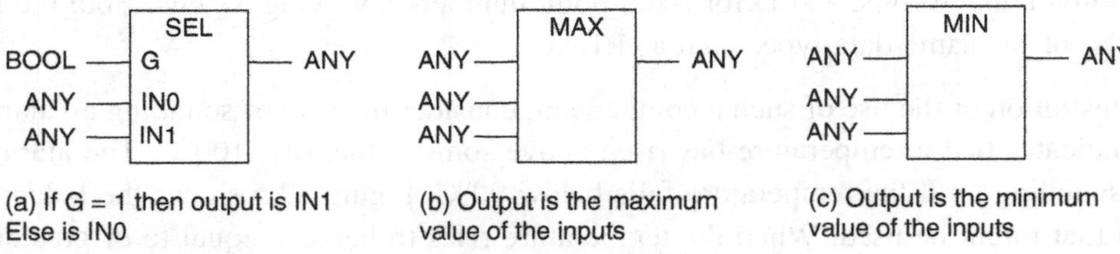

Figure 12.7: IEC symbols: (a) selection, (b) maximum, and (c) minimum.

12.3 Arithmetic Functions

Most PLCs provide BCD-to-binary or integer and integer or binary-to-BCD conversions for use when the input might be a thumbwheel switch or the output to a decimal display. Figure 12.8a shows one form of instructions for use in such situations, Figure 12.8b the form used by Siemens for conversion of BCD to integer and integer to BCD, and Figure 12.8c the form used by Allen-Bradley.

12.3.1 Arithmetic Operations

Some PLCs are equipped to carry out just the arithmetic operations of addition and subtraction, others the four basic arithmetic operations of addition, subtraction, multiplication, and division, and still others can carry out these and various other functions such as the exponential. Addition and subtraction operations are used to alter the value of data held in data registers. For example, this might be to adjust a sensor input reading or perhaps obtain a value by subtracting two sensor values or alter the preset values used by timers and counters. Multiplication might be used to multiply some input before perhaps adding to or subtracting it from another.

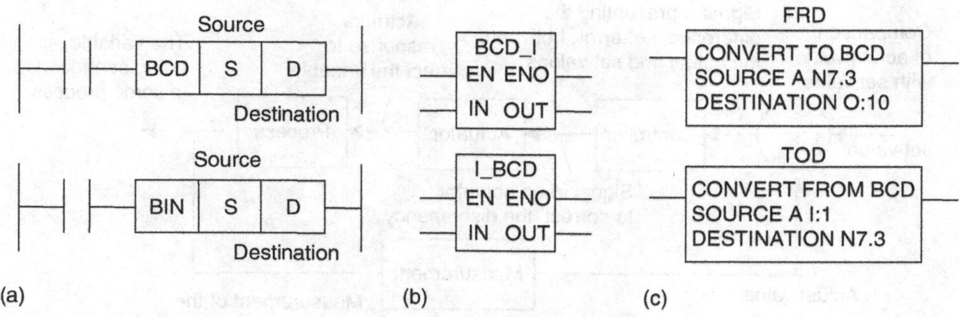

Figure 12.8: Conversion BCD-to-binary and binary-to-BCD: (a) one format, (b) Siemens format, and (c) Allen-Bradley format.

The way PLCs have to be programmed to carry out such operations varies. In its PLCs, Allen-Bradley has such arithmetic operations as add (ADD), subtract (SUB), divide (DIV), multiply (MUL), and square root (SQR). Figure 12.9a shows the format for ADD; the other arithmetic functions have a similar format. The data in source A, which is at N7.1, is added to that in source B, which is at N7.3, and the result is put at the destination N7.5.

Figure 12.9b shows the basic form of the Siemens instructions for arithmetic functions. With integers the functions available are ADD_1 for addition, SUB_1 for subtraction, MUL_1 for multiplication, and DIV_1 for division, with the quotient as the result. The arithmetic functions are executed if there is a 1 at the enable EN input.

12.4 Closed Loop Control

You can control the temperature of a room by switching on an electric fire. The fire will heat the room up to the maximum temperature that is possible, bearing in mind the rate at which the fire heats the room and the rate at which it loses heat. This is termed *open loop control* in that there is no feedback to the fire to modify the rate at which it is heating the room. To control the temperature with feedback, you need a thermostat that can be set to switch the fire on when the room temperature is below the required value and switch it off when it goes above it. There is feedback of temperature information in this system; as such it is termed *closed loop control*.

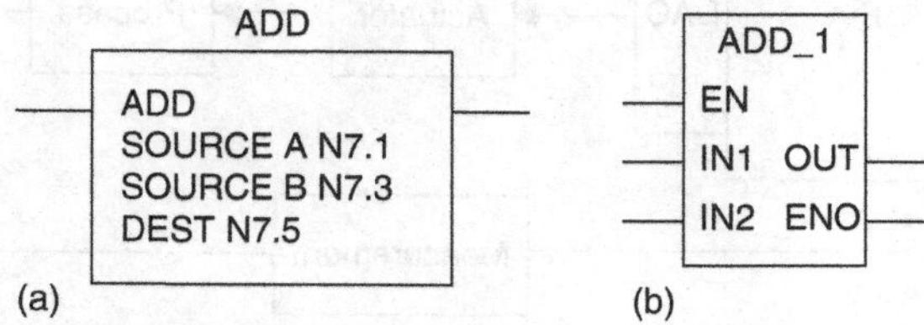

Figure 12.9: ADD: (a) Allen-Bradley format, and (b) Siemens format.

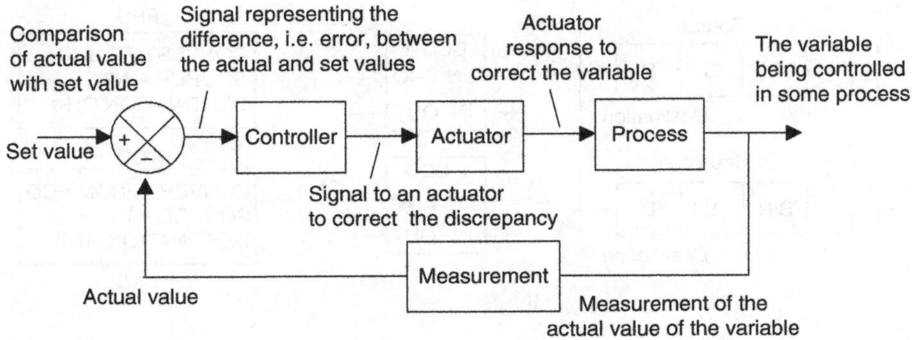

Figure 12.10: Closed loop control.

Closed loop control of some variable, such as the control of the temperature in a room, is achieved by comparing the actual value for the variable with the desired set value and then giving an output, such as switching on a heater, to reduce the difference. Figure 12.10 illustrates this idea by means of a block diagram. The actual value of the variable is compared with the set value and a signal is obtained representing the difference or error. A controller then takes this difference signal and gives an output to an actuator to give a response to correct the discrepancy.

Figure 12.11 shows the arrangement that might be used with a PLC used to exercise the closed loop control. It has been assumed that the actuator and the measured values are analog and thus require conversion to digital; analog-to-digital and digital-to-analog units have thus been shown.

12.4.1 Modes of Control

There are a number of methods by which the controller can react to an error signal:

- *On-off mode,* in which the controller is essentially just a switch that supplies an on/off output signal depending on whether there is an error or not (Figure 12.12a). When there

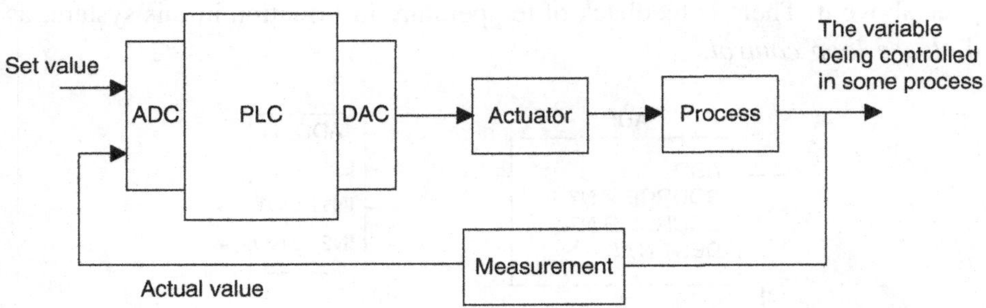

Figure 12.11: A PLC for closed loop control.

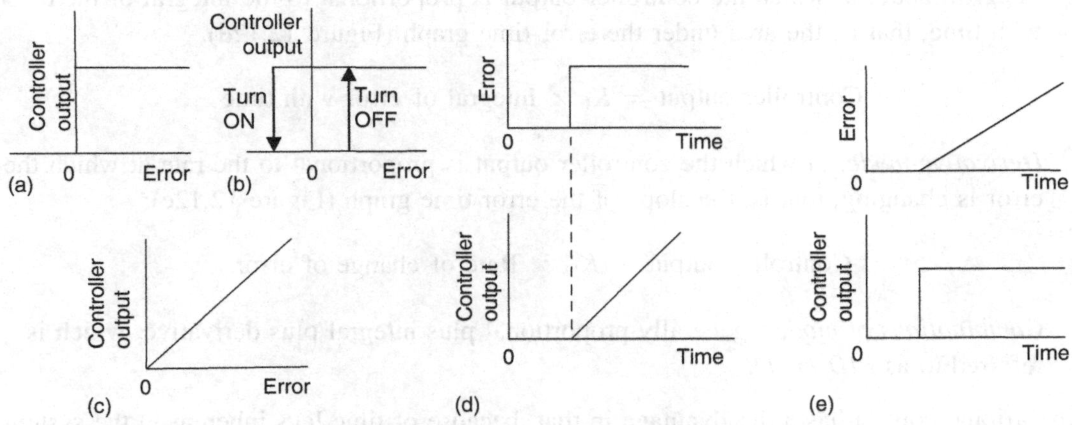

Figure 12.12: Control: (a) on-off, (b) on-off with hysteresis, (c) proportional, (d) integral, and (e) derivative.

is an error, there is a constant output from the controller, regardless of the size of the error. A simple example of such a controller is the bimetallic thermostat (refer back to Figure 2.11) used with the central heating systems of many houses. This is just a form of switch that is off when the temperature is above the required temperature and on when it is below it. Because the control action is discontinuous and there are time lags in the system, oscillations of the controlled variable occur about the set value. For example, when a bimetallic thermostat switch switches on, there is a time delay before the heater begins to have an effect on the temperature, and when the temperature rises to the required temperature, there will be a time delay before the switched-off heater stops heating the system. However, on/off control is not too bad if the system has a large capacitance or inertia, so the effect of changes results in only slow changes in the variable. A dead band or hysteresis is often added to stop the controller from reacting to small error values and consequently constantly switching on and off when the variable is hovering about the set value. The switch-on value of the variable is thus different from the switch-off value (Figure 12.12b).

- *Proportional mode,* in which the controller gives an output to the actuator that is proportional to the difference between the actual value and the set value of the variable, that is, the error (Figure 12.12c). Such a form of control can be given by a PLC with basic arithmetic facilities. The set value and the actual values are likely to be analog and so are converted to digital, and then the actual value is subtracted from the set value and the difference multiplied by some constant, the proportional constant K_P, to give the output, which, after conversion to analog, is the correction signal applied to the actuator:

$$\text{Controller output} = K_P \times \text{Error}$$

- *Integral mode,* in which the controller output is proportional to the integral of the error with time, that is, the area under the error-time graph (Figure 12.12d).

$$\text{Controller output} = K_I \times \text{Integral of error with time}$$

- *Derivative mode,* in which the controller output is proportional to the rate at which the error is changing, that is, the slope of the error-time graph (Figure 12.12e):

$$\text{Controller output} = K_D \times \text{Rate of change of error}$$

- *Combinations of modes*, generally proportional plus integral plus derivative, which is referred to as *PID mode.*

Proportional control has a disadvantage in that, because of time lags inherent in the system, the correcting signal applied to the actuator tends to cause the variable to oscillate about the set value. What is needed is a correcting signal that is reduced as the variable gets close to the set value. This is obtained by *PID control*, the controller giving a correction signal that is computed from a proportional element (the P term), an element that is related to previous values of the variable (the integral I term), and an element related to the rate at which the variable is changing (the derivative D term).

The term *tuning* is used for determining the optimum values of K_P, K_I, and K_D to be used for a particular control system. The value of K_D/K_P is called the *derivative action time* T_D, and the value of K_P/K_I is called the *integral action time* T_I; it is these terms K_P, T_D, and T_I that are generally specified.

12.4.2 Control with a PLC

Figure 12.13 shows a PLC ladder rung that can be used to exercise two-step control. The output is turned on when source A, the actual temperature, is less than source B, the required temperature, that is, the set value.

Many PLCs provide the PID calculation to determine the controller output as a standard function block. All that is then necessary is to pass the desired parameters, that is, the values of K_P, K_I, and K_D, and input/output locations to the routine via the PLC program.

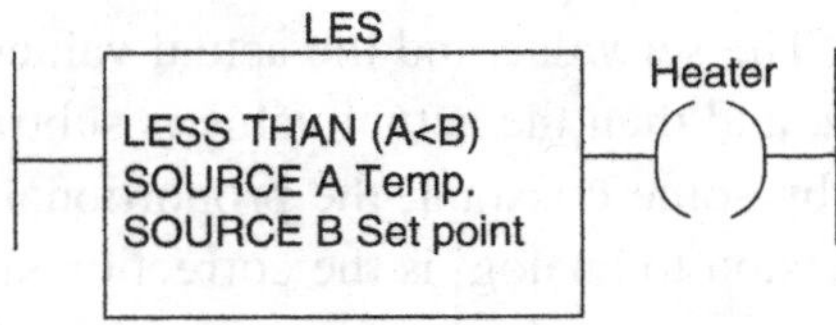

Figure 12.13: Two-step control.

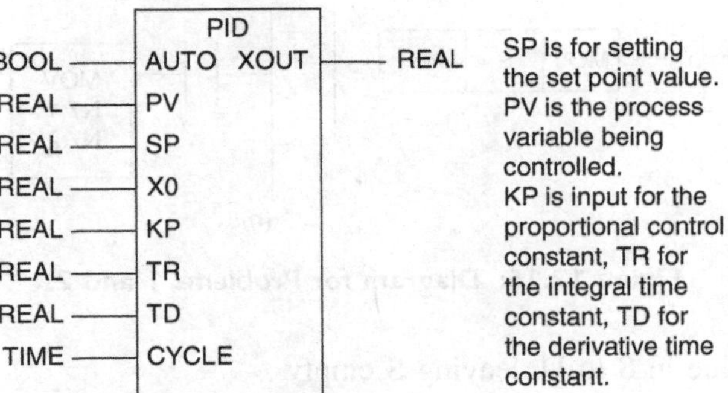

Figure 12.14: IEC 61131-3 standard symbol.

Figure 12.14 shows the IEC 61131-3 standard symbol for the PID control function. When AUTO is set, the function blocks calculate the output value XOUT needed to bring the variable closer to the required set value.

Summary

Data handling consists of operations involving moving or transferring numeric information stored in one memory word location to another word in a different location, comparing data values, and carrying out simple arithmetic operations.

Closed loop control of some variable is achieved by comparing the actual value for the variable with the desired set value and then giving an output to reduce the difference. With proportional control, the controller gives the actuator an output that is proportional to the difference between the actual value and the set value of the variable, that is, the error. With PID control, the controller gives a correction signal that is computed from a proportional element, the P term, the integral term I, is an element giving a signal related to previous values of the variable and is the area under the error-time graph, and an element giving a signal related to the rate at which the variable is changing (the derivative D term).

Problems

Problems 1 through 9 have four answer options: A, B, C, or D. Choose the correct answer from the answer options. Problems 1 and 2 refer to Figure 12.15, which shows two formats used for the move operation.

1. *Decide whether each of these statements is true (T) or false (F).* In Figure 12.15a, the program instruction is to:

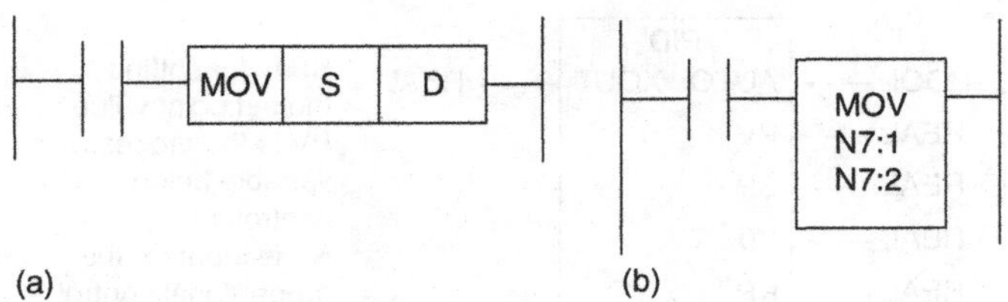

Figure 12.15: Diagram for Problems 1 and 2.

(i) Move the value in S to D, leaving S empty.
(ii) Copy the value in S and put it in D.
A. (i) T (ii) T
B. (i) T (ii) F
C. (i) F (ii) T
D. (i) F (ii) F

2. *Decide whether each of these statements is true (T) or false (F).* In Figure 12.15b, the program instruction is to:
(i) Move the value in N7:1 to N7:2, leaving N7:1 empty.
(ii) Copy the value in N7:1 and put it in N7:2.
A. (i) T (ii) T
B. (i) T (ii) F
C. (i) F (ii) T
D. (i) F (ii) F

Problems 3 and 4 refer to Figure 12.16, which shows two versions of a ladder rung involving a comparison.

3. *Decide whether each of these statements is true (T) or false (F).* In Figure 12.16a, the program instruction is to give an output:
(i) When the accumulated time in timer T450 exceeds a value of 400.
(ii) Until the accumulated time in timer T450 reaches a value of 400.
A. (i) T (ii) T
B. (i) T (ii) F
C. (i) F (ii) T
D. (i) F (ii) F

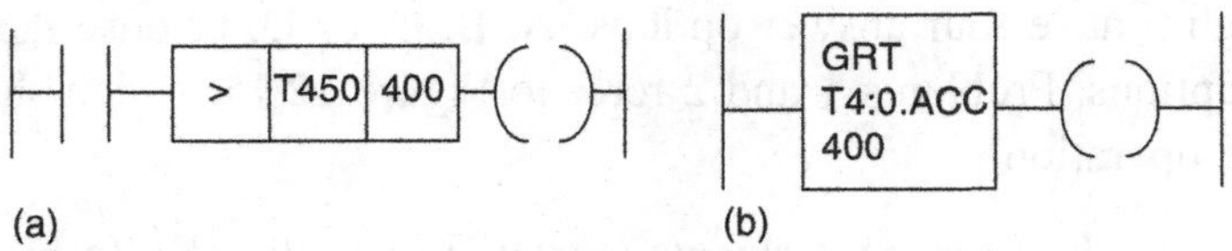

Figure 12.16: Diagram for Problems 3 and 4.

4. *Decide whether each of these statements is true (T) or false (F).* In Figure 12.16b, the program instruction is to give an output:
   (i) When the accumulated time in timer T4:0 exceeds a value of 400.
   (ii) Until the accumulated time in timer T4:0 reaches a value of 400.
   A. (i) T (ii) T
   B. (i) T (ii) F
   C. (i) F (ii) T
   D. (i) F (ii) F

5. *Decide whether each of these statements is true (T) or false (F).* In Figure 12.17, when the input conditions are met, the program instruction is to give an output when the data:
   (i) In N7:10 equals that in N7:20.
   (ii) In N7:10 is less than that in N7:20.
   A. (i) T (ii) T
   B. (i) T (ii) F
   C. (i) F (ii) T
   D. (i) F (ii) F

6. *Decide whether each of these statements is true (T) or false (F).* In Figure 12.18, the program instruction is to give, when the input conditions are met, an output when:
   (i) The data in N7:10 is not equal to that in N7:20.
   (ii) The data in N7:10 is greater or less than that in N7:20.
   A. (i) T (ii) T
   B. (i) T (ii) F
   C. (i) F (ii) T
   D. (i) F (ii) F

7. In Figure 12.19, when the input conditions are met the program instruction is to give:
   A. The sum of the data at sources A and B
   B. The product of the data in sources A and B

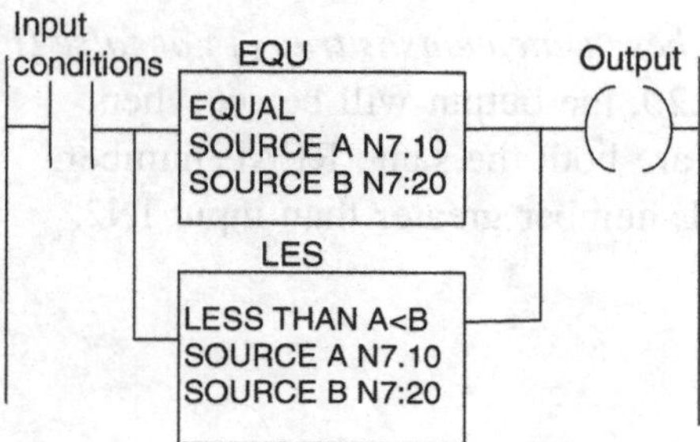

Figure 12.17: Diagram for Problem 5.

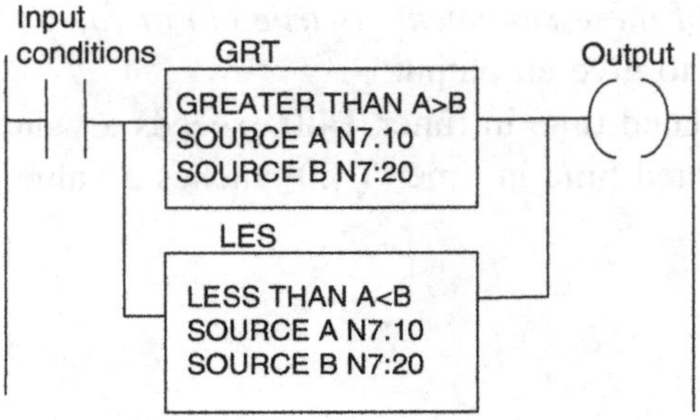

Input conditions
GRT
GREATER THAN A>B
SOURCE A N7.10
SOURCE B N7:20
Output
LES
LESS THAN A<B
SOURCE A N7.10
SOURCE B N7:20

Figure 12.18: Diagram for Problem 6.

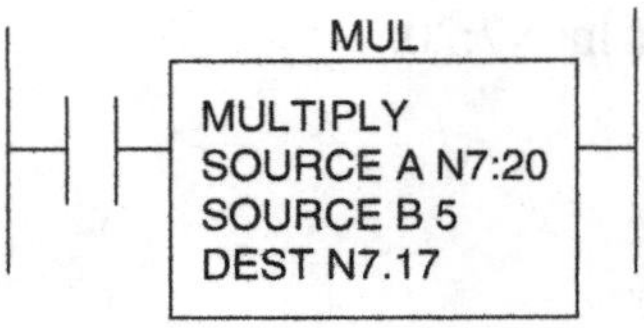

MUL
MULTIPLY
SOURCE A N7:20
SOURCE B 5
DEST N7.17

Figure 12.19: Diagram for Problem 7.

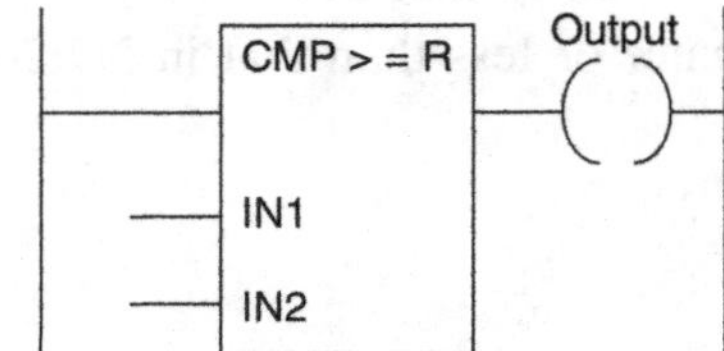

Figure 12.20: Diagram for Problem 8.

C. The difference between the data in sources A and B
D. The value given by dividing the data in source A by that in B

8. *Decide whether each of these statements is true (T) or false (F).* For the Siemens function box shown in Figure 12.20, the output will be set when:
(i) Inputs IN1 and IN2 are both the same REAL number.
(ii) Input IN1 is a REAL number greater than input IN2.
A. (i) T (ii) T
B. (i) T (ii) F
C. (i) F (ii) T
D. (i) F (ii) F

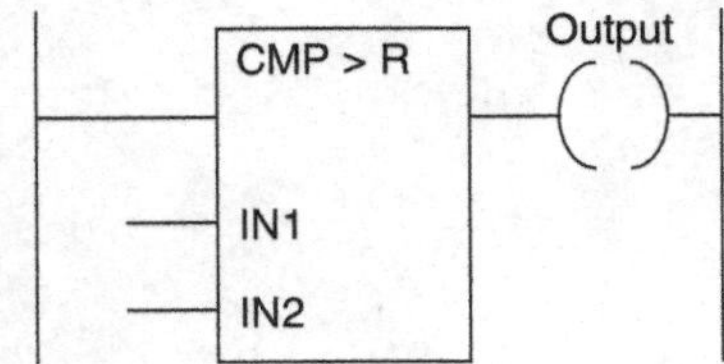

Figure 12.21: Diagram for Problem 9.

9. *Decide whether each of these statements is true (T) or false (F).* For the Siemens function box shown in Figure 12.21, the output will be set when:
   (i) Inputs IN1 and IN2 are both the same REAL number.
   (ii) Input IN1 is a REAL number greater than input IN2.
   A. (i) T (ii) T
   B. (i) T (ii) F
   C. (i) F (ii) T
   D. (i) F (ii) F

10. Devise ladder programs for systems that will carry out the following tasks:
    (a) Switch on a pump when the water level in a tank rises to above 1.2 m and switch it off when it falls below 1.0 m.
    (b) Switch on a pump; then 100 s later, switch on a heater; then a further 30 s later, switch on the circulating motor.
    (c) Switch on a heater when the temperature is less than the set temperature.
    (d) Turn on a lamp when a data source is not giving 100.

11. Describe the operation of an on/off controller and explain how it might be used to control the temperature in a domestic central heating system.

12. Explain the principle of a proportional controller.

Lookup Tasks

13. For a particular PLC model, determine what data-handling functions it has.

CHAPTER 13

Designing Systems

This chapter considers the way programs are designed and how they and a PLC system can be tested and faults found. This involves consideration of both the hardware and the software.

13.1 Program Development

Whatever the language in which a program is to be written, a systematic approach to the problem can improve the chance of high-quality programs being generated in as short a time as possible. A systematic design technique is likely to involve the following steps:

1. A definition of what is required, with the inputs and outputs specified.
2. A definition of the algorithm to be used. An algorithm is a step-by-step sequence that defines a method of solving the problem. This can often be shown by a flowchart or can be written in pseudocode, which involves the use of the words BEGIN, DO, END, IF-THEN-ELSE, and WHILE-DO.
3. The algorithm is then translated into instructions that can be input to the PLC. Because programs are often very long and can end up difficult to write as a long single block and are even more difficult to later follow for fault finding and maintenance, it is advisable to break the program down into areas that are then further subdivided until manageably sized blocks of program occur. This technique is termed *top-down design.*
4. The program is then tested and debugged.
5. The program is documented so that any person using or having to modify the program at a later date understands how the program works.

13.1.1 Flowcharts and Pseudocode

Figure 13.1a shows the symbols used in flowcharts. Each step of an algorithm is represented by one or more of these symbols and linked by lines to represent the program flow (Figure 13.1b). Pseudocode is a way of describing the steps in an algorithm in an informal way.

W. Bolton: Programmable Logic Controllers, Sixth Edition. http://dx.doi.org/10.1016/B978-0-12-802929-9.00013-3

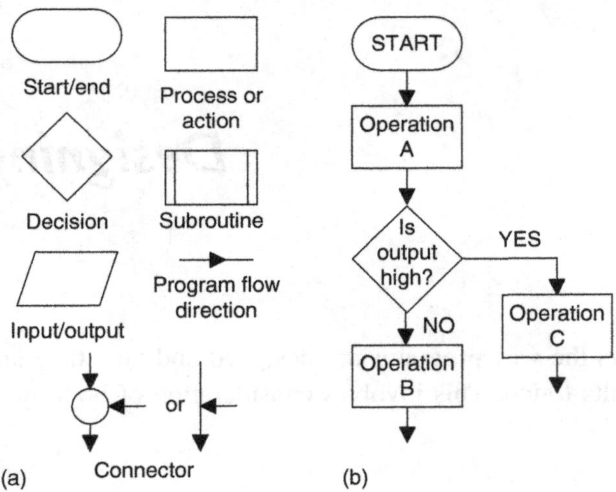

Figure 13.1: (a) Flowchart symbols, and (b) example of a simple flowchart.

Consider how the following program operations can be represented by flowcharts and pseudocode and then programmed using ladder and sequential function chart programming:

- *Sequential*. Consider a sequence in which event A has to be followed by event B. Figure 13.2a shows how this can be represented by a flowchart. In pseudocode this is written as:

```
BEGIN A
    DO A
END A
BEGIN B
    DO B
END B
```

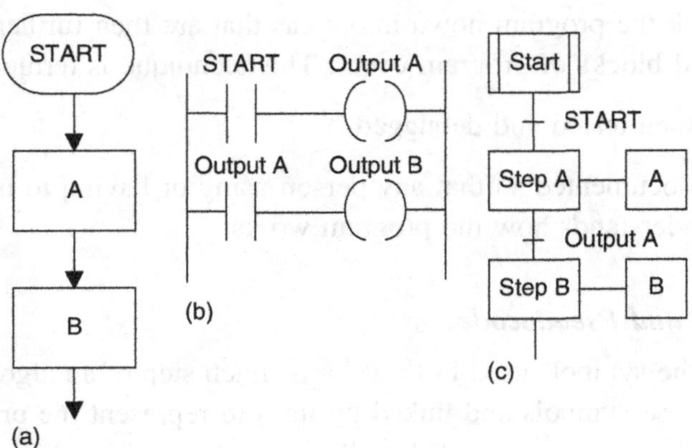

Figure 13.2: Sequence.

A sequence can be translated into a ladder program in the way shown in Figure 13.2b. When the start input occurs, output A happens. When action A happens, it operates the output A relay and results in output B occurring. Figure 13.2c shows the sequential function chart representation of a sequence.

- *Conditional*. Figure 13.3a shows the flowchart for when A or B is to happen if a particular condition X being YES or NO occurs. The pseudocode to describe this situation involves the words IF-THEN-ELSE-ENDIF.

```
IF X
THEN
    BEGIN A
    DO A
    END A
```

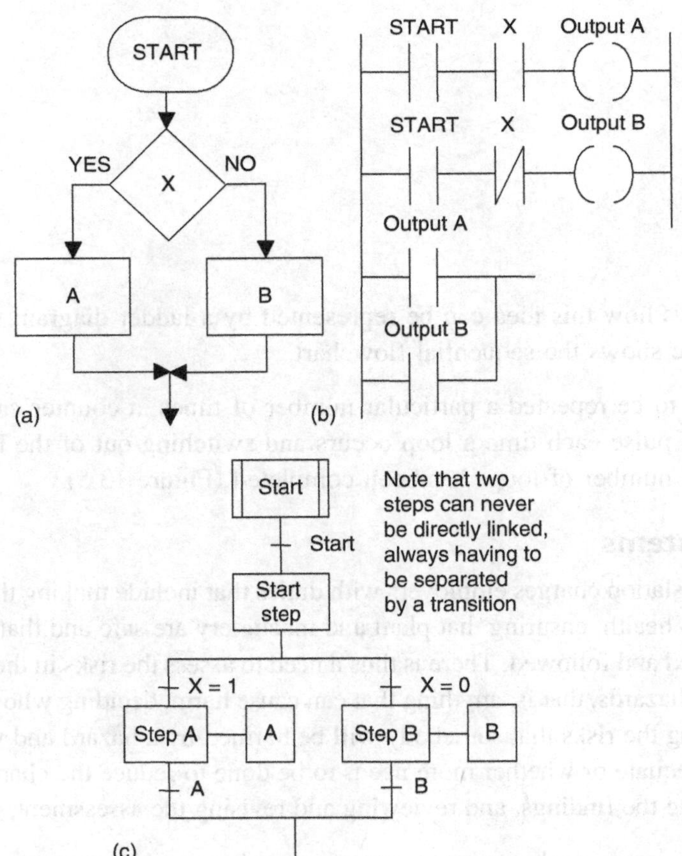

Figure 13.3: *Conditional.*

```
ELSE
    BEGIN B
    DO B
    END B
ENDIF X
```

Such a condition can be represented by the ladder diagram shown in Figure 13.3b. When the start input occurs, the output will be A if there is an input to X; otherwise the output is B. Figure 13.3c shows the sequential function chart for such selective branching.

- *Looping*. A *loop* is a repetition of some element of a program that is repeated as long as some condition prevails. Figure 13.4a shows how this repetition can be represented by a flowchart. As long as condition X is realized, sequence A followed by B occurs and is repeated. When X is no longer realized, the program continues and the looping through A and B ceases. In pseudocode, this can be represented using the words WHILE-DO-ENDWHILE:

```
WHILE X
    BEGIN A
    DO A
    END A
    BEGIN B
    DO B
    END B
ENDWHILE X
```

Figure 13.4b shows how this idea can be represented by a ladder diagram with an internal relay. Figure 13.4c shows the sequential flowchart.

Where a loop has to be repeated a particular number of times, a counter can be used, receiving an input pulse each time a loop occurs and switching out of the loop sequence when the required number of loops has been completed (Figure 13.5).

13.2 Safe Systems

Modern safety legislation charges employers with duties that include making the workplace safe and free of risks to health, ensuring that plant and machinery are safe and that safe systems of work are established and followed. There is thus a need to assess the risks in the workplace. This means looking for hazards, that is, anything that can cause harm, deciding who might be harmed and how, evaluating the risks that somebody will be harmed by a hazard and whether existing precautions are adequate or whether more needs to be done to reduce the chance of harm occurring, recording the findings, and reviewing and revising the assessment, if necessary.

Thus, for example, issues such as emergency stops and access doors on equipment need to be considered, the risks assessed, and safe systems then designed. With regard to access

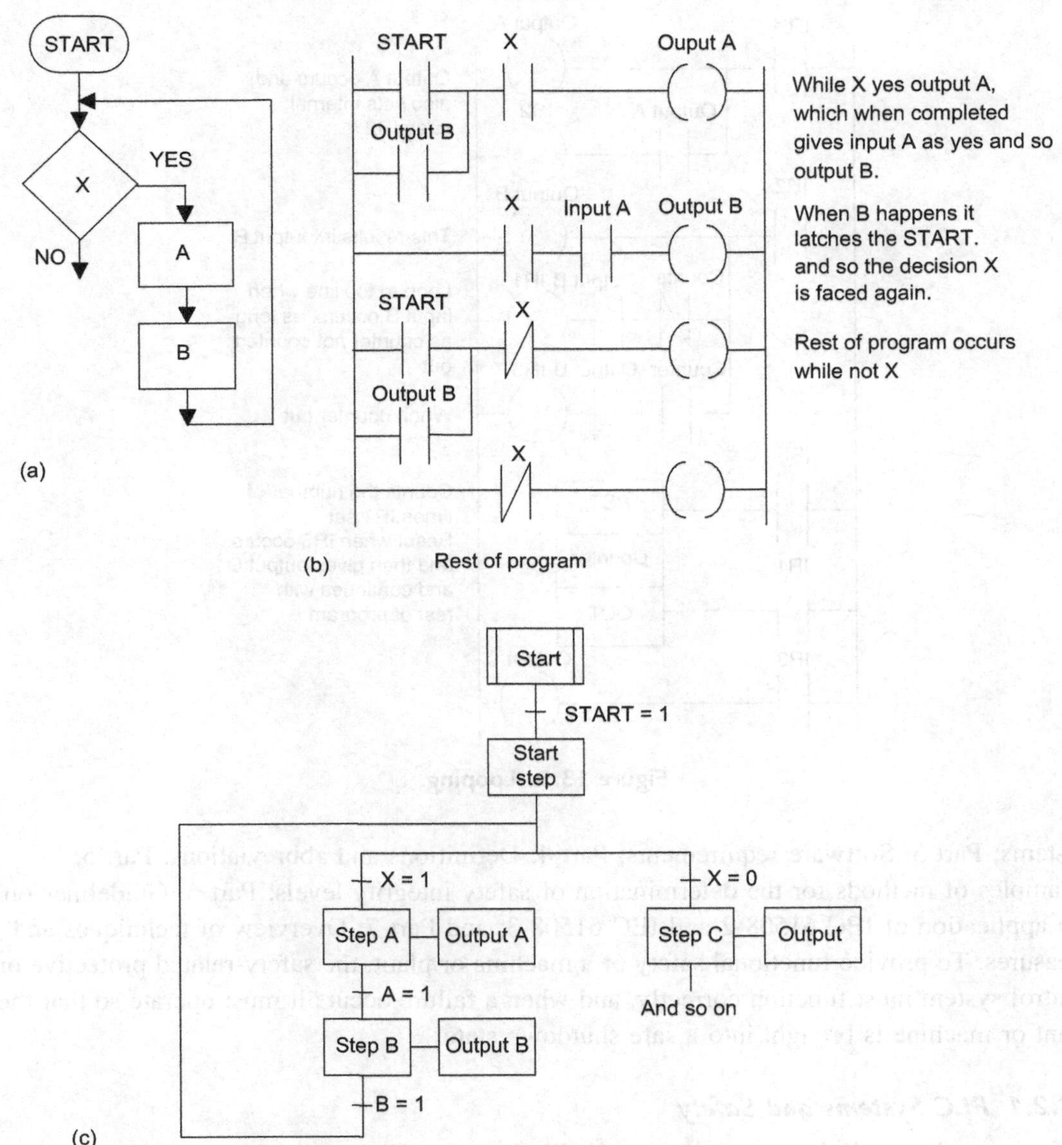

Figure 13.4: Looping.

doors on equipment, switch contacts can be used on such doors so that the system is stopped if the doors are not correctly guarding equipment.

An important standard is IEC 61508: Functional Safety of Electrical/Electronic/ Programmable Electronic Safety-Related Systems. The standard is in seven parts, as follows: Part 1: General requirements; Part 2: Requirements for E/E/PE safety-related

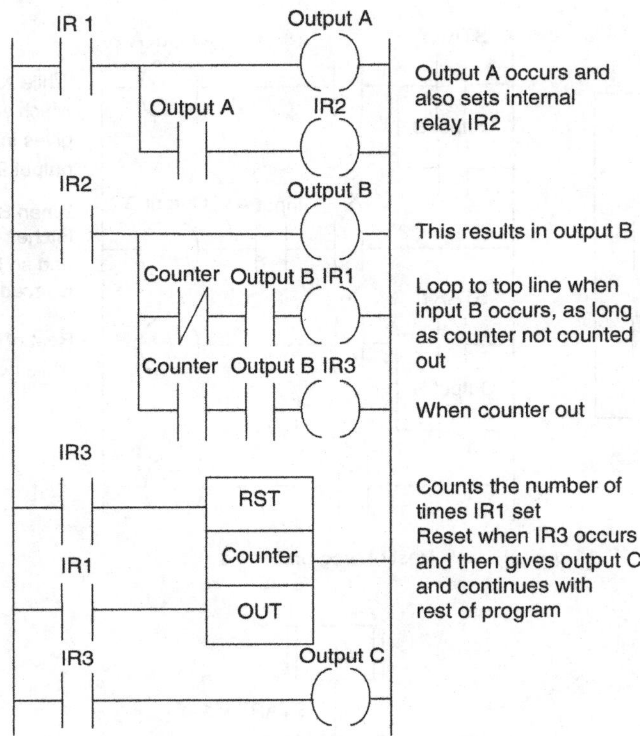
IR 1
Output A
Output A occurs and also sets internal relay IR2
Output A
IR2
IR2
Output B
This results in output B
Counter
Output B
IR1
Loop to top line when input B occurs, as long as counter not counted out
Counter
Output B
IR3
When counter out
IR3
RST
Counter
IR1
OUT
Counts the number of times IR1 set
Reset when IR3 occurs and then gives output C and continues with rest of program
IR3
Output C

Figure 13.5: Looping.

systems; Part 3: Software requirements; Part 4: Definitions and abbreviations; Part 5: Examples of methods for the determination of safety integrity levels; Part 6: Guidelines on the application of IEC 61508-2 and IEC 61508-3; and Part 7: Overview of techniques and measures. To provide functional safety of a machine or plant, the safety-related protective or control system must function correctly, and when a failure occurs it must operate so that the plant or machine is brought into a safe shutdown state.

13.2.1 PLC Systems and Safety

Safety must be a priority in the design of a PLC system. Thus, emergency stop buttons and safety guard switches must be hardwired and not depend on the PLC software for implementation, so that, in a situation where there is a failure of the stop switch or PLC, the system is automatically safe. The system must be *fail-safe*. Thus if failure occurs, the outputs must revert to a fail-safe mode so that no harm can come to anyone. For example, the guards on a machine must not be open or be capable of being opened if the PLC fails.

With a PLC system, a stop signal can be provided by a switch as shown in Figure 13.6. This arrangement is unsafe as an emergency stop because if there is a fault and the switch

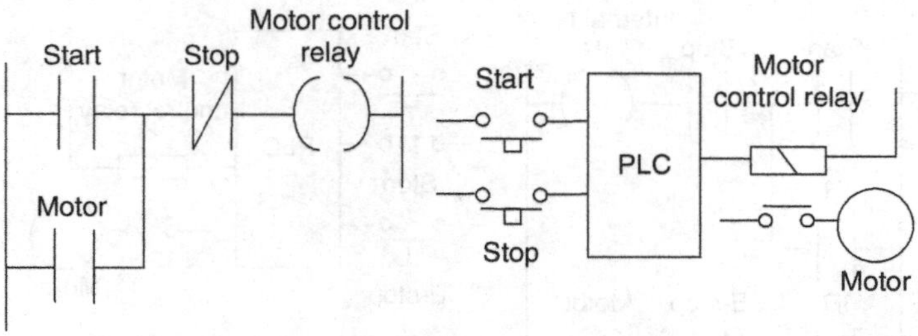

Figure 13.6: An unsafe stop system.

cannot be operated, then no stop signal can be provided. Thus to start we momentarily close the push-button start switch and the motor control internal relay then latches this closure and the output remains on. To stop we have to momentarily open the stop switch; this unlatches the start switch. However, if the stop switch cannot be operated, we cannot stop the system. What we require is a system that will still stop if a failure occurs in the stop switch.

We can achieve this by the arrangement shown in Figure 13.7. The program has the stop switch as open contacts. However, because the hardwired stop switch has normally closed contacts, the program has the signal to close the program contacts. Pressing the stop switch opens the program contacts and stops the system.

For a safe emergency stop system, we need one that will provide a stop signal if there is a fault and the switch cannot be operated. Because there might be problems with a PLC, we also need the emergency stop to operate independently of the PLC. Putting the emergency stop in the input to the PLC gives an unsafe system (Figure 13.8).

Figure 13.9 shows a safer system where the emergency stop switch is hardwired in the output. Pressing the emergency stop button switch stops, say, a running motor. When we release the stop button, the motor will not restart again, because the internal relay contacts have come unlatched.

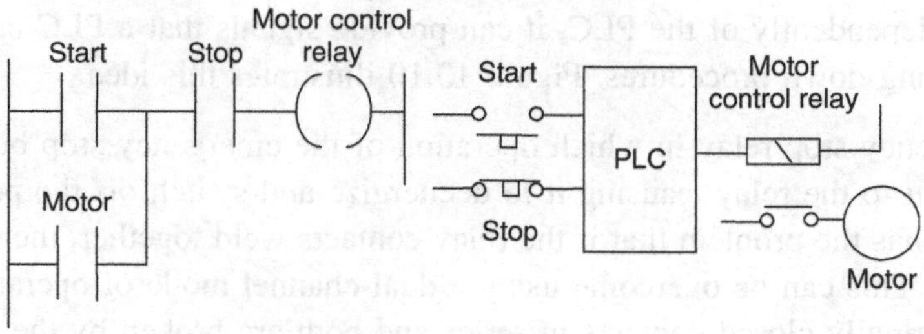

Figure 13.7: A safer stop system.

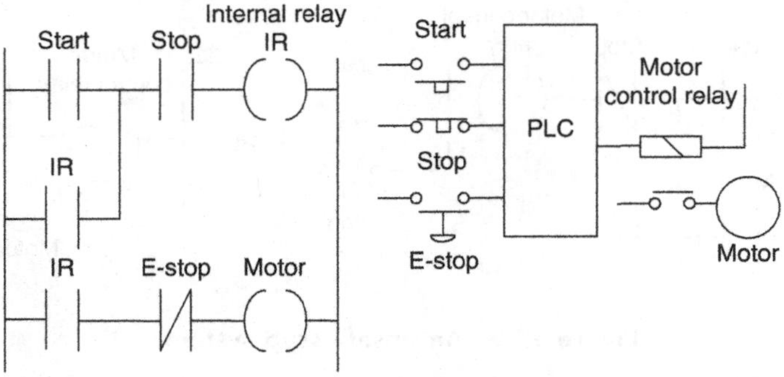

Figure 13.8: An unsafe emergency stop system.

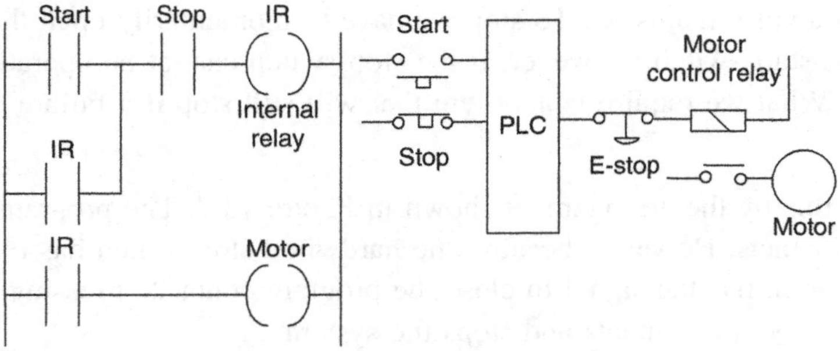

Figure 13.9: A safer emergency stop system.

13.2.2 Emergency Stop Relays

Emergency stop relays are widely used for emergency stop arrangements, such as the PNOZ p1p from Pilz GmbH & Co. This device has LEDs for indicating the status of input and output circuits, the reset circuit and power supply, and faults. However, the base unit can be connected via an interface module so that its status can be read by a PLC. This interface isolates the output from the emergency stop relay from the signal conditioning and input to the PLC by means of optoisolators (refer back to Figure 1.8). Thus, though the emergency stop operates independently of the PLC, it can provide signals that a PLC can use to, say, initiate safe closing-down procedures. Figure 13.10 illustrates this idea.

A simple emergency stop relay in which operation of the emergency stop button breaks the control circuit to the relay, causing it to deenergize and switch off the power (Figure 13.11a), has the problem that if the relay contacts weld together, the emergency stop will not operate. This can be overcome using a dual-channel mode of operation in which there are two normally closed contacts in series and both are broken by the action of the relay deenergizing (Figure 13.11b). Safety can be increased yet further if three contacts in

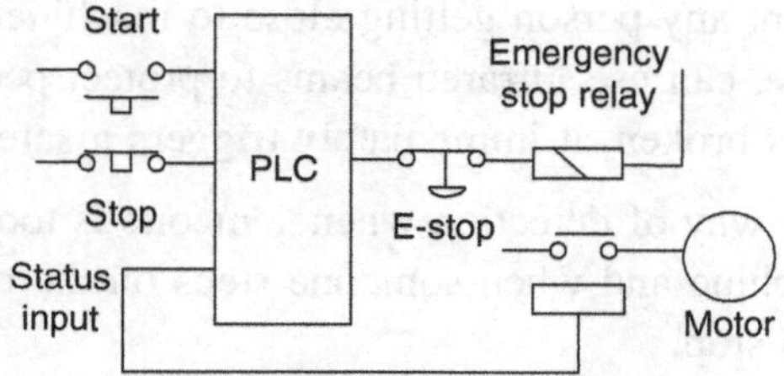

Figure 13.10: Emergency stop relay providing feedback of status.

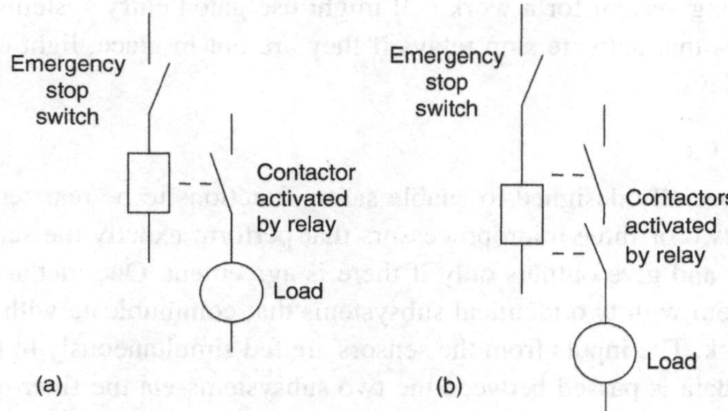

Figure 13.11: Emergency stop relay: (a) single-channel mode, and (b) dual-channel mode.

series are used, one using normally closed contacts and the others normally open contacts. Then one set of contacts has to be deenergized and the other two energized.

13.2.3 Safety Functions

In designing control systems, it is essential that personnel are prevented from coming into contact with machinery while it is active. This might involve:

- *Two-handed engaging* so that both hands must be on switches all the time and the machine will switch off if only one of the switches is being engaged.
- *Protective door monitoring* to prevent access to a machine while it is operating. This can be achieved by the use of safety interlocks such as doors and gates. Limit switches positioned on door and gate latches can be used so that when the door or gate is unlatched, the limit switch is opened and closes down the machinery. However, it is relatively simple for operatives to defeat such limit switches by sticking a device such as a screwdriver in the contacts to force a machine to operate. More sophisticated safety interlocks have thus been devised, such as proximity switches and key locks.

- *Light curtains* to prevent any person getting close to machinery. A danger zone, such as a packaging machine, can use infrared beams to protect people from getting too close. If a light beam is broken, it immediately triggers a safe shutdown command.
- *Safety mats* are another way of detecting when someone is too close to a machine. They are placed round a machine and when someone steps on the mat, a contact is closed, causing the machine to stop.
- *Emergency stop relays*, to enable machinery to be stopped in the event of an emergency (see Section 13.2.2).

Thus a safe-operating system for a work cell might use gated entry systems, such as guards on machines that activate stop relays if they are not in place, light curtains, and emergency stop relays.

13.2.4 Safety PLCs

Safety PLCs are specially designed to enable safety functions to be realized. In a safety PLC there can be two or three microprocessors that perform exactly the same logic, check against each other, and give outputs only if there is agreement. One method that is used is a two-channel system with two identical subsystems that communicate with each other via a fiber-optic cable link. The inputs from the sensors are fed simultaneously to both subsystems. During operation, data is passed between the two subsystems via the fiber-optic cable. They operate in synchronism with the same program and compare input and output signals, the results of logic operations, counters, and the like, and automatically go into a safe-stop condition if there are different outputs or internal faults or failures. For safety-related digital outputs, actuators are switched on or off from both subsystems. This means that one subsystem alone can shut down equipment.

13.3 Commissioning

Commissioning of a PLC system involves:

1. Checking that all the cable connections between the PLC and the plant being controlled are complete, safe, to the required specification, and meet local standards.
2. Checking that the incoming power supply matches the voltage setting for which the PLC is set.
3. Checking that all protective devices are set to their appropriate trip settings.
4. Checking that emergency stop buttons work.
5. Checking that all input/output devices are connected to the correct input/output points and giving the correct signals.
6. Loading and testing the software.

13.3.1 Testing Inputs and Outputs

Input devices, such as switches, can be manipulated to give the open and closed contact conditions and the corresponding LED on the input module observed. It should be illuminated when the input is closed and not illuminated when it is open. Failure of an LED to illuminate could be because the input device is not correctly operating, there are incorrect wiring connections to the input module, the input device is not correctly powered, or the LED or input module is defective. For output devices that can be safely started, push buttons might have been installed so that each output could be tested.

Another method that can be used to test inputs and outputs is *forcing*. This involves software, rather than mechanical switching on or off, being used with instructions from the display on the screen being used to turn off or on inputs/outputs. Forcing thus overwrites an input or output condition. This permits the operation of inputs and outputs to be tested, also whether the inputs and outputs are correctly wired to the PLC.

For example, with the Siemens S7-1200 system, under the Watch and Force menu, the Force Table entry is clicked and then the address of the item being forced is entered and then the Force value of TRUE or FALSE. With the Mitsubishi FX PLC keys are pressed to select the Forced I/O facility and within the resulting Set/Reset window the Input/Output address to be forced is entered and the value of on or off entered. Thus an output or an input is forced to the selected forced value and the consequences of that action observed. Figure 13.12(a) illustrates this when an output Q0.1 is forced on. As a consequence output

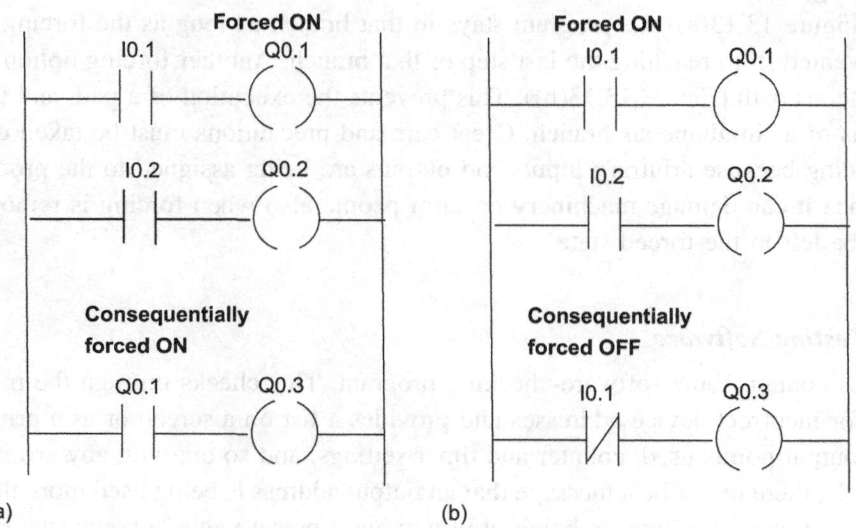

Figure 13.12: Ladder program, (a) forcing an output, (b) forcing an input.

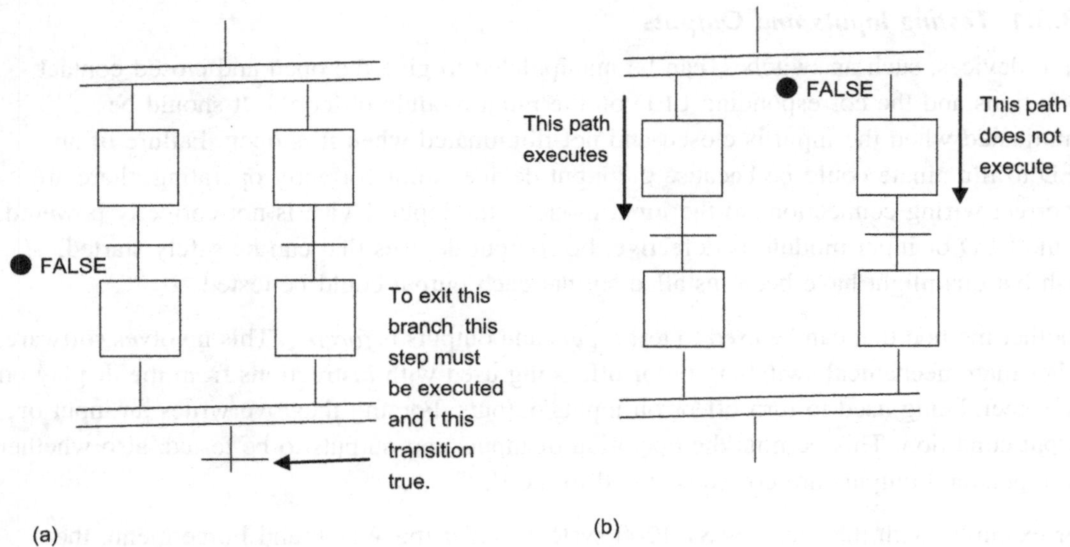
FALSE
To exit this branch this step must be executed and t this transition true.
(a)
FALSE
This path executes
This path does not execute
(b)

Figure 13.13: Sequential function chart program, (a) forcing a transition, (b) forcing a path.

Q0.3 is switched on. Figure 13.12(b) shows the effects of an input I0.1 being forced on. Forcing can only be stopped by clicking the Stop Forcing icon or using a stop forcing instruction. Figure 13.13 shows how forcing can be used with a sequential function chart program. A transition can be forced and will override the conditions of the transition each time the program reaches that transition. If a transition is forced in a simultaneous branch to be false (Figure 13.13(a)) the program stays in that branch as long as the forcing is active, being prevented from reaching the last step of that branch. Another forcing option is to force a simultaneous path (Figure 13.13(b)). This prevents the execution of a path and thus one or more paths of a simultaneous branch. Great care and precautions must be taken during the use of forcing because arbitrary inputs and outputs are being assigned to the process. As a consequence it can damage machinery or harm people also when forcing is removed items may still be left in the forced state.

13.3.2 Testing Software

Most PLCs contain some software-checking program. This checks through the installed program for incorrect device addresses and provides a list on a screen or as a printout of all the input/output points used, counter and timer settings, and so on, with any errors detected. For example, there might be a message that an output address is being used more than once in a program, a timer or counter is being used without a preset value, a counter is being used without a reset, or the like.

13.3.3 Simulation

Many PLCs are fitted with a simulation unit that reads and writes information directly into the input/output memory and so simulates the actions of the inputs and outputs. The installed program can thus be run and inputs and outputs simulated so that they, and all preset values, can be checked. To carry out this type of operation, the terminal has to be placed in the correct mode. For Mitsubishi this is termed the *monitor mode*, for Siemens the *test mode*, and for Telemecanique the *debug mode*.

With a Mitsubishi in monitor mode, Figure 13.14 shows how inputs appear when open and closed and how output looks when not energized and energized. The display shows a selected part of the ladder program and what happens as the program proceeds. Thus at some stage in a program the screen might appear in the form shown in Figure 13.15a. For rung 12, with

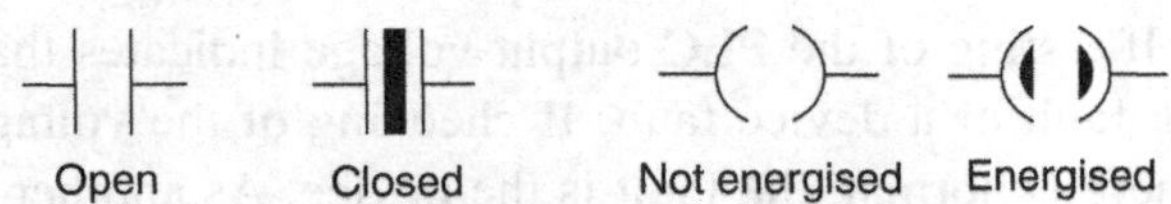

Figure 13.14: Monitor mode symbols.

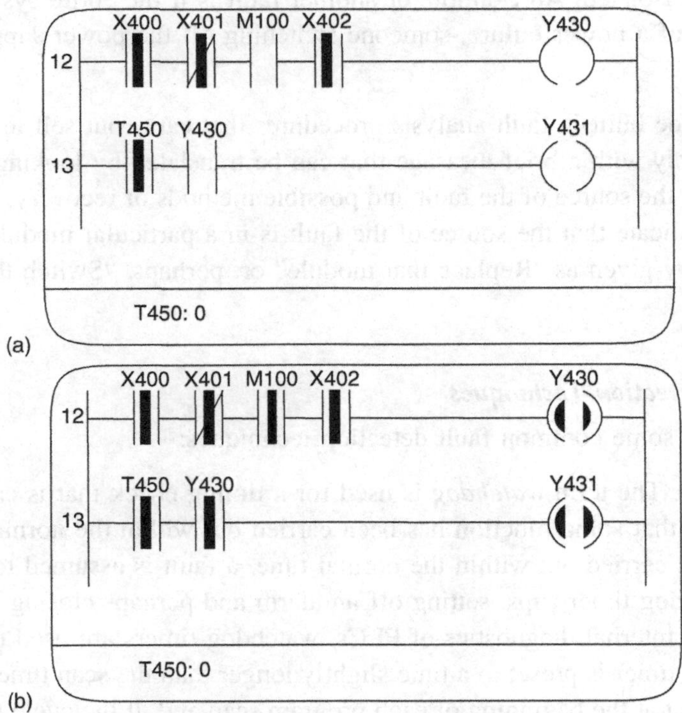

Figure 13.15: Ladder program monitoring.

inputs to X400, X401, and X402 but not M100, there is no output from Y430. For rung 13, the timer T450 contacts are closed, the display at the bottom of the screen indicating that there is no time left to run on T450. Because Y430 is not energized, the Y430 contacts are open, so there is no output from Y431. If we now force an input to M100, the screen display changes to that shown in Figure 13.5b. Now Y430, and consequently Y431, come on.

13.4 Fault Finding

With any PLC-controlled plant, by far the greater percentage of faults are likely to be with sensors, actuators, and wiring rather than within the PLC itself. Of the faults within the PLC, most are likely to be in the input/output channels or power supply rather than in the CPU.

As an illustration of a fault, consider a single output device failing to turn on, even though the output LED is on. If testing of the PLC output voltage indicates that it is normal, the fault might be a wiring fault or a device fault. If checking of the voltage at the device indicates the voltage there is normal, the fault is the device. As another illustration, consider all the inputs failing. This might be as a result of a short circuit or earth fault with an input. A possible procedure to isolate the fault is to disconnect the inputs one by one until the faulty input is isolated. An example of another fault is if the entire system stops. This might be a result of a power failure, someone switching off the power supply, or a circuit breaker tripping.

Many PLCs provide built-in fault analysis procedures that carry out self-testing and display fault codes, possibly with a brief message that can be translated by looking up the code in a list, which gives the source of the fault and possible methods of recovery. For example, the fault code may indicate that the source of the fault is in a particular module, with the method of recovery given as "Replace that module" or, perhaps, "Switch the power off and then on."

13.4.1 Fault Detection Techniques

The following are some common fault detection techniques:

- *Timing checks*. The term *watchdog* is used for a timing check that is carried out by the PLC to check that some function has been carried out within the normal time. If the function is not carried out within the normal time, a fault is assumed to have occurred and the watchdog timer trips, setting off an alarm and perhaps closing down the PLC. As part of the internal diagnostics of PLCs, watchdog timers are used to detect faults. The watchdog timer is preset to a time slightly longer than the scan time would normally be. It is then set at the beginning of each program scan and, if the cycle time is normal, it does not time out and is reset at the end of a cycle, ready for the next cycle. However, if

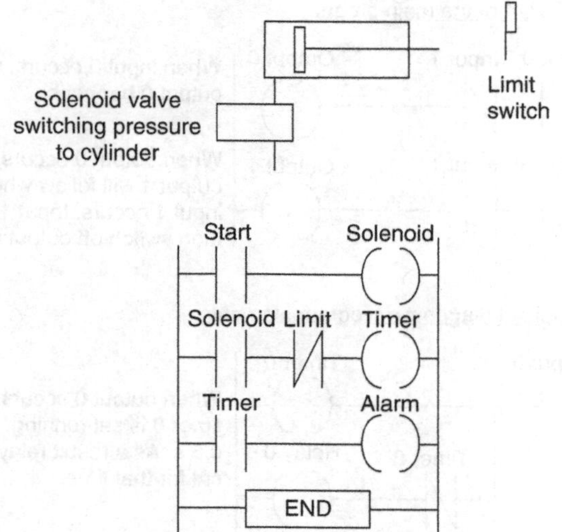

Figure 13.16: Watchdog timer.

the cycle time is longer than it would normally be, the watchdog timer times out and indicates that the system has a fault.

Within a program, additional ladder rungs are often included so that when a function starts, a timer is started. If the function is completed before the time runs out, the program continues, but if not, the program uses the jump command to move to a special set of rungs, which triggers an alarm and perhaps stops the system. Figure 13.16 shows an example of a watchdog timer that might be used with the movement of a piston in a cylinder. When the start switch is closed, the solenoid of a valve is energized and causes the piston in the cylinder to start moving. It also starts the timer. When the piston is fully extended, it opens a limit switch and stops the timer. If the time taken for the piston to move and switch off the timer is greater than the preset value used for the timer, the timer sets off the alarm.

- *Last output set*. This technique involves the use of status lamps to indicate the last output that has been set during a process that has come to a halt. Such lamps are built into the program so that as each output occurs, a lamp comes on. The lamps that are on thus indicate which outputs are occurring. The program has to be designed to turn off previous status lamps and turn on a new status lamp as each new output is turned on. Figure 13.17 illustrates this concept.

  Such a technique can be cumbersome in a large system with many outputs. In such a case, the outputs might be grouped into sets and a status lamp used for each set. A selector switch can then be used within a group to select each output in turn to determine whether it is on. Figure 13.18 illustrates this idea.

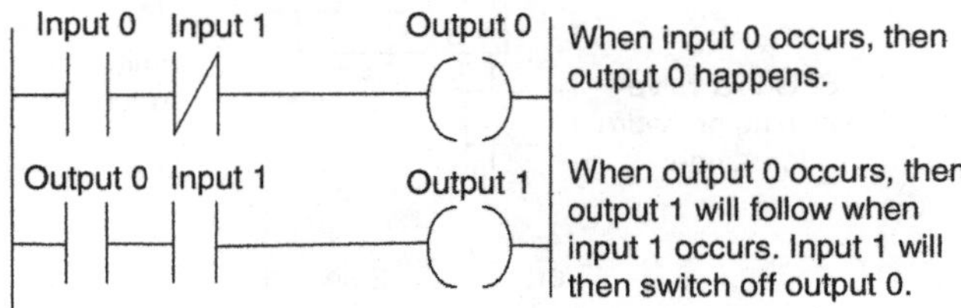

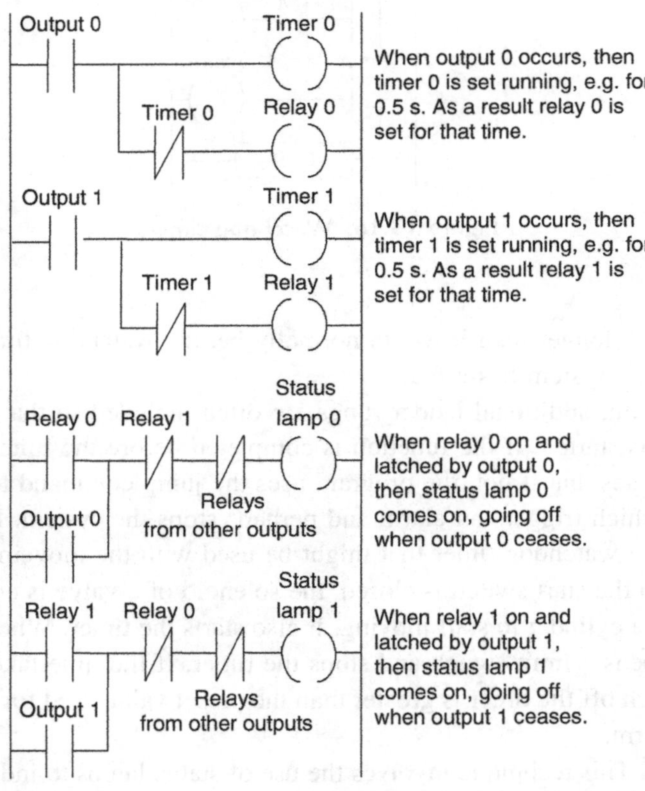

Figure 13.17: Last output set diagnostic program.

As an illustration of the use of this program to indicate which action occurred last, Figure 13.19 shows the program that might be used with a pneumatic system operating cylinders in a sequence. The program indicates at which point in the sequence a fault occurred, such as a piston sticking, and would be added to the main program used to sequence the cylinders. Each of the cylinder movements has a light-emitting diode associated with it, with the last cylinder movement indicated by its LED being illuminated.

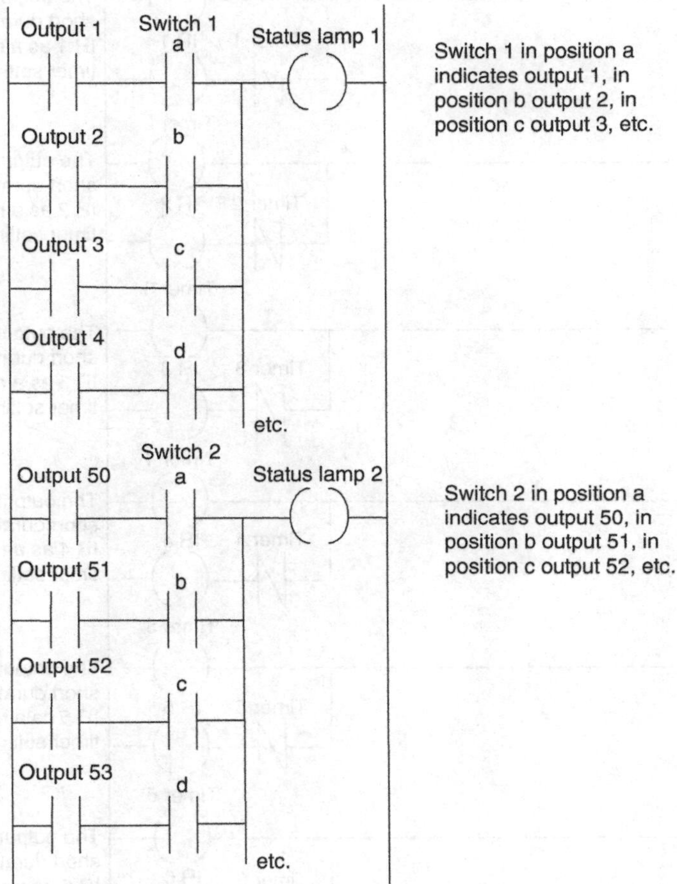

Figure 13.18: Single status lamp for a group of outputs.

- *Replication.* Where there is concern regarding safety in the case of a fault developing, checks may be constantly used to detect faults. One technique is replication checks, which involve duplicating, that is, replicating, the PLC system. This could mean that the system repeats every operation twice and, if it gets the same result, it is assumed that there is no fault. This procedure can detect transient faults. A more expensive alternative is to have duplicate PLC systems and compare the results given by the two systems. In the absence of a fault, the two results should be the same.
- *Expected value checks.* Software errors can be detected by checking whether an expected value is obtained when a specific input occurs. If the expected value is not obtained, a fault is assumed to be occurring.

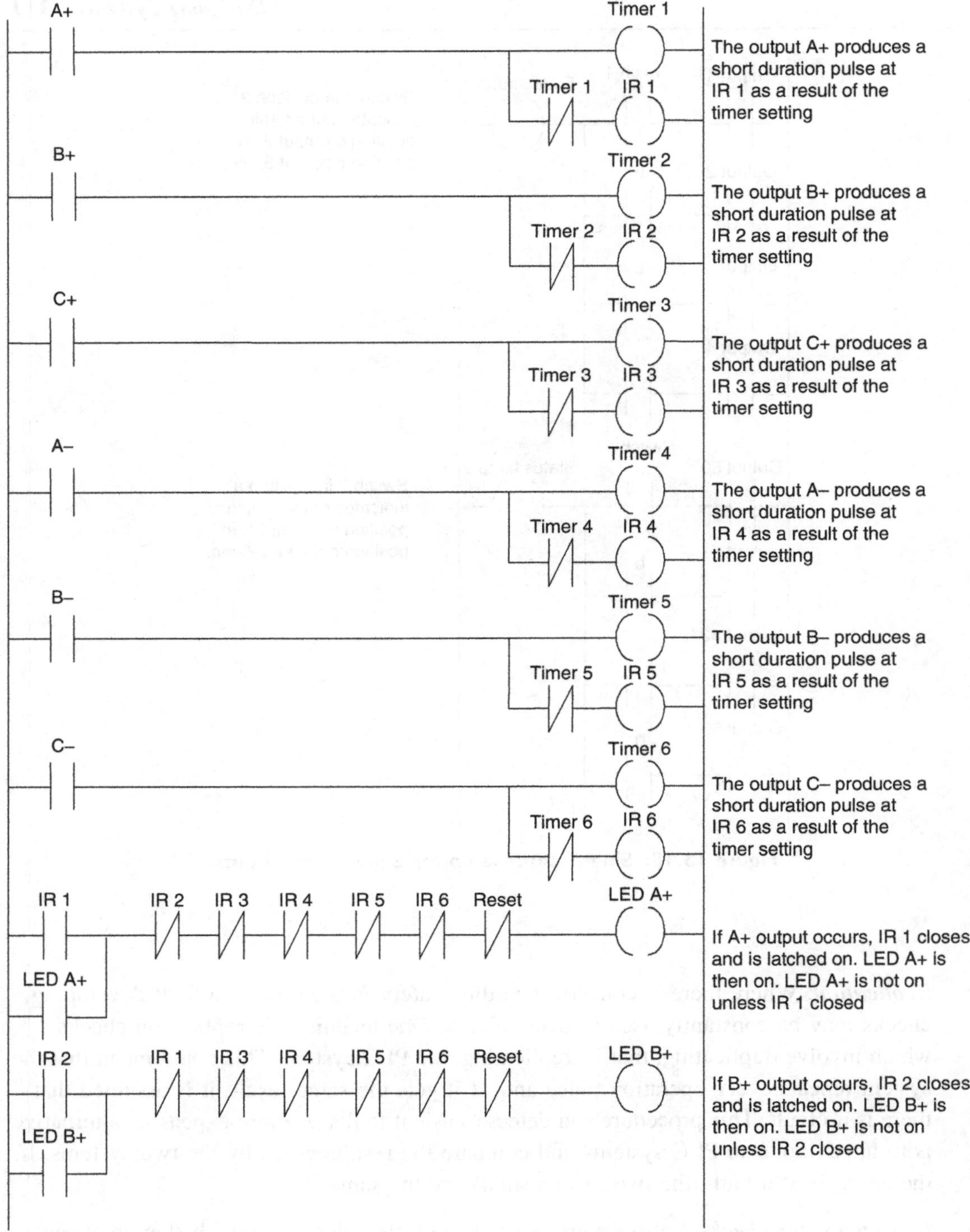

Figure 13.19: Diagnostic program for last cylinder action.

(Continued)

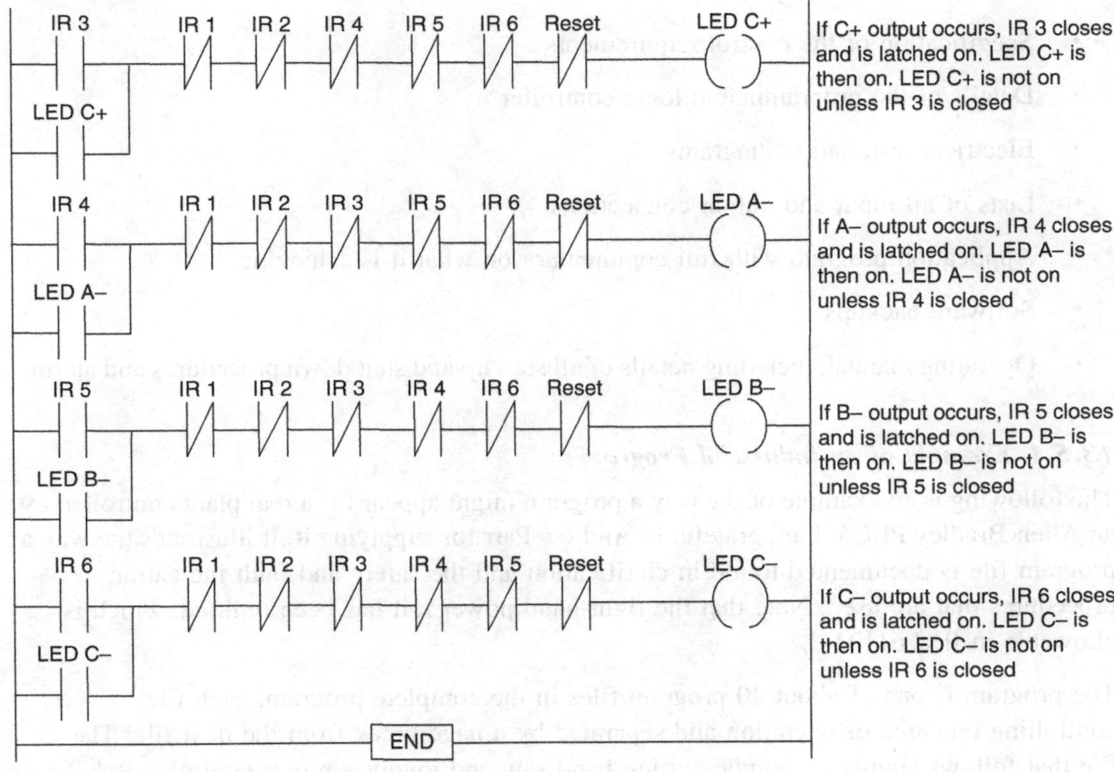

Figure 13.19—Cont'd

13.4.2 Program Storage

Applications programs may be loaded into battery-backed RAM in a PLC. A failure of the battery supply means a complete loss of the stored programs. An alternative to storing applications programs in battery-backed RAM is to use EPROM. This form of memory is secure against the loss of power. Against the possibility of memory failure occurring in the PLC and loss of the stored application program, a backup copy of each application program should be kept. If the program has been developed using a computer, the backup may be on a CD or a hard disk. Otherwise the backup may be on an EPROM cartridge. The program can then again be downloaded into the PLC without it having to be rewritten.

13.5 System Documentation

The documentation is the main guide used by everyday users, including for troubleshooting and fault finding with PLCs. It thus needs to be complete and in a form that is easy to follow. The documentation for a PLC installation should include the following:

- A description of the plant
- Specification of the control requirements
- Details of the programmable logic controller
- Electrical installation diagrams
- Lists of all input and output connections
- Application program with full commentary on what it is achieving
- Software backups
- Operating manual, including details of all start up and shut down procedures and alarms

13.5.1 Example of an Industrial Program

The following is an example of the way a program might appear for a real plant controlled by an Allen-Bradley PLC5; I am grateful to Andrew Parr for supplying it. It illustrates the way a program file is documented to aid in clarification and the safety and fault indication procedures that are used. Note that the right-hand power rail has been omitted, which is allowable in IEC 61131-3.

The program is one of about 40 program files in the complete program, each file controlling one area of operation and separated by a page break from the next file. The file that follows controls a bundle-cutting band saw and involves motor controls, desk lamps, and a small state transition sequence.

Note the rung cross-references, such as [38], below B3/497 in rung 2. These are used to show that B3/497 originates, for example, in rung 38 in the current program file. Also note that all instructions are tagged with descriptions and the file is broken down into page sections. The software allows you to go straight to a function via the page titles.

All the motor starter rungs work in the same way. The PLC energizes the contactor and then one second later looks for the auxiliary relay (labeled as Aux in the program file) coming back to say the contactor has energized. If there is a fault that causes the contactor to deenergize, such as a loss of supply, or a trip or open circuit coil, it causes the PLC to signal a fault and deenergize the contactor output so that the machine does not spring into life when the fault is cleared.

The saw normally sits raised clear of the bundle. To cut the bundle, the blade motor has to be started and the lower push-button pressed (at rung 8). The saw falls under gravity at a fast or slow speed that is set by hydraulic valves. To raise the saw, a hydraulic pump is started to pump oil into the saw support cylinders. At any time the saw can be raised, such as to clear swarf, to what is termed the *pause state*. Otherwise, cutting continues until the bottom limit is reached. The saw then is raised to the top limit for the next bundle. A cut can be aborted by pressing the raise button for two seconds. While a bundle is being cut, it is held by clamp solenoids.

The final three rungs of the program set the length to be cut. There are two photocells about 20 mm apart on a moveable carriage. These are positioned at the required length. The operator runs the bundle in until the first is blocked and the second is clear. These control the long/correct/short desk lamps.

```
                              Bundle Cutting Saw
                         ***Saw Cutting...Saw Motor
                              Stacking Machine
File #14 Saw Proj: FLATS3           Page:00001                    21:08 12/05/02
-----------------------------------------------------------------------------------
   | Saw_Motor
   |  Tripped    Saw_Motor  Saw_ESR                            Saw_Motor
   | 1=Tripped  Start_Fault Healthy                            Available
   |  I:032        B3       I:031                                 B3
 0 +---]/[---------]/[---------] [-------------------------------( )-------
   |   10         517        17                                  516
   |              [2]
   |
   |                                    Saw_Blade
   |  Saw_Motor   Saw_Motor Saw_Motor  Tension_LS    Saw_Motor
   |   Start_PB    Stop_PB  Available  1=Healthy     Contactor
   |    I:030       I:030      B3        I:032         O:034            >
 1 +--+---] [-----+---] [--------] [--------] [------+----( )--------------->
   |  |    00     |   01      516         03         |    10            >
   |  |           |           [0]                    |
   |  |Saw_Motor  |                                   |
   |  |Contactor  |                                   |
   |  |  O:034    |                                   |
   |  +---] [-----+                                   +--------------------
   |      [1]
   |                                      <-------------------+----------
   |                                       Saw_Motor          |
   |                                       Start_Motor        |
   |                                       +-TON-----------+  |
   |                                       +Timer On Delay +-(EN)--+
   |                                       |Timer:   T4:109|
   |                                       |Base (SEC): 0.01+-(DN)
   |                                       |Preset:     100|
   |                                       |Accum:        0|
   |                                       +---------------+
   |
   |  Saw_Motor    Saw_Motor  Saw_Alarms                        Saw_Motor
   | Start_Fault  Running_Aux   Accept                         Start_Fault
   |   T4:109       I:032        B3                                 B3
 2 +-+---] [---------]/[-----+-----]/[---------------------------( )------
   | |   DN          11      |    497                              517
   | |   [1]                 |    [38]
   | |Saw_Motor              |
   | |   B3                  |
   | +---] [-----------------+
   |    517
   |    [2]
   |
```

```
                              Bundle Cutting Saw
                                ...Coolant Pump
                               Stacking Machine
File #14 Saw Proj: FLATS3          Page:00002                    21:08 12/05/02
--------------------------------------------------------------------------------
  |Coolant_Pump  Coolant_Pump  Saw_ESR                           Coolant_Pump
  | 1=Tripped     Start_Fault  Healthy                             Available
  |   I:032           B3        I:031                                  B3
3 +-----]/[-----------]/[-----------] [---------------------------------( )----
  |    12            519         17                                   518
  |                  [6]                   Test_Run
  |Coolant_Pump                          Coolant_Pump
  |Select_SW      OneShot                  TOF_Timer
  |  I:030           B3                   +--TOF--------------+
4 +-----] [-----------]ONS[---------------+ Timer Off Delay   +--(EN)------
  |    02            520                  | Timer:     T4:110 |
  |                                       | Base (SEC): 1.0   +--(DN)
  |                                       | Preset:         4 |
  |                                       | Accum:          4 |
  |                                       +-------------------+
  |Coolant_Pump  Saw_Motor                Coolant_Pump
  |Select_SW     Running_Aux                 Contactor
  |   I:030         I:032                      O:034
5 +--+----] [-----------] [--------+-----------+---------( )------------------+---------
  |  |   02           11           |           |          11                  |
  |  |Test_Run                     |           |Coolant_Pump                  |
  |  |Coolant_Pump                 |           |Start_Fault                   |
  |  |TOF_Timer                    |           |TON Timer                     |
  |  |T4:110                       |           |+--TON --------------+        |
  |  +----] [----------------------+           ++Timer On Delay      +-(EN)-+
  |       DN                                    |Timer:      T4:111|
  |       [4]                                   |Base (SEC): 0.01  +--(DN)
  |                                             |Preset:        100|
  |                                             |Accum:           0|
  |                                             +------------------+
  |Coolant Pump
  |Start_Fault  Coolant_Pump  Saw_Alarms                           Coolant_Pump
  |TON_Timer     Running_Aux    Accept                              Start_Fault
  |  T4:111         I:032         B3                                    B3
6 +--+----] [-----------] [-------+-----]/[--------------------------------( )----
  |  |    DN           13         |    497                               519
  |  |    [5]                     |    [38]
  |  |Coolant_Pump                |
  |  |Start_Fault                 |
  |  |     B3                     |
  |  +----] [---------------------+
  |   [6]519
  |
  | Saw_Motor    Coolant_Pump                                     Saw_Motor &
  | Running_Aux  Running_Aux                                       Coolant_OK
  |   I:032         I:032                                              B3
7 +-----] [-----+-------] [-------+----------------------------------( )----
  |    11       |      13         |                                  496
  |             |Coolant_Pump     |
  |             |Select_SW        |
  |             +------]/[--------+
  |                    02
```

```
                         Bundle Cutting Saw
                          ...Coolant Pump
                          Stacking Machine
File #14 Saw Proj: FLATS3      Page:00003                  21:08 12/05/02
---------------------------------------------------------------------------
Blank page for future modification
```

```
                         Bundle Cutting Saw
                 ...Saw Cut Sequence...Transitions
                          Stacking Machine
File #14 Saw Proj: FLATS3        Page:00004              21:08 12/05/02
-------------------------------------------------------------------------
     |
     |State_0                               Saw_Hyd    Saw_Blade
     |Ready_for Saw_Motor_& Saw_Hyd_Pump Permit_SW Tension_LS Saw_Lower  Trans_A
     |Start     Coolant_OK Healthy       1=Permit  1=Healthy PushButton Seq_Start
     |   B3          B3         B3          I:031      I:032      I:030      B3
   8 +----] [-----------] [----------] [-----------] [----------] [-----------] [-------( )----
     |  500         496        499          14         03         04        506
     |  [15]        [7]        [24]
     |
     |                                                                      Trans_B
     | State_1       Saw                                                   Cut_Done
     | Cutting    End_Cut_LS                                               or_Fault
     |   B3          I:032                                                     B3
   9 +---] [------+-------] [-----+-------------------------------------------( )----
     |  501      |   00         |                                            507
     |  [16]     |              |
     |           |Saw_Motor_&   |
     |           | Coolant_OK   |
     |           |    B3        |
     |           +-----]/[------+
     |               496
     |               [7]
     |
     | State_2
     | Raise_to    Saw_Top_LS                                             Trans_C
     |Top_Limit    Struck_TON                                            At_Top_LS
     |   B3          T4:112                                                 B3
  10 +----] [------------] [-----------------------------------------------( )----
     |  502           DN                                                   508
     |  [17]         [20]
     |
     | State_1     Saw_Raise                                              Trans_D
     | Cutting     Pushbutton                                            Pause_Req
     |   B3          I:030                                                  B3
  11 +----] [------------] [-----------------------------------------------( )----
     |  501           03                                                   509
     |  [16]
     |
     | State_3     Saw_Lower                                              Trans_E
     |  Paused     Pushbutton                                            Pause_End
     |   B3          I:030                                                  B3
  12 +----] [------------] [-----------------------------------------------( )---
     |  503           04                                                   510
     |  [18]
     |
     |
```

```
                              Bundle Cutting Saw
                      ...Saw Cut Sequence...Transitions
                               Stacking Machine
File #14 Saw Proj: FLATS3          Page:00005                  21:08 12/05/02
-----------------------------------------------------------------------------
   |
   |              Raise_PB                                         Trans_F
   | State_3    Raise_to_Top                                      Pause_End
   | Paused      TON_Timer                                        Go_To_Top
   |   B3         T4:108                                             B3
13 +-----] [------+--------] [--------+---------------------------------( )-------
   |   503     |     DN        |                                    511
   |  [18]     |    [19]       |
   |           |  Saw_Motor_&  |
   |           |  Coolant_OK   |
   |           |      B3       |
   |           +--------]/[--------+
   |                  496
   |                  [7]
   |                                                               Trans_G
   | State_3    Saw_Top_LS                                       Hit_Top_LS
   | Paused     Struck_TON                                      While_Paused
   |   B3         T4:112                                             B3
14 +-----] [-------------] [-------------------------------------------( )-------
   |   503          DN                                              495
   |  [18]         [20]
   |
   |
```

```
                              Bundle Cutting Saw
                                  ...States
                               Stacking Machine
File #14 Saw Proj: FLATS3      Page:00006                       21:08 12/05/02
-------------------------------------------------------------------------------
    |
    |          State_2                                               State_0
    | State_1  Raise_to   State_3                                    Ready_for
    | Cutting  Top_Limit  Paused                                      Start
    |   B3        B3        B3                                          B3
15 +----]/[---------]/[---------]/[-------------------------------------( )-------
    |  501       502       503                                         500
    |  [16]      [17]      [18]
    |
    |                Trans_B
    |    Trans_A     Cut_Done   Trans_D    Saw_ESR                    State_1
    |    Seq_Start   or_Fault   Pause_Req  Healthy                    Cutting
    |       B3          B3         B3      I:031                        B3
16 +--+------] [------+----]/[---------]/[---------] [--------------------( )--------
    | |     506       |    507        509       17                      501
    | |     [8]       |    [9]        [11]
    | | Trans_E       |
    | | Pause_End     |
    | |     B3        |
    | +------] [------+
    | |     510       |
    | |     [12]      |
    | | State_1       |
    | | Cutting       |
    | |    B3         |
    | +----] [--------+
    |     501
    |     [16]
    |
    |    Trans_B                                                      State_2
    |    Cut_Done     Trans_C    Saw_ESR                              Raise_to
    |    or_Fault     At_Top_LS  Healthy                              Top_Limit
    |      B3            B3      I:031                                  B3
17 +--+-----] [------+-----]/[---------] [---------------------------------( )--------
    | |    507       |    508        17                                 502
    | |    [9]       |    [10]
    | | Trans_F      |
    | | Pause_End    |
    | | Go_To_Top    |
    | |    B3        |
    | +------] [-----+
    | |    511       |
    | |    [13]      |
    | | State_2      |
    | | Raise_to     |
    | | Top_Limit    |
    | |    B3        |
    | +------] [-----+
    |      502
    |      [17]
```

```
                              Bundle Cutting Saw
                                  ...States
                               Stacking Machine
File #14 Saw Proj: FLATS3        Page:00007                      21:08 12/05/02
-------------------------------------------------------------------------------
   |
   |                                Trans_F     Trans_G
   |  Trans_D      Trans_E      Pause_End    Hit_Top_LS    Saw_ESR            State_3
   | Pause_Req    Pause_End    Go_To_Top    While_Paused   Healthy            Paused
   |    B3           B3           B3            B3          I:031               B3
18 +-+--] [-----+-----]/[---------]/[-----------]/[----------] [---------------( )----
   | |  509     |    510          511           495          17                503
   | |  [11]    |
   | |State_3   |
   | | Paused   |
   | |   B3     |
   | +--] [-----+
   |    503
   |    [18]
   |
   |
```

```
                         Bundle Cutting Saw
                           .... .Timers
                          Stacking Machine
File #14 Saw Proj: FLATS3       Page:00008                    21:08 12/05/02
---------------------------------------------------------------------------
    |
    |If Raise PB is pressed for more than 2 secs go right to top limit switch
    |
    |                                             Raise_PB
    | State_3    Saw_Raise                     Raise_to_Top
    |  Paused   PushButton                       TON_Timer
    |    B3       I:030                     +-- TON --------------+
 19 +-------] [---------] [-----------------+Timer On Delay       +--(EN)------
    |    503       03                       |Timer:       T4:108 |
    |                                       |Base (SEC): 0.01+--(DN)
    |                                       |Preset:         200 |
    |                                       |Accum:            0 |
    |                                       +---------------------+
    |T4:112 ensures saw carriage goes past top limit to help avoid creeping
    |off the top position
    |Saw_Top_LS                          Saw_Top_LS
    | 1=Struck                           Struck_TON
    |  I:032                           +--- TON -----------------+
 20 +----] [---------------------------++ Time On Delay         +--(EN)--+---
    |    01                            || Timer:      T4:112 |          |
    |                                  || Base (SEC): 0.01 +---(DN)    |
    |                                  || Preset:          100 |          |
    |                                  || Accum:           101 |          |
    |                                  |+--------------------+          |
    |                                  |Saw_Top_LS                     |
    |                                  |Struck_TOF                     |
    |                                  | 1=At_Top                      |
    |                                  |+-- TOF ------------+          |
    |                                  ++ Timer Off Delay   +----(EN)--+
    |                                   | Timer:     T4:113 |
    |                                   | Base (SEC): 0.01 +----(DN)
    |                                   | Preset:         300 |
    |                                   | Accum:            0 |
    |                                   +-------------------+
    |
    |Permissive for bundle delivery/despatch
    |  Saw_Top_LS                                                   Saw_Not
    |   1=Struck                                                   Operating
    |    I:032                                                        B3
 21 +-+----] [-------+-----------------------------------------------( )----
    | |     01       |                                                524
    | |   Saw_Hyd    |
    | |  Permit_SW   |
    | |  1=Permit    |
    | |    I:031     |
    | +----]/[-------+
    |      14
    |
```

```
                              Bundle Cutting Saw
                        .... .Solenoids and Hydraulic Pump
                               Stacking Machine
File #14 Saw Proj: FLATS3         Page:00009                      21:08 12/05/02
--------------------------------------------------------------------------------
    | The saw lowers at slow or fast speed under gravity.
    | It is raised by starting the pump which drives the saw up to the top
    | limit or for a time for a pause.
    |
    |                State_0
    | Saw_Lower    Ready_for                                        Saw_Lower
    |PushButton      Start                                           Fast_SOV
    |   I:030         B3                                              O:033
 22 +----] [-------+-----] [-----+------------------------------------------( )------
    |    04         |   500      |                                          11
    |               |   [15]     |
    |               | State_1    |
    |               | Cutting    |
    |               |    B3      |
    |               +-----] [-----+
    |  State_1          501                                           Saw_Lower
    |  Cutting          [16]                                           Healthy
    |    B3                                                             O:033
 23 +----] [-----------------------------------------------------------------( )------
    |   501                                                                 10
    |
    |Saw_Hyd_Pump  Saw_Hyd_Pump                                     Saw_Hyd_Pump
    |  1=Tripped   Start_Fault                                         Healthy
    |   I:032          B3                                                B3
 24 + ----]/[--------------]/[-----------------------------------------( )-----
    |    14           498                                               499
    |                 [26]
    |  Saw_Raise                                Saw_Lower   Saw_Lower
    |  PushButton                               Slow_SOV    Fast_SOV
    |    I:030                                   O:033       O:033          >
 25 +-+-----] [-----------------------------+-------]/[-----------]/[------------>
    | |    03                                |        10          11            >
    | | State_2                              |       [23]        [22]
    | | Raise_to                             |
    | | Top_Limit                            |
    | |     B3                               |
    | +-----] [------------------------------+
    | |    502                               |
    | |    [17]                              |
    | | State_0                   Saw_Hyd    |
    | | Ready_for Saw_Top_LS     Permit_SW   |
    | | Start     Struck_TON     1=Permit    |
    | |    B3       T4:112        I:031      |
    | +-----] [---------]/[---------] [--------+
    |    500         DN           14
    |    [15]       [20]
    |              Saw_Hyd_Pump  Saw_ESR   Saw_Hyd-Pump
    |                Healthy     Healthy    Contactor
    |              <     B3        I:031       O:034
    |              <-----] [----------] [-----+------( )----------------+-----
    |              <                          >                         >
    |
```

```
                         Bundle Cutting Saw
                    .... .Solenoids and Hydraulic Pump
                          Stacking Machine
File #14 Saw Proj: FLATS3         Page:00010                   21:08 12/05/02
-------------------------------------------------------------------------------
    |                  <                   >                       >
    |                  <     499     17    |   12                  |
    |                                      |                       |
    |                                      | Saw_Hyd_Pump          |
    |                                      | Start_Fault           |
    |                                      | TON_Timer             |
    |                                      |+--- TON -------------+ |
    |                                      ++Timer On Delay      +-(EN)-+
    |                                       |Timer:       T4:114|
    |                                       |Base (SEC):  0.01+-(DN)
    |                                       |Preset:        100|
    |                                       |Accum:           0|
    |                                       +------------------+
    |
    |Saw_Hyd_Pump
    | Start_Fault   Saw_Hyd_Pump    Saw_Alarms                 Saw_Hyd_Pump
    |  TON_Timer     Running_Aux      Accept                    Start_Fault
    |   T4:114        I:032            B3                           B3
 26 +--+-----] [------------]/[-------+------]/[-----------------------( )-------
    |  |     DN            15         |    497                          498
    |  |    [25]                      |   [38]
    |  |                              |
    |  |Saw_Hyd_Pump                  |
    |  | Start_Fault                  |
    |  |     B3                       |
    |  +-----] [----------------------+
    |       498
    |      [26]
    |
```

```
                          Bundle Cutting Saw
                         ...Blade Tensioning
                          Stacking Machine
File #14 Saw Proj: FLATS3        Page:00011                  21:08 12/05/02
---------------------------------------------------------------------------
   |Saw tension is changed via two hydraulic soleniods.
   |The TOF timer on the pump reduces start commands on the pump.
   |
   |Saw_Tension
   |Motor_Tripped  TensionPump  Saw_ESR                      Tension_Pump
   | 1=Tripped     Start_Fault  Healthy                        Available
   |   I:032           B3        I:031                            B3
27 +-----]/[------------]/[--------------] [----------------------( )--------
   |    05            513         17                              512
   |                  [30]
   |Saw_Tension     TensionPump                       TensionPump
   |Increase_PB      Available                        Run_Cmd_TOF
   |   I:030            B3                            +---TOF-------------+
28 +-+--] [----------+-----] [------------------------+Timer Off Delay+--(EN)-----
   | |   02          |     512                        |Timer:    T4:115|
   | |               |     [27]                       |Base (SEC): 1.0+--(DN)
   | | Saw_Tension   |                                |Preset:        5|
   | | Decrease_PB   |                                |Accum:         5|
   | |   I:030       |                                +----------------+
   | +-----] [-------+
   |       06
   | Tension_Pump                            Tension_Pump
   |  Run_Cmd_TOF                             Contactor
   |   T4:115                                   O:034
29+ -------] [-------------------------------+-----( )--------------------+------
   |      DN                                 |     13                     |
   |                                         |Tension_Pump                |
   |                                         | Start_Fault                |
   |                                         |  TON Timer                 |
   |                                         |+--TON----------------+     |
   |                                         ++ Timer On Delay      +---(EN)--+
   |                                          | Timer:      T4:116  |
   |                                          | Base (SEC): 0.01    +---(DN)
   |                                          | Preset:        100  |
   |                                          | Accum:           0  |
   |                                          +---------------------+
   | Tension_Pump
   |  Start_Fault  Saw_Tension  Saw_Alarms                      Tension_Pump
   |   TON_Timer    Pump_Aux      Accept                         Start_Fault
   |    T4:116        I:032         B3                                B3
30 +-+----] [-------------]/[-------+------]/[------------------------( )--------
   | |    DN          06          |     497                           513
   | |   [29]                     |    [38]
   | |Tension_Pump                |
   | |Start_Fault                 |
   | |    B3                      |
   | +-----] [--------------------+
   |      513
   |      [6]
```

```
                          Bundle Cutting Saw
                          ...Blade Tensioning
                           Stacking Machine
File #14 Saw Proj: FLATS3       Page:00012                 21:08 12/05/02
--------------------------------------------------------------------------------
     |
     |Saw_Tension    Saw_Tension                                   Saw_Tension
     |Increase_PB    Decrease_SOV                                  Increase_SOV
     |   I:030          O:033                                         O:033
  31 +----] [--------------]/[----------------------------------------( )-----
     |   05              13                                            12
     |                  [32]
     |
     |Saw_Tension    Saw_Tension                                   Saw_Tension
     |Decrease_PB    Increase_SOV                                  Decrease_SOV
     |   I:030          O:033                                         O:033
  32 +----] [--------------]/[----------------------------------------( )-----
     |   06              12                                            13
     |                  [31]
     |
```

```
                               Bundle Cutting Saw
                                 ...Saw Clamps
                                Stacking Machine
File #14 Saw Proj: FLATS3         Page:00013                    21:08 12/05/02
-------------------------------------------------------------------------------
   |T4:118 & 119 operate the clamp/unclamp solenoids for a fixed time.
   |
   |   Saw_Clamp  Saw_Unclamp  Saw_Unclamp        Saw_Clamp  Saw_Clamp
   |   PushButon   Solenoid    PushButton         TON_Timer  Solenoid
   |     I:034      O:006        I:034              T4:118    O:006
33 +-+------] [------+----]/[------------]/[--------+--------]/[--------( )-------+---
   | |      00       |   14           01            |        DN        13         |
   | |               |  [34]                        |       [33]                  |
   | |   Saw_Clamp   |                              |                             |
   | |   Solenoid    |                              |Saw_Clamp                    |
   | |    O:006      |                              |TON_Timer                    |
   | +------] [------+                              |+--TON-----------+           |
   |        13                                      ++Timer On Delay  +--(EN)-+
   |       [33]                                      |Timer:   T4:118 |
   |                                                 |Base (SEC): 1.0 +--(DN)
   |                                                 |Preset:       2 |
   |                                                 |Accum:        0 |
   |                                                 +----------------+
   |  Saw_UnClamp  Saw_Clamp   Saw_Clamp      Saw_UnClamp  Saw_UnClamp
   |  PushButton   Solenoid    PushButton      TON_Timer    Solenoid
   |     I:034      O:006        I:034           T4:119      O:006
34 +-+-------] [------+----]/[------------]/[--------+--------]/[--------( )--------+---
   | |      01        |   13           00            |        DN        14          |
   | |                |  [33]                        |       [34]                   |
   | | Saw_UnClamp    |                              |                              |
   | |  Solenoid      |                              |Saw_UnClamp                   |
   | |   O.006        |                              |TON_Timer                     |
   | +------] [-------+                              | +--TON-----------+           |
   |        14                                       ++Timer On Delay  +---(EN)-+
   |       [34]                                       | Timer:   T4:119 |
   |                                                  | Base (SEC): 1.0 +--(DN)
   |                                                  | Preset:       2 |
   |                                                  | Accum:        0 |
   |                                                  +-----------------+
   | Saw_Clamp                                                     Saw_Clamps
   | Solenoid                                                      Last_Clamped
   |   O.006                                                           B3
35 +------] [-------------------------------------------------------(L)---------
   |      13                                                          488
   |     [33]
   |
   | Saw_UnClamp                                                   Saw_Clamps
   | Solenoid                                                      Last_Clamped
   |   O.006                                                           B3
36 +------] [-------------------------------------------------------(U)---------
   |      14                                                          488
   |     [34]
   |
```

```
                          Bundle Cutting Saw
                            ...Saw Clamps
                           Stacking Machine
File #14 Saw Proj: FLATS3    Page:00014                 21:08 12/05/02
-----------------------------------------------------------------------
     |
     |                                                   Saw_Clamp
     | Saw_Clamps                                      Loading_Valve
     |  Solenoid                                         Required
     |   0.006                                              B3
  37 +- +----] [-------+--------------------------------------( )-----
     |  |   13         |                                     481
     |  |  [33]        |
     |  |Saw_UnClamp   |
     |  |Solenoid      |
     |  |   O:006      |
     |  +-----] [------+
     |      14
     |     [34]
     |
     |Disch_Desk
     | Lamp_Test                                          Saw_Alarms
     |PushButton                                            Accept
     |   I:031                                                B3
  38 +------] [-----------------------------------------------( )-----
     |     12                                                497
     |
     |
```

```
                        Bundle Cutting Saw
                        ...Saw Desk Lamps
                        Stacking Machine
File #14 Saw Proj: FLATS3     Page:00015                 21:08 12/05/02
-------------------------------------------------------------------------
   |     Saw_ESR                                           Saw_Intlock
   |     Healthy                                           Healthy_Lamp
   |      I:031                                               O:030
39 +--+----] [--------+--------------------------------------------( )------
   |  |    17         |                                             00
   |  | Disch_Desk    |
   |  |  Lamp_Test    |
   |  | PushButton    |
   |  |    I:031      |
   |  +-----] [-------+
   |        12
   |
   |     Saw_ESR                                           Saw_Intlock
   |     Healthy                                            Fault_Lamp
   |      I:031                                               O:030
40 +--+----]/[--------+--------------------------------------------( )------
   |  |    17         |                                             01
   |  | Disch_Desk    |
   |  |  Lamp_Test    |
   |  | PushButton    |
   |  |    I:031      |
   |  +-----] [-------+
   |        12
   |
   | Saw_Hyd_Pump                                          Saw_Hyd_Pump
   |     Healthy                                           Healthy_Lamp
   |       B3                                                 O:030
41 +--+----] [--------+--------------------------------------------( )------
   |  |    499        |                                             02
   |  |   [24]        |
   |  | Disch_Desk    |
   |  |  Lamp_Test    |
   |  | PushButton    |
   |  |    I:031      |
   |  +-----] [-------+
   |        12
   |
   | Saw_Hyd_Pump                                          Saw_Hyd_Pump
   |     Healthy                                           Running_Lamp
   |      I:032                                               O:030
42 +--+----] [--------+--------------------------------------------( )------
   |  |    15         |                                             03
   |  | Disch_Desk    |
   |  |  Lamp_Test    |
   |  | PushButton    |
   |  |    I:031      |
   |  +-----] [-------+
   |        12
   |
```

```
                              Bundle Cutting Saw
                             ...Saw Desk Lamps
                              Stacking Machine
File #14 Saw Proj: FLATS3        Page:00016                    21:08 12/05/02
-------------------------------------------------------------------------------
      |                                                            Saw_Motor
      |  Saw_Motor                                                  Healthy
      |  Available                                                 Desk_Lamp
      |     B3                                                      O:030
  43 +--+----] [--------+--------------------------------------------( )-----
      |  |   516        |                                              04
      |  |   [0]        |
      |  | Disch_Desk   |
      |  | Lamp_Test    |
      |  | PushButton   |
      |  |   I:031      |
      |  +-----] [------+
      |        12
      |                                                            Saw_Motor
      |  Saw_motor                                                Saw_Intlock
      |  Running_Aux                                               Desk_Lamp
      |    I:032                                                     O:030
  44 +--+----] [--------+--------------------------------------------( )-----
      |  |    11        |                                              05
      |  | Disch_Desk   |
      |  | Lamp_Test    |
      |  | PushButton   |
      |  |   I:031      |
      |  +-----] [------+
      |        12
      |
      |  Coolant_Pump                                             Coolant_Pump
      |   Available                                               Healthy_Lamp
      |     B3                                                       O:030
  45 +--+----] [--------+--------------------------------------------( )-----
      |  |   518        |                                              06
      |  |   [3]        |
      |  | Disch_Desk   |
      |  | Lamp_Test    |
      |  | PushButton   |
      |  |   I:031      |
      |  +-----] [------+
      |        12
      |
      |  Coolant_Pump                                              Saw_Hyd_Pump
      |  Running_Aux                                               Running_Lamp
      |    I:032                                                     O:030
  46 +--+----] [--------+--------------------------------------------( )-----
      |  |    13        |                                              07
      |  | Disch_Desk   |
      |  | Lamp_Test    |
      |  | PushButton   |
      |  |   I:031      |
      |  +-----] [------+
      |        12
      |
```

```
                                  Bundle Cutting Saw
                                  ...Saw Desk Lamps
                                  Stacking Machine
File #14 Saw Proj: FLATS3          Page:00017                       21:08 12/05/02
---------------------------------------------------------------------------------
    | Saw_Top_LS                                                    Saw_at_Top
    |  1=Struck                                                      Desk_Lamp
    |   I:032                                                         O:030
 47 +-+----] [-------+-----------------------------------------------( )--------
    | |    01        |                                                 10
    | |Disch_Desk    |
    | |Lamp_Test     |
    | |Push Button   |
    | |  I:031       |
    | +-----] [------+
    |      12
    |Saw_Hyd_Pump     State_3                                        Saw_Raising
    |Running_Aux      Passed                                          Desk_Lamp
    |  I:032            B3                                              O:030
 48 +-+----] [--------------]/[-------+------------------------------( )--------
    | |    15          503            |                                11
    | | State_3        [18]           |
    | | Paused       Fast_Flash       |
    | |   B3            B3            |
    | +----] [--------------] [-------+
    | |   503           14            |
    | |   [18]        [2:34]          |
    | |Disch_Desk                     |
    | |Lamp_Test                      |
    | |Push Button                    |
    | |  I:031                        |
    | +----] [------------------------+
    |     12
    | Saw_Lower       State_3                                       Saw_Lowering
    | Slow_SOV        Passed                                          Desk_Lamp
    |   O:033           B3                                              O:030
 49 +-+----] [--------------]/[-------+------------------------------( )--------
    | |    10          503            |                                12
    | |   [23]         [18]           |
    | |Saw_Lower      State_3         |
    | |Fast_SOV       Paused          |
    | |  O:033          B3            |
    | +----] [--------------]/[-------+
    | |    11          503            |
    | |   [22]         [18]           |
    | | State_3                       |
    | | Paused       Fast_Flash       |
    | |   B3            B3            |
    | +----] [--------------] [-------+
    | |   503           14            |
    | |   [18]        [2:34]          |
    | |Disch_Desk                     |
    | |Lamp_Test                      |
    | |Push_Button                    |
    | |  I:031                        |
    | +----] [------------------------+
    |     12
```

```
                              Bundle Cutting Saw
                               ...Saw Desk Lamps
                               Stacking Machine
File #14 Saw Proj: FLATS3         Page:00018                    21:08 12/05/02
--------------------------------------------------------------------------------
   | State_2
   | Raise_to                                                        End_Cut
   | Top_Limit                                                      Desk_Lamp
   |     B3                                                           O:030
50 +--+----] [--------+----------------------------------------------( )-------
   |  |    502        |                                               13
   |  |   [17]        |
   |  | Disch_Desk    |
   |  | Lamp_Test     |
   |  | PushButton    |
   |  |    I:031      |
   |  +------] [------+
   |        12
   |  Saw_Blade                                                     Saw_Blade
   | Tension_LS                                                    Tension_OK
   | 1=Healthy                                                      Desk_Lamp
   |   I:032                                                          O:030
51 +--+----] [--------+----------------------------------------------( )-------
   |  |    03         |                                               14
   |  | Disch_Desk    |
   |  | Lamp_Test     |
   |  | PushButton    |
   |  |    I:031      |
   |  +------] [------+
   |        12
   |                                                              Tension_Pump
   | Saw_Tension                                                     Running
   |   Pump_Aux                                                     Desk_Lamp
   |    I:030                                                         O:030
52 +--+----] [--------+----------------------------------------------( )-------
   |  |    06         |                                               15
   |  | Disch_Desk    |
   |  | Lamp_Test     |
   |  | PushButton    |
   |  |    I:031      |
   |  +------] [------+
   |        12
   |                                                                 Bundle
   | Saw_Clamps                                                      Clamped
   | Last_Clamped                                                   Desk_Lamp
   |     B3                                                           O:031
53 +--+----] [--------+----------------------------------------------( )-------
   |  |    488        |                                               10
   |  |   [36]        |
   |  | Disch_Desk    |
   |  | Lamp_Test     |
   |  | PushButton    |
   |  |    I:031      |
   |  +------] [------+
   |        12
```

```
                              Bundle Cutting Saw
                              ...Saw Desk Lamps
                              Stacking Machine
File #14 Saw Proj: FLATS3        Page:00019                    21:08 12/05/02
-----------------------------------------------------------------------------
   |                                                                 Bundle
   | Saw_Clamps                                                    UnClamped
   |  Raised_LS                                                    Desk_Lamp
   |    O:034                                                        O:031
54 +--+----] [-----------------------+-------------------------------( )---
   |  |     10                       |                                11
   |  |                              |
   |  |                              |
   |  |Saw_Clamps                    |
   |  |Last_Clamped                  |
   |  |     B3                       |
   |  +------]/[---------------------+
   |  |     488                      |
   |  |     [36]                     |
   |  |                              |
   |  | Saw_Unclamp                  |
   |  |  Solenoid        Fast_Flash  |
   |  |   O:006             B3       |
   |  +------] [-----------] [-------+
   |  |     14              14       |
   |  |     [34]          [2:34]     |
   |  | Disch_Desk                   |
   |  | Lamp_Test                    |
   |  | Push_Button                  |
   |  |    I:031                     |
   |  +------] [---------------------+
   |        12
   |
```

```
                              Bundle Cutting Saw
                    ...Saw Cutting Length Lamps (from PECs)
                              Stacking Machine
File #14 Saw Proj: FLATS3      Page:00020                       21:08 12/05/02
-------------------------------------------------------------------------------
     |
     | PECs (photocells) operate at +/- 10 mm from set length
     |
     | West_Saw_Cut        East_Saw_Cut
     | Photocell           Photocell                                Length_Short
     | 1=Mat_Present       1-Mat_Present                             Desk_Lamp
     |     I:054              I:054                                    O:031
 55 +--+----]/[--------------]/[----+----------------------------------( )------
     |  |   01                 02    |                                   13
     |  |                            |
     |  | Disch_Desk                 |
     |  | Lamp_Test                  |
     |  | PushButton                 |
     |  |     I:031                  |
     |  +-----] [--------------------+
     |        12
     |
     | West_Saw_Cut        East_Saw_Cut
     | Photocell           Photocell                              Length_Correct
     | 1=Mat_Present       1-Mat_Present                             Desk_Lamp
     |     I:054              I:054                                    O:031
 56 +--+----] [--------------]/[----+----------------------------------( )------
     |  |   01                 02    |                                   14
     |  |                            |
     |  | Disch_Desk                 |
     |  | Lamp_Test                  |
     |  | PushButton                 |
     |  |     I:031                  |
     |  +-----] [--------------------+
     |        12
     |
     | West_Saw_Cut        East_Saw_Cut
     | Photocell           Photocell                                Length_Long
     | 1=Mat_Present       1-Mat_Present                             Desk_Lamp
     |     I:054              I:054                                    O:031
 57 +--+----] [--------------] [----+----------------------------------( )------
     |  |   01                 02    |                                   15
     |  |                            |
     |  | Disch_Desk                 |
     |  | Lamp_Test                  |
     |  | PushButton                 |
     |  |     I:031                  |
     |  +-----] [--------------------+
     |        12
     |
     |
 58 +-------------------------------------------------------------------[END]------
     |
     |
```

Summary

A systematic approach to the writing of programs can improve the chances of high-quality programs being generated in as short a time as possible. This is likely to involve the following: a definition of what is required, a definition of the algorithm to be used, translation of the algorithm into instructions for the PLC, testing and debugging of the program, and documentation to ensure that any person using the program can understand how it works.

Modern safety legislation charges employers with duties that include making the workplace safe and without risks to health and ensuring that plant and machinery are safe and that safe systems of work are set and followed. An important standard relevant to PLCs is IEC 61508: Functional Safety of Electrical/Electronic/Programmable Electronic Safety-Related Systems. Emergency stop buttons and safety guard switches must be hardwired and not depend on the PLC software for implementation so that, in the situation where there is a failure of the stop switch or PLC, the system is automatically safe. The system must be fail-safe. This can involve two-handed engaging, protective door monitoring, light curtains, safety mats, and emergency stop relays.

Commissioning of a PLC system involves checking that all the cable connections between the PLC and the plant being controlled are complete, safe, and to the required specification and local standards, checking that the incoming power supply matches the voltage setting for which the PLC is set, checking that all protective devices are set to their appropriate trip settings, checking that emergency stop buttons work, checking that all input/output devices are connected to the correct input/output points and giving the correct signals, loading the software, and testing the software. Fault detection techniques include watchdog timing checks, indicators to show last output set, replication, and expected value checks.

The documentation for a PLC installation should include a description of the plant, specification of the control requirements, details of the programmable logic controller, electrical installation diagrams, lists of all input and output connections, the application program with full commentary on what it is achieving, software backups, and an operating manual, including details of all startup and shutdown procedures and alarms.

Problems

Questions 1 through 7 have four answer options: A, B, C, or D. Choose the correct answer from the answer options.

1. The ladder program elements given in Figure 13.20 can be described by a basic algorithm of the type:

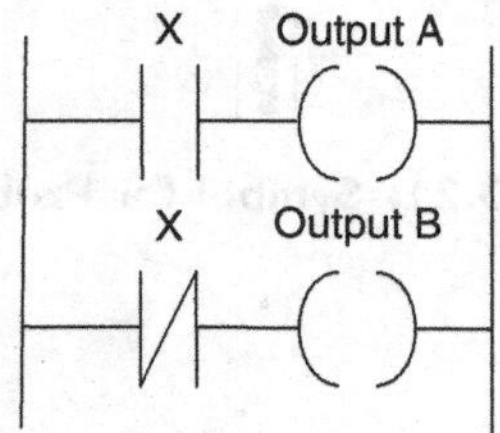

Figure 13.20: Program elements for Problem 1.

A. DO-THEN-DO-ENDDO
B. IF-THEN-ELSE-ENDIF
C. WHILE-DO-ENDWHILE
D. Not described by A, B, or C

2. *Decide whether each of these statements is true (T) or false (F).* The term *forcing,* when applied to a PLC input/output, means using a program to:
   (i) Turn on or off inputs/outputs.
   (ii) Check that all inputs/outputs give correct responses when selected.
   A. (i) T (ii) T
   B. (i) T (ii) F
   C. (i) F (ii) T
   D. (i) F (ii) F

3. *Decide whether each of these statements is true (T) or false (F).* The term *watchdog,* when applied to a PLC, means a checking mechanism that:
   (i) Excessive currents are not occurring.
   (ii) Functions are carried out within prescribed time limits.
   A. (i) T (ii) T
   B. (i) T (ii) F
   C. (i) F (ii) T
   D. (i) F (ii) F

4. *Decide whether each of these statements is true (T) or false (F).* When a PLC is in monitor/test/debug mode, it:
   (i) Enables the operation of a program to be simulated.
   (ii) Carries out a fault check.
   A. (i) T (ii) T
   B. (i) T (ii) F
   C. (i) F (ii) T
   D. (i) F (ii) F

Figure 13.21: Symbol for Problem 5.

5. When a PLC is in monitor/test/debug mode and the symbol shown in Figure 13.21 occurs, it means that an input is:
   A. Defective
   B. Correctly operating
   C. On
   D. Off
6. *Decide whether each of these statements is true (T) or false (F).* Failure of an input sensor or its wiring, rather than failure of an LED or in the PLC input channel, will show as:
   (i) The input LED not coming on.
   (ii) Forcing of that input, making the input LED come on.
   A. (i) T (ii) T
   B. (i) T (ii) F
   C. (i) F (ii) T
   D. (i) F (ii) F
7. A single output device fails to turn on when the output LED is on. The voltage at the output is tested and found normal, but the voltage at the device is found to be absent. The fault is:
   A. Faulty wiring
   B. A faulty output device
   C. A fault in the PLC
   D. A fault in the program
8. Explain how, using forcing, the failure of an input sensor or its wiring can be detected.
9. Suggest possible causes of a complete stoppage of the control operation and the PLC with the power-on lamp off.
10. Suggest possible causes of an output LED being on but the output device failing to turn on.
11. Devise a timing watchdog program to be used to switch off a machine if faults occur in any of the systems controlling its actions.

12. Design the program for a pneumatic system for control by a PLC with the cylinder sequence A+, B+, B–, A– and an LED indicating, in the presence of a fault such as a sticking cylinder, at which point in the cycle the fault occurred. Explain the action of all elements in the system.

Lookup Tasks

13. Look up the safety standard IEC61508. You will also find summaries of the main implications of it on the Web.
14. Find out details of light curtain systems that are commercially available.
15. Find out details of electronic safety relays that are commercially available.

CHAPTER 14

Programs

This chapter extends the examples given in previous chapters to show programs developed to complete specific tasks. These include tasks that involve temperature control and a number involving pneumatic valves.

14.1 Temperature Control

Consider the task of using a PLC as an on/off controller for a heater in the control of temperature in some enclosure. The heater is to be switched on when the temperature falls below the required temperature and switched off when the temperature is at or above the required temperature. The basic algorithm might be considered to be:

```
IF temperature below set value
THEN
    DO switch on heater
ELSE
    DO switch off heater
ENDIF
```

What is required to give the input to a PLC is a sensor-signal conditioner arrangement that will give no input when the temperature is below the required temperature and an input when it is above. One such device is the bimetallic thermostat (Figure 14.1; see Section 2.1.5). A bimetallic strip consists of two different metals strips of the same length bonded together. Because the metals have different coefficients of expansion, when the temperature increases the composite strip bends into a curve with the higher coefficient metal on the outside of the curve. The movement may be used to open or close electrical contacts and so operate as a temperature-dependent switch. If the actual temperature is above the required temperature, the bimetallic strip is in an off position and there is no circuit voltage output. If the actual temperature is below the required temperature, the bimetallic strip moves into the on position and the voltage is switched on, so there is an output. The thermostat output is thus just off or on and the input to the PLC is no voltage or a voltage. Figure 14.2 shows the PLC connections and a ladder program that can be used.

W. Bolton: Programmable Logic Controllers, Sixth Edition. http://dx.doi.org/10.1016/B978-0-12-802929-9.00014-5

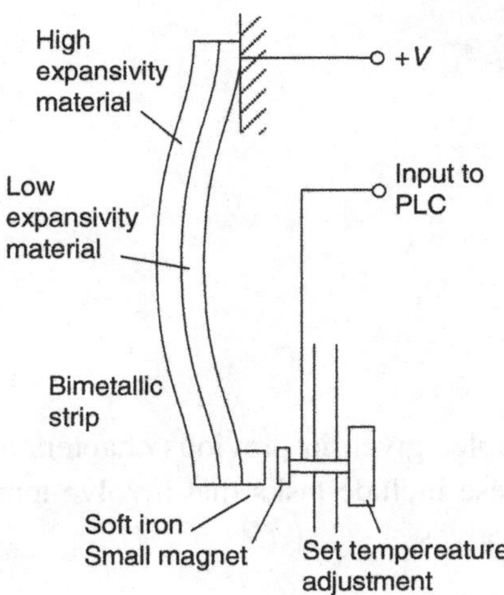

Figure 14.1: Bimetallic thermostat with voltage applied to the PLC input to switch it on when the temperature is below the set value and no voltage applied when above it.

An alternative to the bimetallic thermostat is to use a thermocouple, a thermistor, or an integrated chip (see Section 2.1.5) as a temperature sensor. When connected in an appropriate circuit, the sensor will give a suitable voltage signal related to the temperature. For example, with a thermistor, its resistance changes when the temperature changes, and this can be converted into a voltage signal by using a potential divider circuit. Figure 14.3 shows a possible solution involving a potential divider circuit to convert the resistance change into a voltage change. Suppose we use a 4.7 kΩ bead thermistor. This has a resistance of 4.7 kΩ

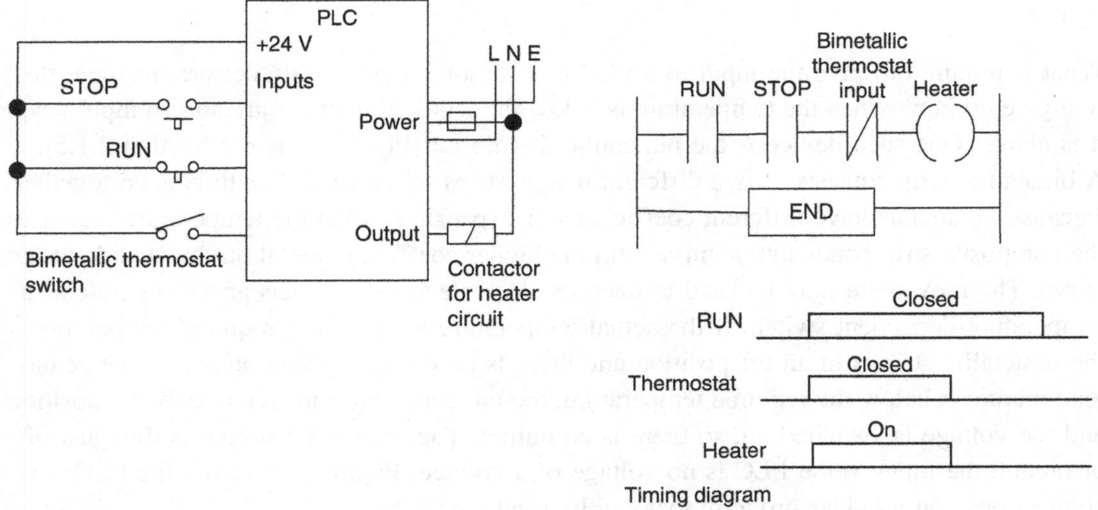

Figure 14.2: PLC arrangement and program with the bimetallic thermostat.

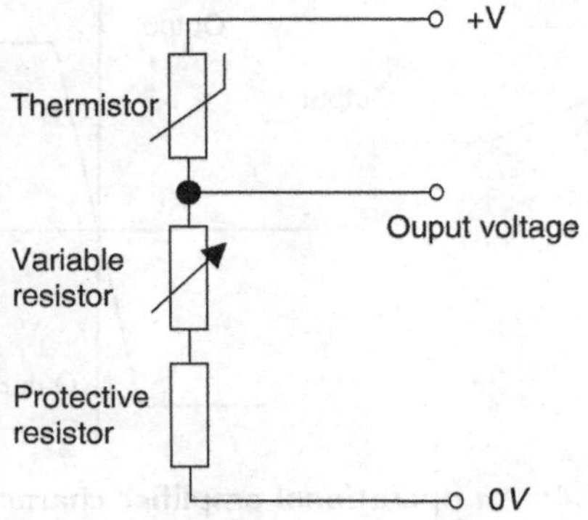

Figure 14.3: A thermistor with a potential divider circuit.

at 25°C, 15.28 kΩ at 0°C, and 0.33 kΩ at 100°C. The variable resistor might be 0 to 10 kΩ. It enables the sensitivity of the arrangement to be altered. However, if the variable resistor was set to zero resistance, without a protective resistor, we could possibly have a large current passed through the thermistor. The maximum power that the thermistor can withstand is specified as 250 mW. Thus, with a 6 V supply, the variable resistor set to zero resistance, the protective resistance of R, and the thermistor at 100°C, the current I through the thermistor is given by $V = IR$ as $6 = I(0 + R + 330)$, and so $I = 6/(R + 330)$. The power dissipated by the thermistor is $I^2 \times 330$, so if we want this to be significantly below the maximum possible—say, 100 mW—then we have $0.100 = I^2 \times 330$, and so R needs to be about 15 Ω.

This voltage output from the thermistor can be compared, using an operational amplifier, with the voltage set for the required temperature so that a high-output signal is given when the temperature is above the required temperature and a low-output signal when it is below, or vice versa. Operational amplifiers are widely used to give on/off signals based on the relative value of two input signals—a set-point voltage and a sensor voltage (see Section 4.2.2). One signal is connected to the noninverting terminal and the other to the inverting terminal. The operational amplifier determines whether the signal to the inverting terminal is above or below that of the noninverting terminal. Figure 14.4 shows the output/input relationship for the operational amplifier. By reversing the reference and input voltage connections, the output polarity is changed.

Thus we can arrange that when the temperature falls from above the required temperature to below it, the output signal switches from a high to a low value. The PLC can then be programmed to give an output when there is a low input, and this output can be used to switch on the heater and switch off the heater when the input goes high.

Figure 14.5 shows the arrangement that might be used and the basic elements of a ladder program. The input from the operational amplifier has been connected to the input port. This

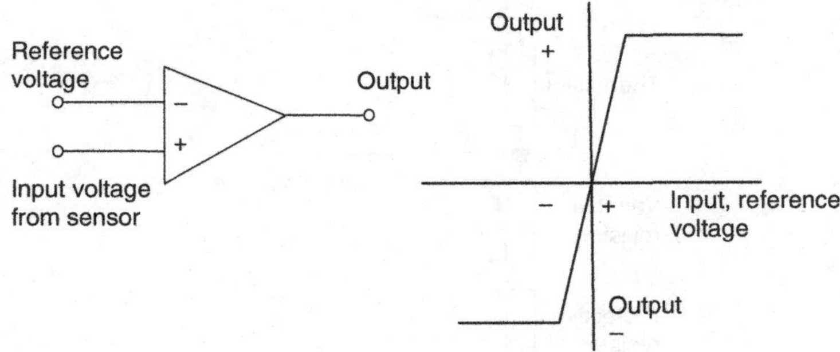

Figure 14.4: An operational amplifier characteristic.

input has contacts that are normally closed. When the input goes high, the contacts open. The output to the heater is taken from the output port. Thus there is an output when the input from the sensor is low and no output when the input is high.

Consider a more complex temperature control task involving a domestic central heating system (Figure 14.6). The central heating boiler is to be thermostatically controlled and will supply hot water to the radiator system in the house as well as to the hot water tank to provide hot water from the taps in the house. Pump motors have to be switched on to direct the hot water from the boiler to either or both the radiator and hot water systems, according to whether the temperature sensors for the room temperature and the hot water tank indicate that the radiators or tank need heating. The entire system is to be controlled by a clock so that it operates for only certain hours of the day. Figure 14.7 shows how this might be programmed.

The inputs might be from sensors feeding through operational amplifiers so that a high output is given when the sensor temperature is above the set value and so there is a need to switch

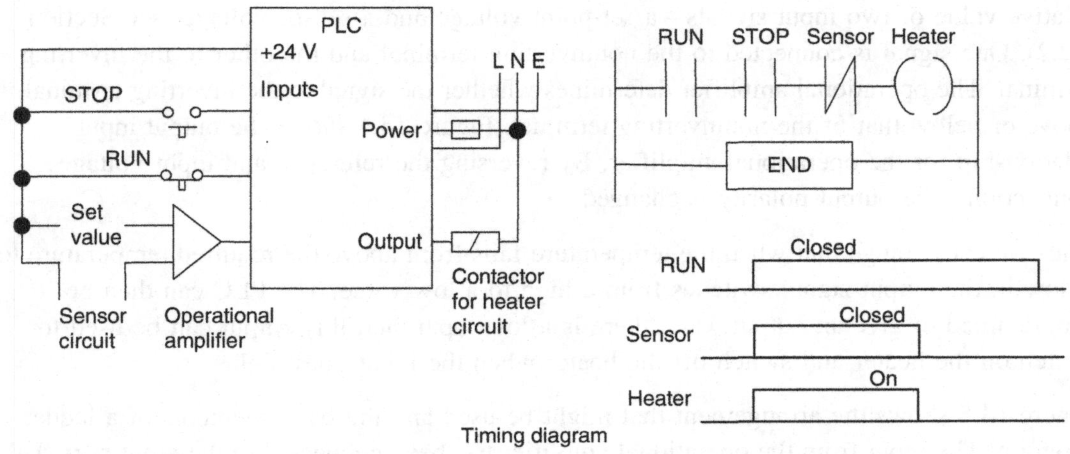

Figure 14.5: Temperature control.

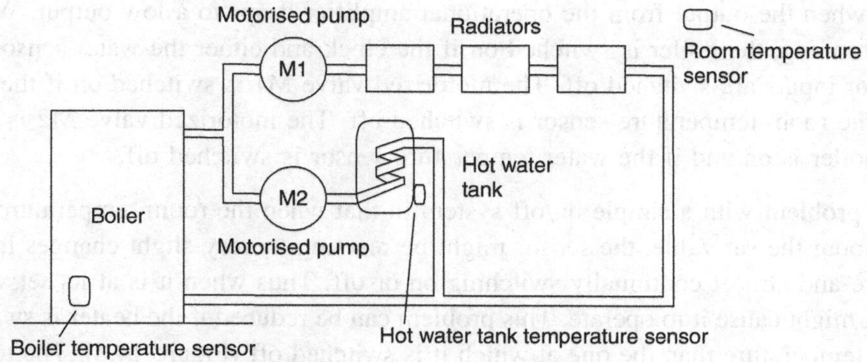

Figure 14.6: Central heating system.

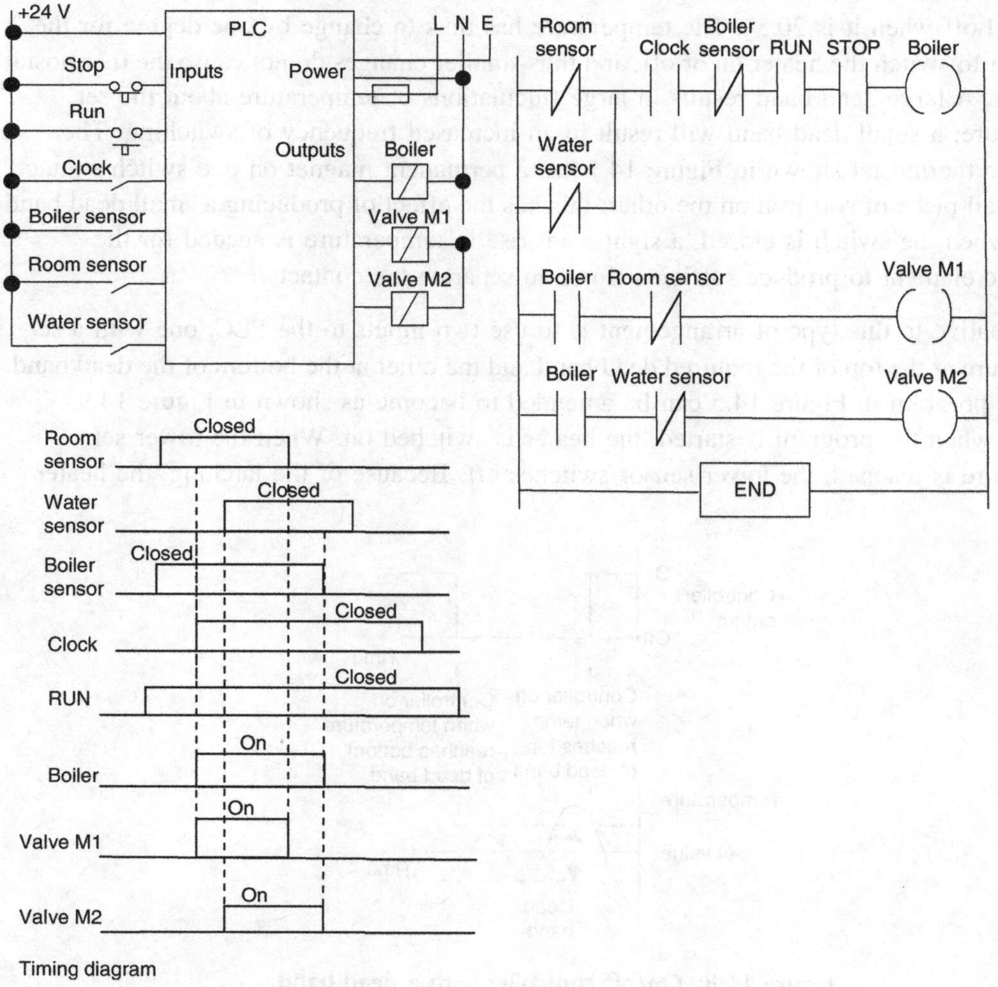

Figure 14.7: A central heating system.

things off when the output from the operational amplifier drops to a low output. When the system is running, the boiler is switched on if the clock and either the water sensor or the room sensor inputs are switched off. The motorized valve M1 is switched on if the boiler is on and if the room temperature sensor is switched off. The motorized valve M2 is switched on if the boiler is on and if the water temperature sensor is switched off.

There is a problem with a simple on/off system in that when the room temperature is hovering about the set value, the sensor might be reacting to very slight changes in temperature and almost continually switching on or off. Thus when it is at its set value, a slight draft might cause it to operate. This problem can be reduced if the heater is switched on at a lower temperature than the one at which it is switched off (Figure 14.8). The term *dead band* is used for the values between the on and off values. For example, if the set value on a thermostat is 20°C, a dead band might mean that it switches on when the temperature falls to 19.5° and off when it is 20.5°. The temperature has thus to change by one degree for the controller to switch the heater on or off, and thus smaller changes do not cause the thermostat to switch. A large dead band results in large fluctuations of temperature about the set temperature; a small dead band will result in an increased frequency of switching. The bimetallic thermostat shown in Figure 14.1 has a permanent magnet on one switch contact and a small piece of soft iron on the other; this has the effect of producing a small dead band in that, when the switch is closed, a significant rise in temperature is needed for the bimetallic element to produce sufficient force to separate the contacts.

An alternative to this type of arrangement is to use two inputs to the PLC, one with a set temperature at the top of the required dead band and the other at the bottom of the dead band. Thus the program in Figure 14.5 can be amended to become as shown in Figure 14.9. Initially, when the program is started, the heater is switched on. When the lower set temperature is reached, the lower sensor switches off. Because of the latching, the heater

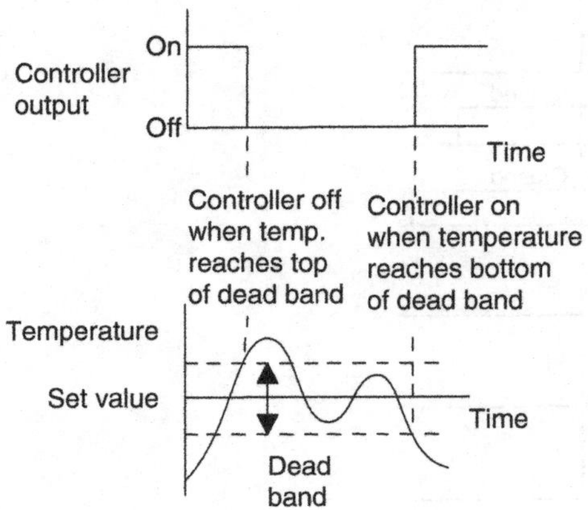

Figure 14.8: On/off controller with a dead band.

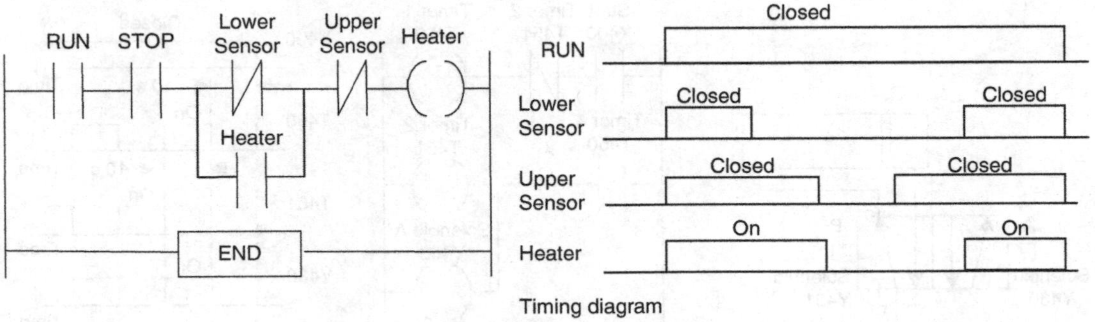

Figure 14.9: Dead-band program.

remains on. When the upper set temperature is reached, the upper sensor switches off, and consequently the heater is switched off. When the temperature drops, it has to drop to the lower set temperature before the heater is switched on again.

In the previous discussion a simple on/off form of temperature control has been used, with perhaps a comparator op-amp giving a 1 output when the temperature is above the set temperature and 0 when it is below. The output to the heating system is then just on or off. A more elaborate system is to use proportional control with the output to the heating system being a signal proportional to the difference in temperature between that occurring and the set value. The program might then carry out the following tasks:

1. Read the actual temperature input after conversion from analog to digital by an ADC.
2. Input the set point temperature.
3. Subtract the actual temperature from the set point temperature.
4. Multiply the result by the proportional constant.
5. Use the result to control the value of the output to the heater.

14.2 Valve Sequencing

Consider tasks involving directional control valves (see Section 2.2.2 for an introductory discussion).

14.2.1 Cyclic Movement

Consider the task of obtaining cyclic movement of a piston in a cylinder. This might be to periodically push workpieces into position in a machine tool with another similar but out-of-phase arrangement used to remove completed workpieces. Figure 14.10 shows the valve and piston arrangement that might be used, a possible ladder program, and a chart indicating the timing of each output.

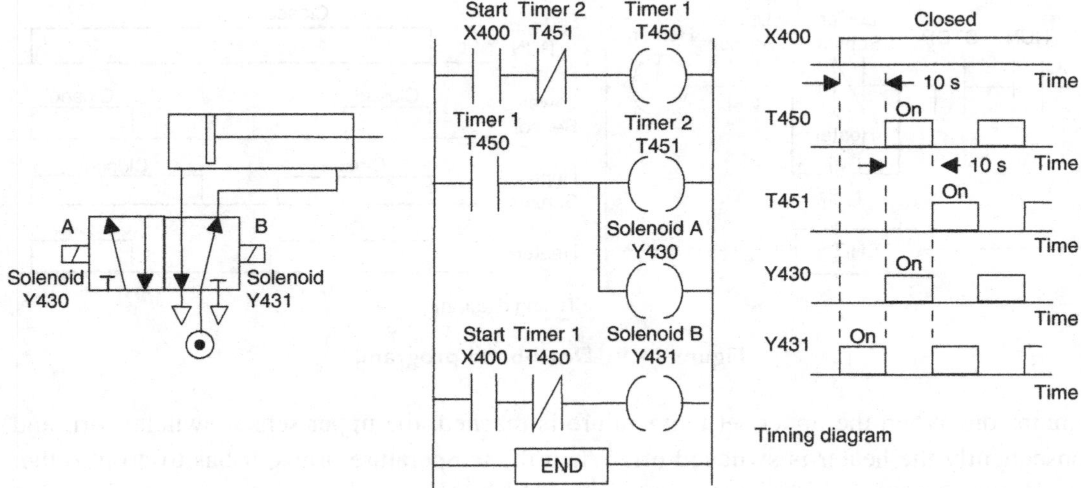

Figure 14.10: Cyclic movement of a piston.

Consider both timers set for 10 s. When the start contacts X400 are closed, timer T450 starts. There is also an output from Y431. Output Y431 is one of the solenoids used to actuate the valve. When it is energized it causes pressure supply P to be applied to the right-hand end of the cylinder and the left-hand side to be connected to the vent to the atmosphere. The piston thus moves to the left. After 10 s, the normally open T450 contacts close and the normally closed T450 contacts open. This stops output Y431, starts timer T451, and energizes output Y430. As a result, pressure supply P is applied to the left-hand side of the piston and the right-hand side is connected to the vent to the atmosphere. The piston now moves to the right. After 10 s, the normally closed T451 contacts are opened. This causes the normally closed T450 contacts to close, and so Y431 is energized. Thus the sequence repeats itself.

14.2.2 Sequencing

Consider another task involving three pistons A, B, and C that have to be actuated in this sequence: A to the right, A to the left, B to the right, B to the left, C to the right, C to the left. (Such a sequence is often written A+, A−, B+, B−, C+, C−.) Figure 14.11 illustrates the valves that might be used; Figures 14.12 and 14.13 show ladder programs involving timers that might be used. An alternative would involve the use of a shift register.

X400/I0.0 is the start switch. When it is closed there is an output from Y430/Q2.0, and timer T450/T0 starts. The start switch is latched by the output. Piston A moves to the right. After the set time, K = 4, the normally closed timer T450/internal relay F0.0 contacts open and the normally open timer T450/internal relay F0.0 contacts close. This switches off Y430/Q2.0,

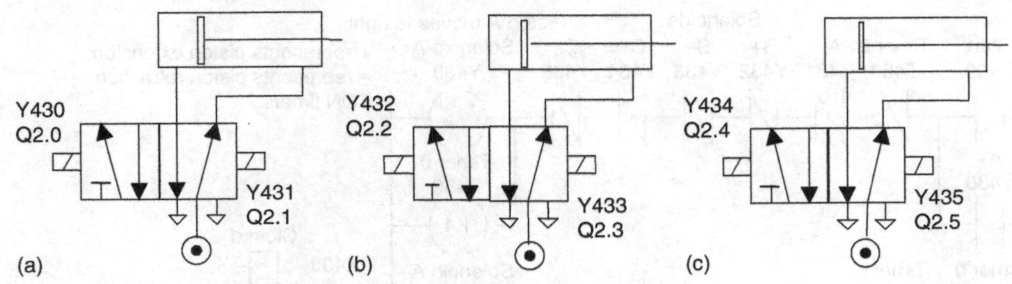

Figure 14.11: The valves.

energizes Y431/Q2.1, and starts timer T451/T1. Piston A moves left. In rung 2, the T450/internal relay F0.0 contacts are latched and so the output remains on until the set time has been reached. When this occurs, the normally closed timer T451/internal relay F0.1 contacts open and the normally open T451/internal relay F0.1 contacts close. This switches off Y431/Q2.1, energizes Y432/Q2.2, and starts timer T452/T2. Piston B moves right. Each succeeding rung activates the next solenoid. Thus, in sequence, each of the outputs is energized.

The program instruction list, in the Mitsubishi format, for the preceding program is as follows:

```
LD    X400   (*Start switch*)
OR    Y430
ANI   T450
ANI   Y431
ANI   Y432
ANI   Y433
ANI   Y434
ANI   Y435
OUT   Y430   (*Piston A moves to right*)
OUT   T450   (*Timer T450 starts*)
LD    T450
OR    Y431
ANI   T451
OUT   Y431   (*Piston A moves to left*)
OUT   T451   (*Timer T451 starts*)
LD    T451
OR    Y432
ANI   T452
OUT   Y432   (*Piston B moves to right*)
OUT   T452   (*Timer T452 starts*)
LD    T452
OR    Y433
ANI   T453
OUT   Y433   (*Piston B moves to left*)
```

(Continued on Pg. 344)

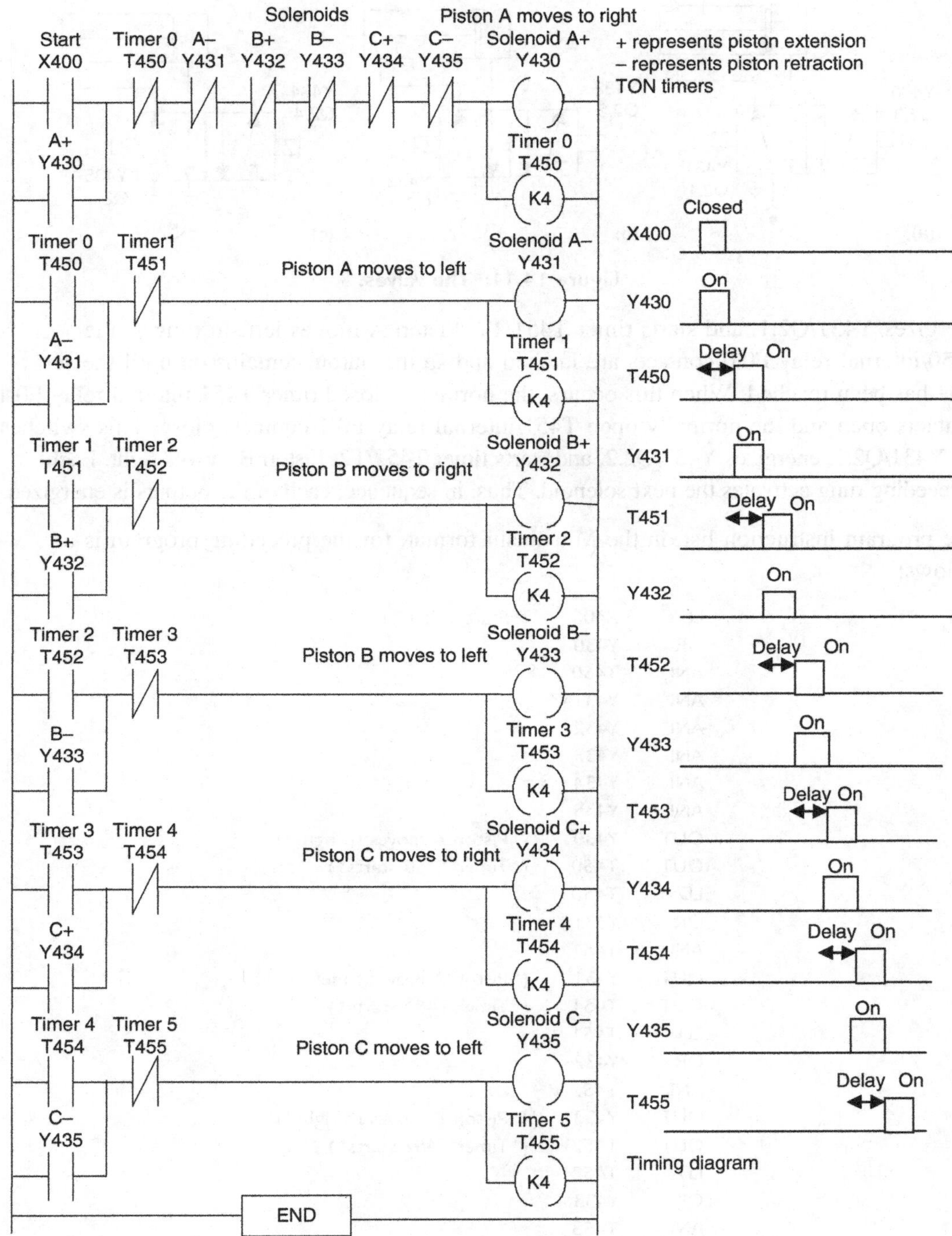

Figure 14.12: Mitsubishi format program.

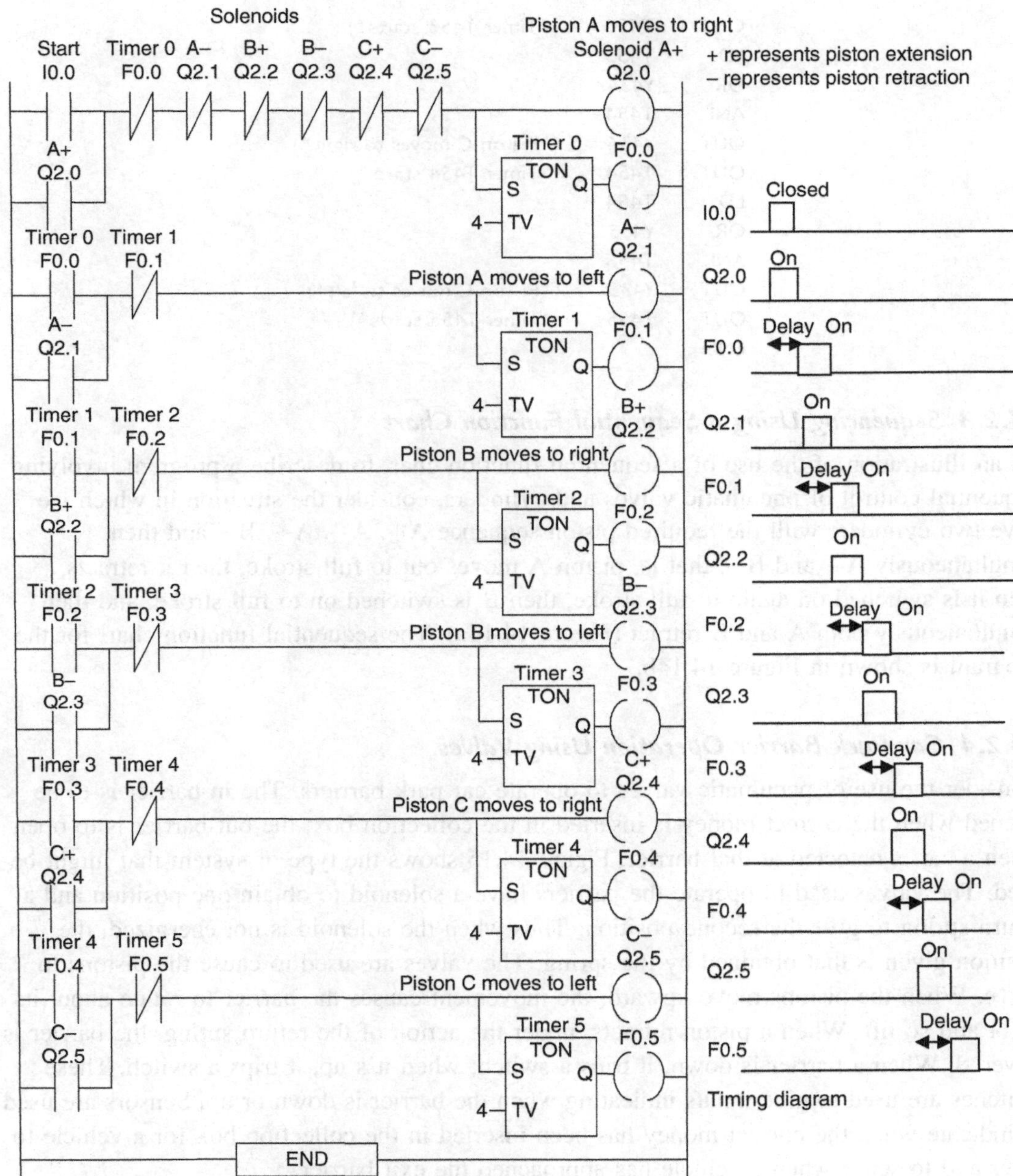

Figure 14.13: Siemens format program.

```
OUT    T453    (*Timer T453 starts*)
LD     T453
OR     Y434
ANI    T454
OUT    Y434    (*Piston C moves to right*)
OUT    T454    (*Timer T454 starts*)
LD     T454
OR     Y435
ANI    T455
OUT    Y435    (*Piston C moves to left*)
OUT    T455    (*Timer T455 starts*)
END
```

14.2.3 Sequencing Using a Sequential Function Chart

As an illustration of the use of a sequential function chart to describe a program involving sequential control of pneumatic valves and cylinders, consider the situation in which we have two cylinders with the required piston sequence A+. A−, A+, B+ and then simultaneously A− and B−, that is, piston A moves out to full stroke, then it retracts, then it is switched on again to full stroke, then B is switched on to full stroke, and then simultaneously both A and B retract (Figure 14.14a). The sequential function chart for the program is shown in Figure 14.14b.

14.2.4 Car Park Barrier Operation Using Valves

Consider the use of pneumatic valves to operate car park barriers. The in-barrier is to be opened when the correct money is inserted in the collection box; the out-barrier is to open when a car is detected at that barrier. Figure 14.15 shows the type of system that might be used. The valves used to operate the barriers have a solenoid to obtain one position and a return spring to give the second position. Thus when the solenoid is not energized, the position given is that obtained by the spring. The valves are used to cause the pistons to move. When the pistons move upward, the movement causes the barrier to rotate about its pivot and so lift. When a piston retracts, under the action of the return spring, the barrier is lowered. When a barrier is down, it trips a switch; when it's up, it trips a switch. These switches are used to give inputs indicating when the barrier is down or up. Sensors are used to indicate when the correct money has been inserted in the collection box for a vehicle to enter and to sense when a vehicle has approached the exit barrier.

Figure 14.16 shows the form a ladder program could take: (a) the Mitsubishi program and (b) the Siemens program. The output Y430/Q2.0 to solenoid 1 to raise the entrance barrier is given when the output from the coin box sensor gives the X400/I0.0 input. The Y430/Q2.0 is latched and remains on until the internal relay M100/F0.1 opens. The output will also not

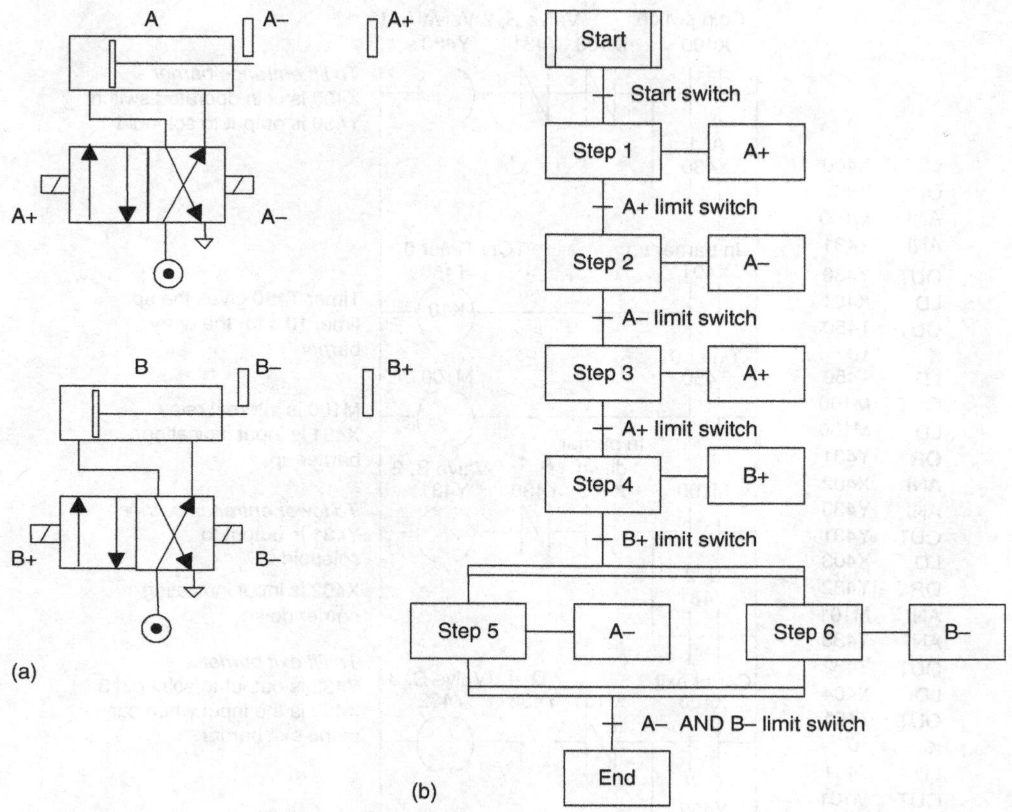

Figure 14.14: Piston sequence B+. B–, B+, A+ and then, simultaneously, A– and B–.

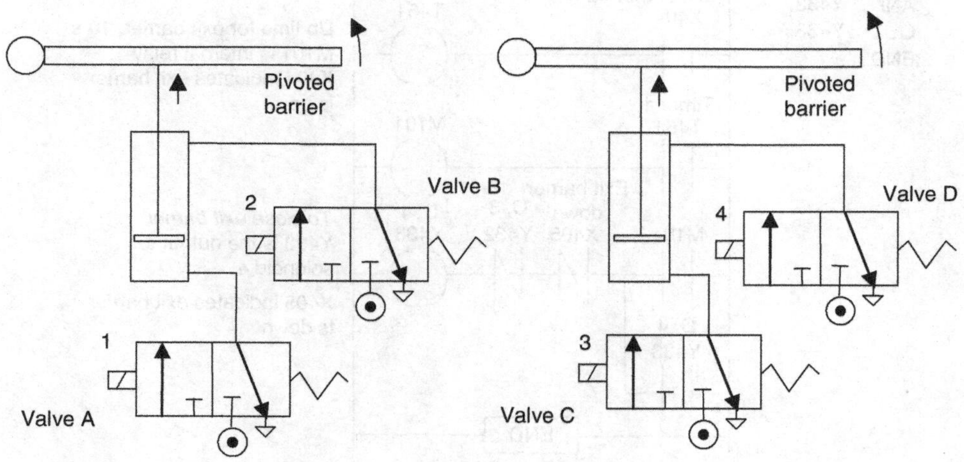

Figure 14.15: Valve/piston system.

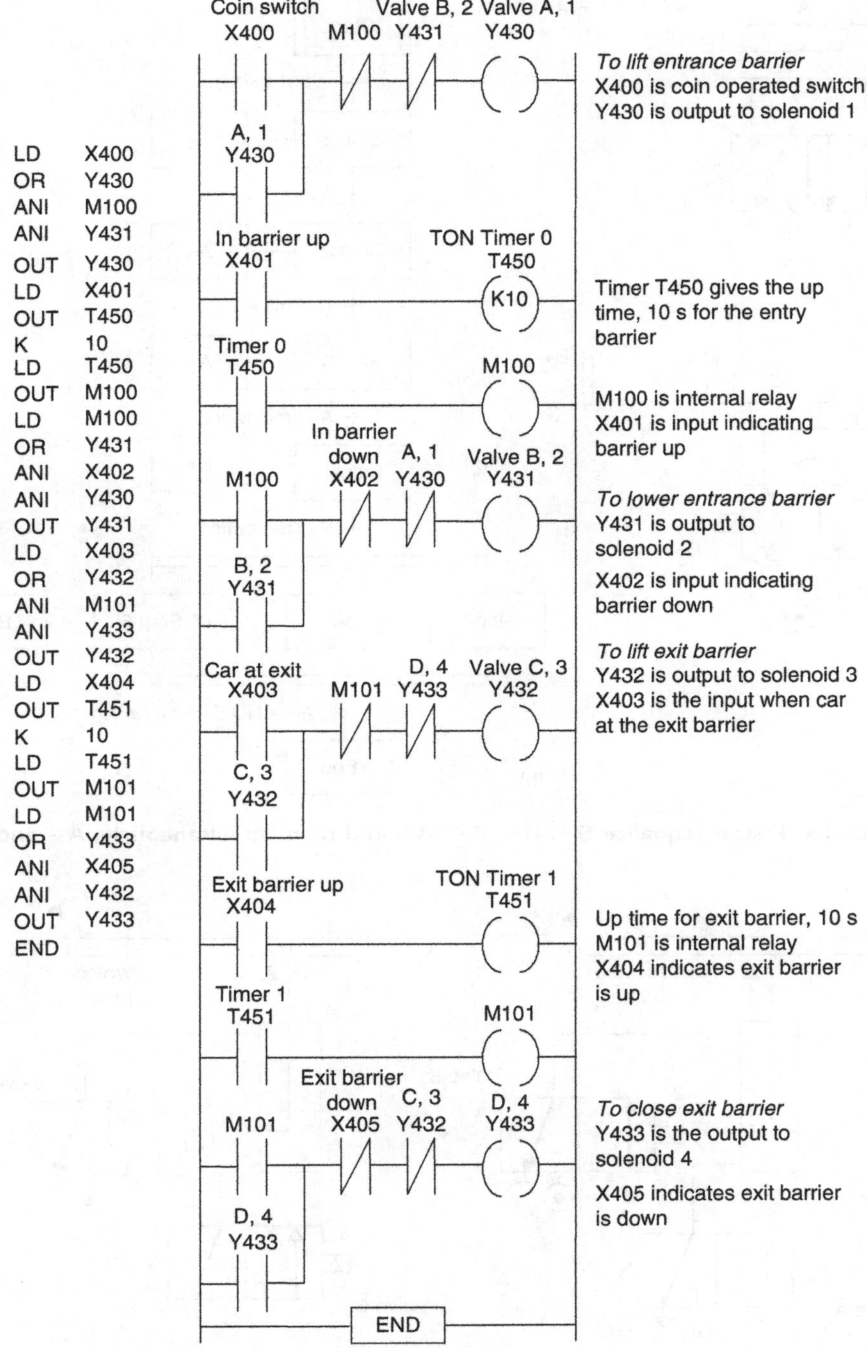
LD X400
OR Y430
ANI M100
ANI Y431
OUT Y430
LD X401
OUT T450
K 10
LD T450
OUT M100
LD M100
OR Y431
ANI X402
ANI Y430
OUT Y431
LD X403
OR Y432
ANI M101
ANI Y433
OUT Y432
LD X404
OUT T451
K 10
LD T451
OUT M101
LD M101
OR Y433
ANI X405
ANI Y432
OUT Y433
END
Coin switch
X400
M100
Valve B, 2
Y431
Valve A, 1
Y430
A, 1
Y430
To lift entrance barrier
X400 is coin operated switch
Y430 is output to solenoid 1
In barrier up
X401
TON Timer 0
T450
K10
Timer T450 gives the up time, 10 s for the entry barrier
Timer 0
T450
M100
M100 is internal relay
X401 is input indicating barrier up
In barrier down
X402
A, 1
Y430
Valve B, 2
Y431
M100
B, 2
Y431
To lower entrance barrier
Y431 is output to solenoid 2
X402 is input indicating barrier down
Car at exit
X403
M101
D, 4
Y433
Valve C, 3
Y432
C, 3
Y432
To lift exit barrier
Y432 is output to solenoid 3
X403 is the input when car at the exit barrier
Exit barrier up
X404
TON Timer 1
T451
Up time for exit barrier, 10 s
M101 is internal relay
X404 indicates exit barrier is up
Timer 1
T451
M101
Exit barrier down
X405
C, 3
Y432
D, 4
Y433
M101
D, 4
Y433
To close exit barrier
Y433 is the output to solenoid 4
X405 indicates exit barrier is down
END

Figure 14.16(a): Car barrier program, Mitsubishi format.

```
A    I0.0
O    Q2.0
AN   F0.1
AN   Q2.1
=    Q2.0
A    I0.1
LKT  I0.2
SR   T0
A    T0
=    Q2.0
A    F0.1
O    Q2.1
AN   I0.2
AN   Q2.0
=    Q2.1
A    I0.3
O    Q2.2
AN   F0.2
AN   Q2.3
=    Q2.2
A    I0.4
LKT  I0.2
SR   T1
A    T1
=    F0.2
A    F0.2
O    Q2.3
AN   I0.5
AN   Q2.2
=    Q2.3
END
```

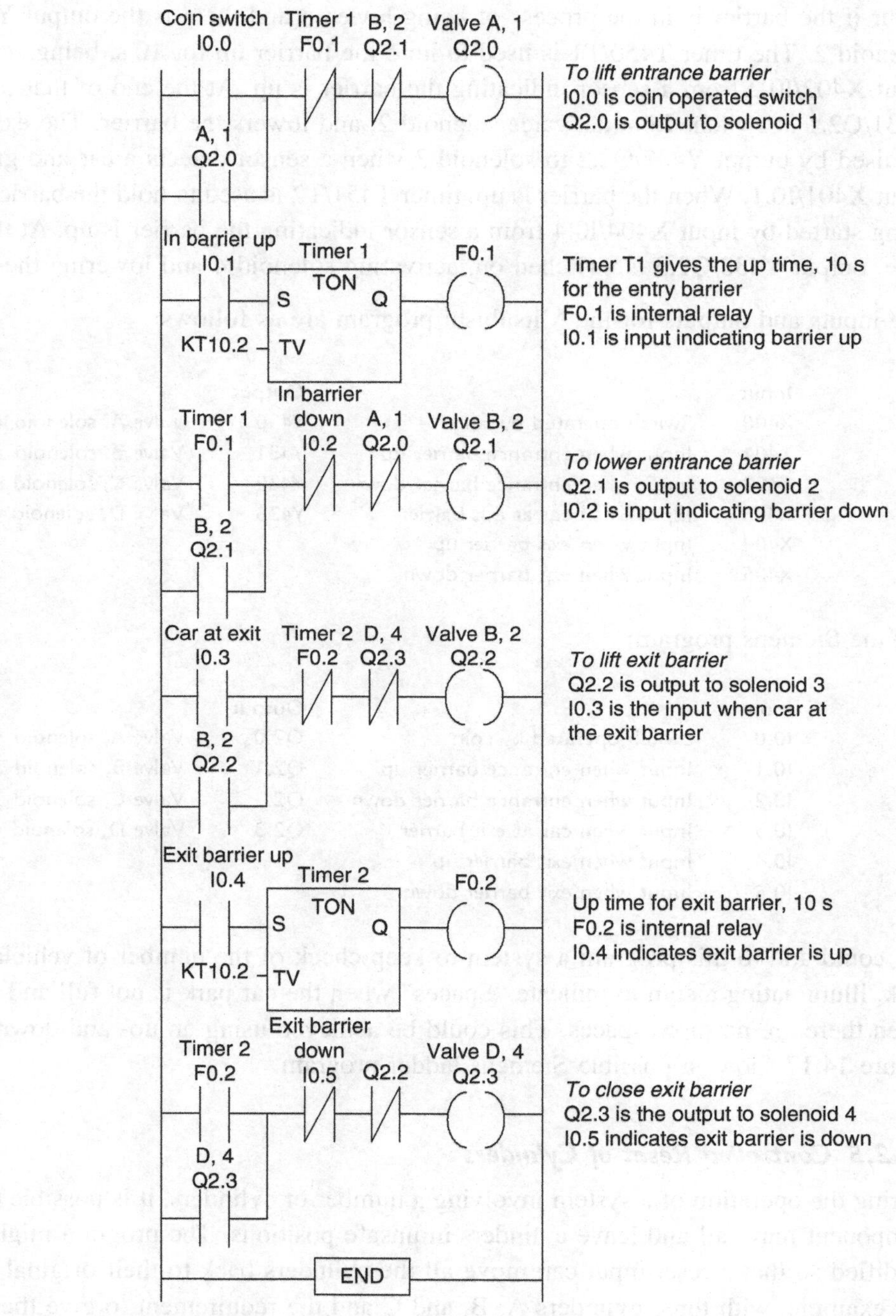

Figure 14.16(b): Car barrier program, Siemens format.

occur if the barrier is in the process of being lowered and there is the output Y431/Q2.1 to solenoid 2. The timer T450/T1 is used to hold the barrier up for 10 s, being started by input X402/I0.2 from a sensor indicating the barrier is up. At the end of that time, output Y431/Q2.1 is switched on, activates solenoid 2, and lowers the barrier. The exit barrier is raised by output Y432/Q2.2 to solenoid 3 when a sensor detects a car and gives input X401/I0.1. When the barrier is up, timer T451/T2 is used to hold the barrier up for 10 s, being started by input X404/I0.4 from a sensor indicating the barrier is up. At the end of the time, output Y433/Q2.3 is switched on, activating solenoid 4 and lowering the barrier.

The inputs and outputs for the Mitsubishi program are as follows:

Input		Output	
X400	Switch operated by coin	Y430	Valve A, solenoid 1
X401	Input when entrance barrier up	Y431	Valve B, solenoid 2
X402	Input when entrance barrier down	Y432	Valve C, solenoid 3
X403	Input when car at exit barrier	Y433	Valve D, solenoid 4
X404	Input when exit barrier up		
X405	Input when exit barrier down		

For the Siemens program:

Input		Output	
I0.0	Switch operated by coin	Q2.0	Valve A, solenoid 1
I0.1	Input when entrance barrier up	Q2.1	Valve B, solenoid 2
I0.2	Input when entrance barrier down	Q2.2	Valve C, solenoid 3
I0.3	Input when car at exit barrier	Q2.3	Valve D, solenoid 4
I0.4	Input when exit barrier up		
I0.5	Input when exit barrier down		

We could add to this program a system to keep check of the number of vehicles in the car park, illuminating a sign to indicate "Spaces" when the car park is not full and a sign "Full" when there are no more spaces. This could be achieved using an up- and down-counter. Figure 14.17 shows a possible Siemens ladder program.

14.2.5 Controlled Reset of Cylinders

During the operation of a system involving a number of cylinders, it is possible that a system component may fail and leave cylinders in unsafe positions. The program might thus be modified so that a reset input can move all the cylinders back to their original positions. For example, with three cylinders A, B, and C and the requirement to give the sequence A+ B+ C+ C− B− A−, we can incorporate a RESET input that will return all the

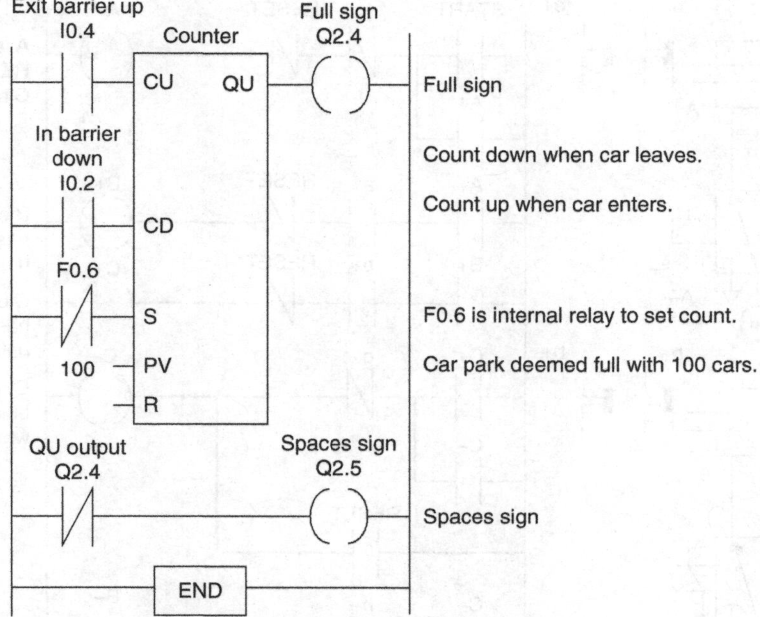

Figure 14.17: Car park with spaces or full.

cylinders to their unextended positions, that is, A−, B− and C−. Figure 14.18 shows such a program.

When a fault occurs, a possibly safer closing-down operation is to close the cylinders down in a particular sequence. Figure 14.19 shows a possible program for closing down the cylinders in the sequence C− B− A−.

14.3 Conveyor Belt Control

Consider a program that is used to count the number of items put onto a conveyor belt from work cells and give an alert when the number reaches 100. This program might be part of a bigger program used to control a production unit. A proximity sensor can be used to sense when an item is put on the conveyor so that a 1 signal is produced each time. The program might thus be as shown in Figure 14.20, which uses the Allen-Bradley format.

A further possibility in this conveyor belt problem is that too many items must not be put on the belt at any one time. A program that might achieve this goal involves instituting a time delay after an item is put on the belt and before the next item can be loaded onto the belt. Figure 14.21 shows the program elements for that specification. When an item passes the proximity sensor, the on-delay timer is started, and only when that is completed can another item be loaded.

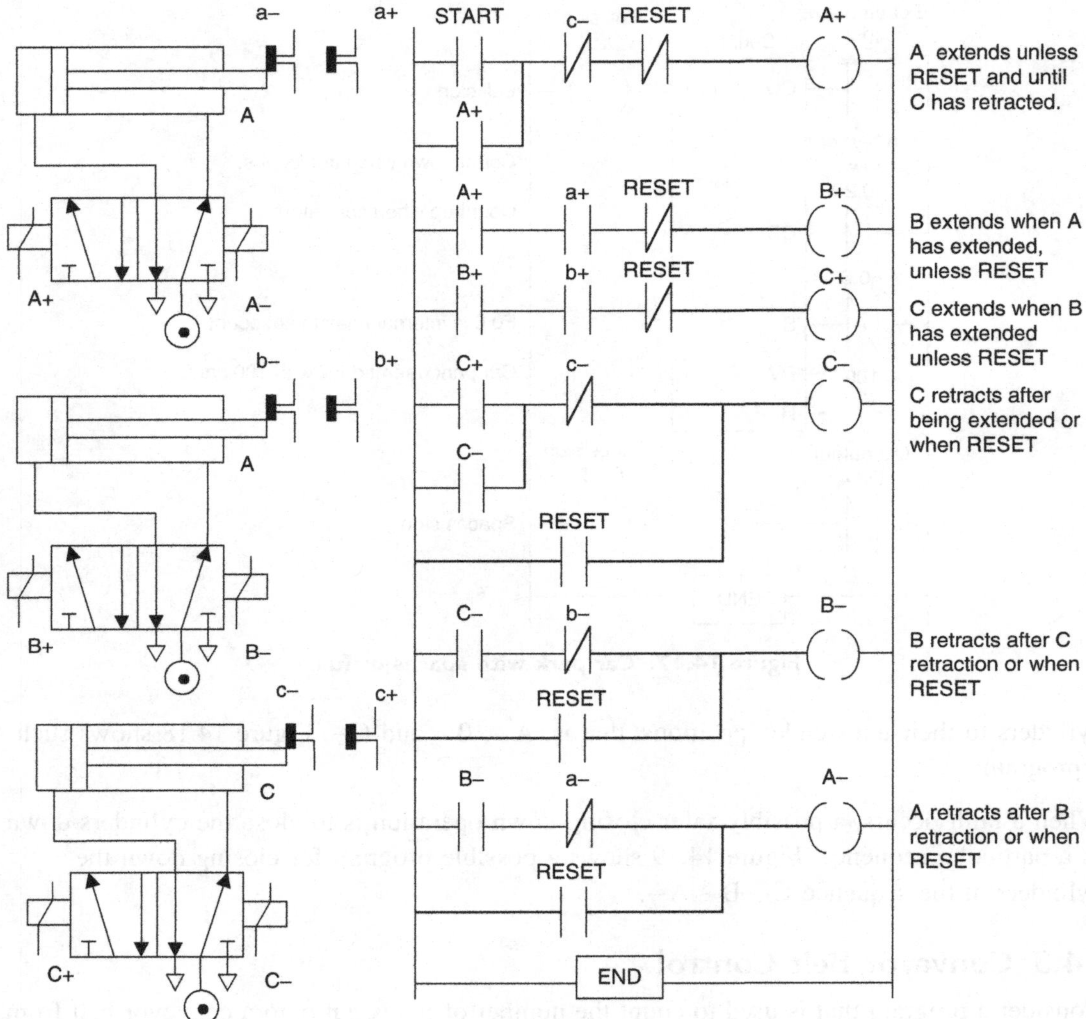

Figure 14.18: All cylinders simultaneously reset.

14.3.1 Bottle Packing

Consider a production-line problem involving a conveyor being used to transport bottles to a packaging unit, with the items being loaded onto the conveyor, checked to ensure they are full and capped, and then the correct number of bottles (four) being packed in a container. The required control actions are thus: If a bottle is not full, the conveyor is stopped; the capping machine is activated when a bottle is at the required position, the conveyor being stopped during this time; count four bottles and activate the packing machine, with the

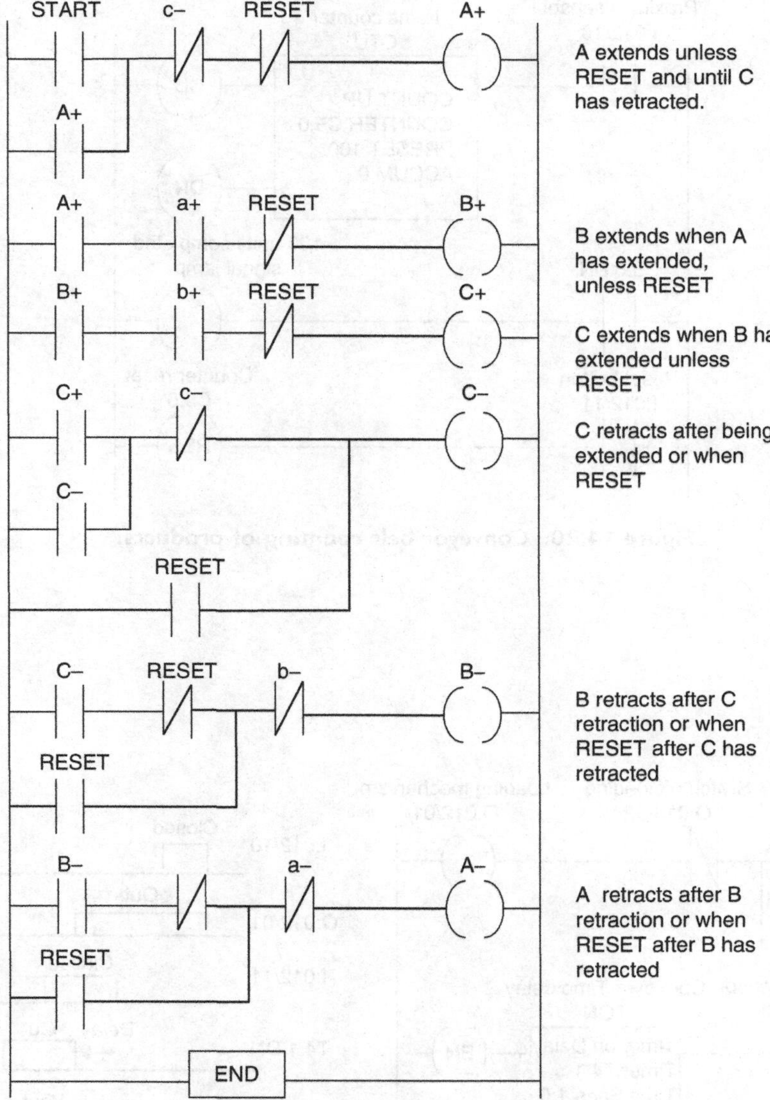

Figure 14.19: Reset in order C– B– A–.

conveyor stopped if another bottle comes to the packing point at that time; and sound an alarm when the conveyor is stopped.

The detection of whether a bottle is full could be done with a photoelectric sensor that could then be used to activate a switch (X402/I0.2 input). The presence of a bottle for the capping machine could also be by means of a photoelectric sensor (X403/I0.3 input). The input to the counter to detect the four bottles could be also from a photoelectric sensor

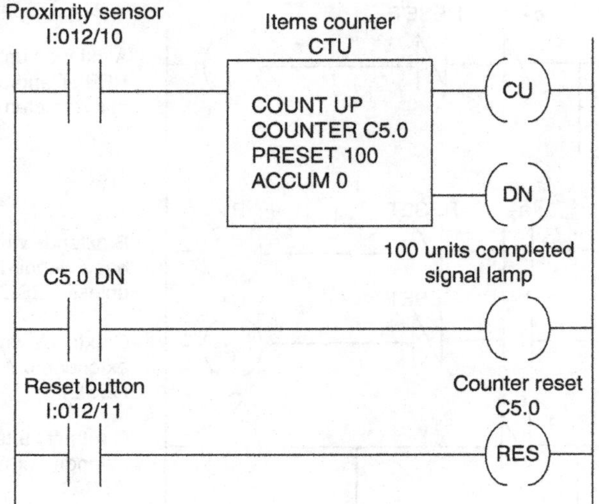

Figure 14.20: Conveyor belt counting of products.

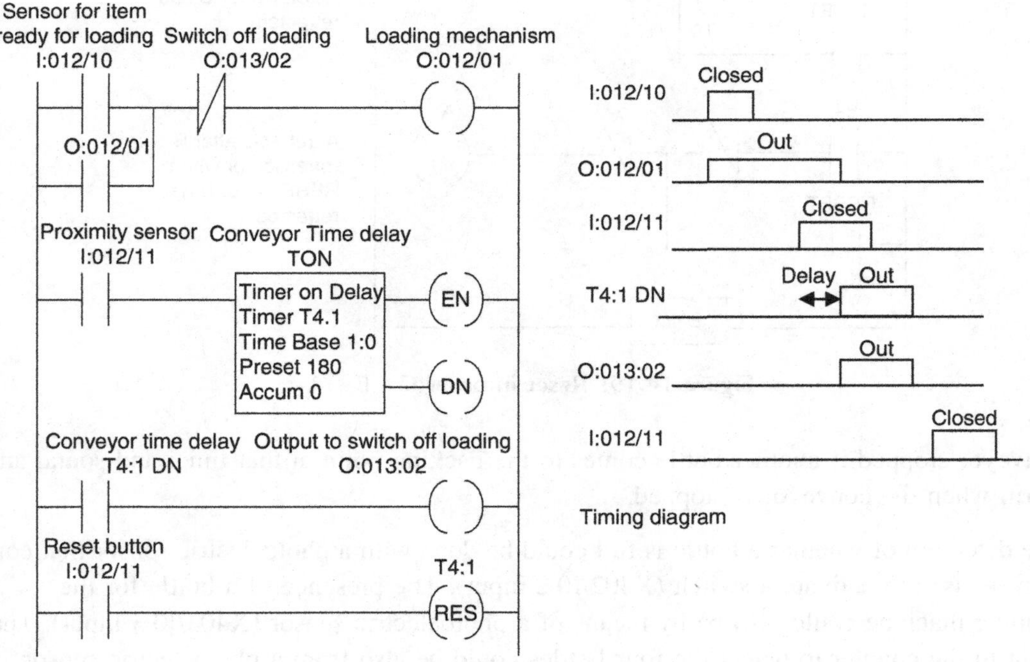

Figure 14.21: Conveyor belt time delays.

(X404/I0.4 input). The other inputs could be start (X400/I0.0 input) and stop (X401/I0.1 input) switches for the conveyor and a signal (X405/I0.5 input) from the packaging machine as to when it is operating and has received four bottles and so is not ready for any further caps. Figures 14.22 and 14.23 show a possible ladder program that could be used in Mitsubishi format and in Siemens format, respectively.

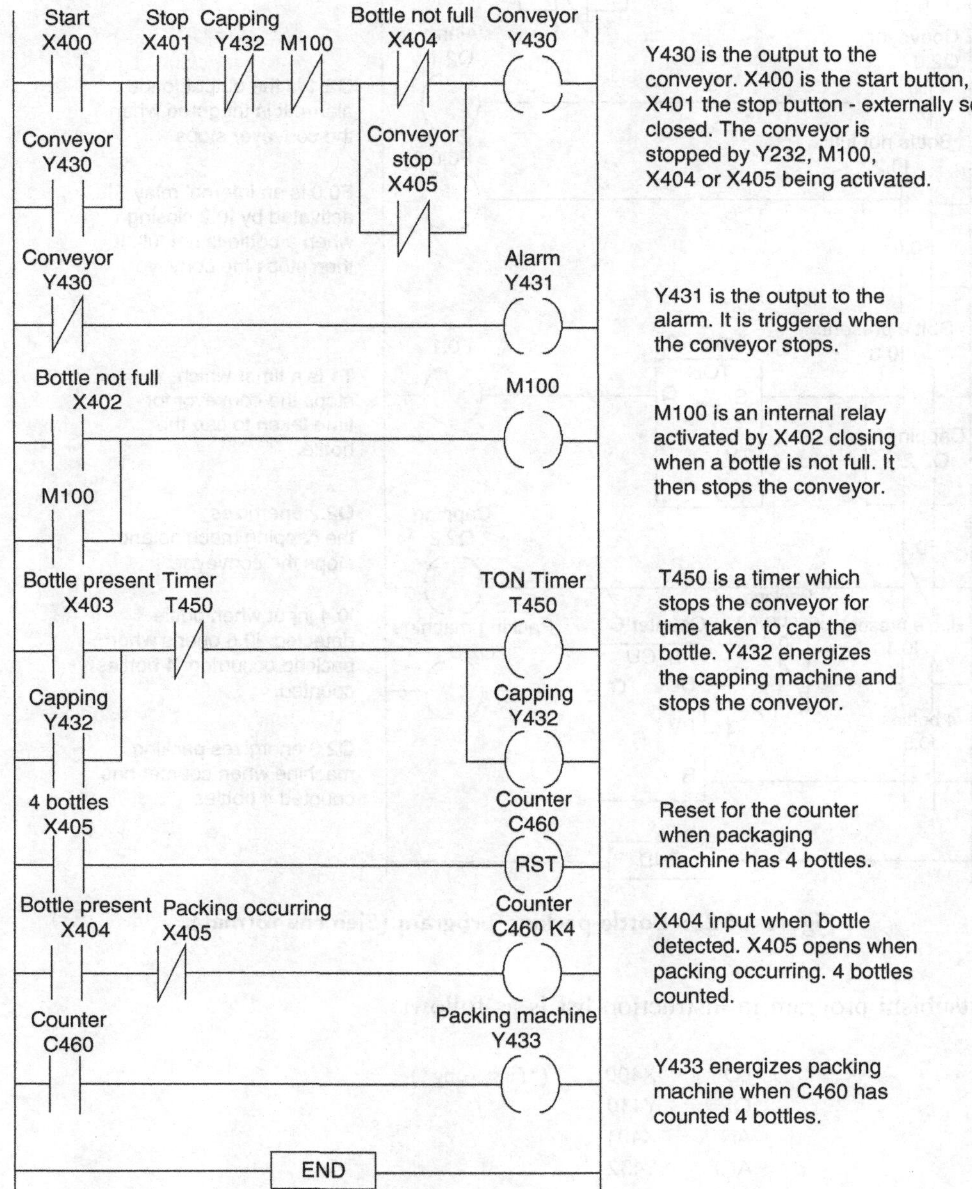
Start X400
Stop X401
Capping Y432
M100
Bottle not full X404
Conveyor Y430
Y430 is the output to the conveyor. X400 is the start button, X401 the stop button - externally set closed. The conveyor is stopped by Y232, M100, X404 or X405 being activated.
Conveyor Y430
Conveyor stop X405
Conveyor Y430
Alarm Y431
Y431 is the output to the alarm. It is triggered when the conveyor stops.
Bottle not full X402
M100
M100 is an internal relay activated by X402 closing when a bottle is not full. It then stops the conveyor.
M100
Bottle present X403
Timer T450
TON Timer T450
T450 is a timer which stops the conveyor for time taken to cap the bottle. Y432 energizes the capping machine and stops the conveyor.
Capping Y432
Capping Y432
4 bottles X405
Counter C460
RST
Reset for the counter when packaging machine has 4 bottles.
Bottle present X404
Packing occurring X405
Counter C460 K4
X404 input when bottle detected. X405 opens when packing occurring. 4 bottles counted.
Counter C460
Packing machine Y433
Y433 energizes packing machine when C460 has counted 4 bottles.
END

Figure 14.22: Bottle-packing program (Mitsubishi format).

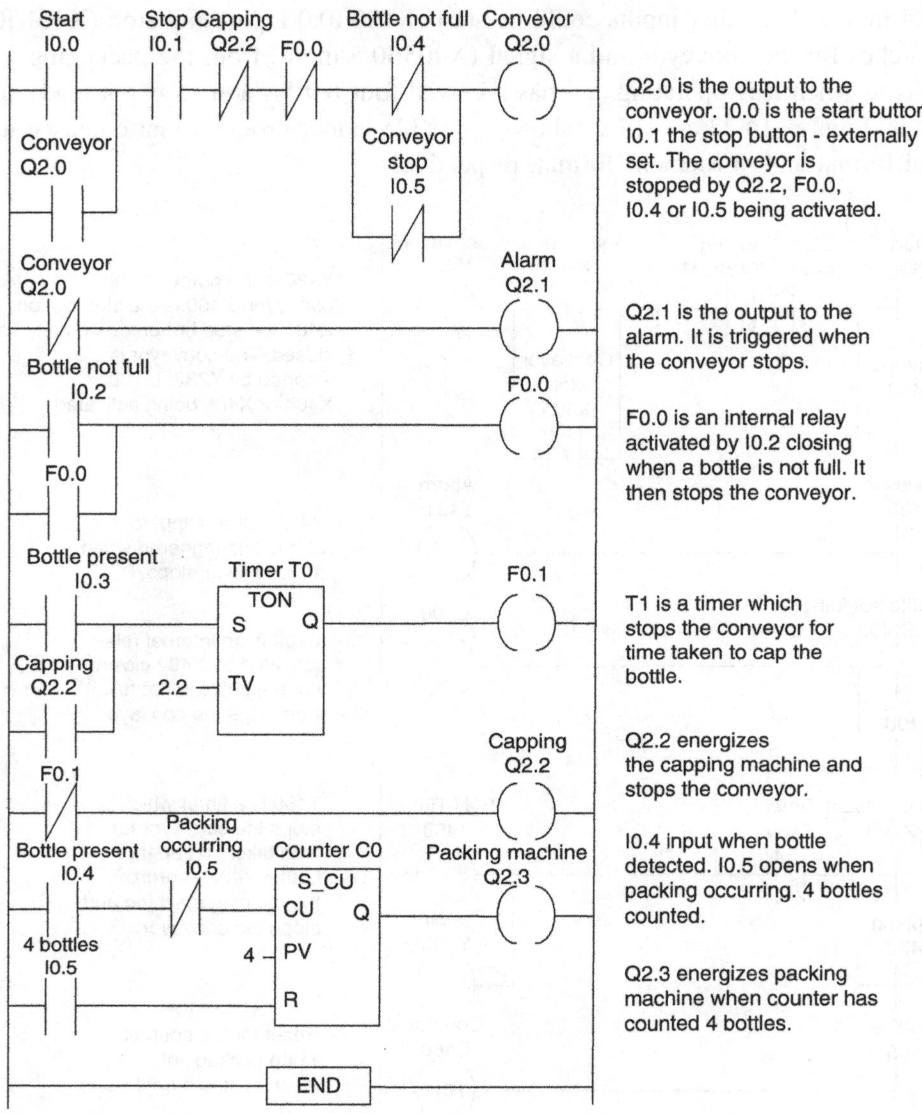

Figure 14.23: Bottle-packing program (Siemens format).

The Mitsubishi program in instruction list is as follows:

```
LD    X400    (*First rung*)
OR    Y430
AN    X401
ANI   Y432
ANI   M100
LDI   X404
```

```
ORI     X405
ANB
OUT     Y430
LDI     Y430    (*Second rung*)
OUT     Y431
LD      X402    (*Third rung*)
OR      M100
OUT     M100
LD      X403    (*Fourth rung*)
OR      Y432
ANI     T450
OUT     T450
K       2       (*2 s allowed for capping*)
OUT     Y432
LD      X405    (*Fifth rung*)
RST     C460
LD      X404    (*Sixth rung*)
ANI     X405
OUT     C460
K       4       (*Four bottles counted*)
LD      C460    (*Seventh rung*)
OUT     Y433
END             (*End rung*)
```

The Siemens program in instruction list is:

```
A       I0.0    (*First rung*)
O       Q2.0
A       I0.1
AN      Q2.2
AN      F0.0
(AN     I0.4
ON      I0.5
)
=       Q2.0
AN      Q2.0    (*Second rung*)
=       Q2.1
A       I0.2    (*Third rung*)
O       F0.0
=       F0.0
A       I0.3    (*Fourth rung*)
O       Q2.2
LKT     2.2     (*2 s allowed for capping*)
SR      T0
A       T0
=       F0.1
AN      F0.1    (*Fifth rung*)
=       Q2.2
```

```
A       I0.4    (*Sixth rung*)
AN      I0.5
CU      C0
LKC     4       (*Four bottles counted*)
A       I0.5
R       C0
=       Q2.3
END             (*End rung*)
```

14.4 Control of a Process

The following is an illustration of the use of a sequential flow chart for programming. The process (Figure 14.24a) involves two fluids filling two containers: When the containers are full, their contents are then emptied into a mixing chamber, from which the mixture is then discharged. The whole process is then repeated. Figure 14.24b shows the type of valve that might be used in such a process. It is solenoid operated to give flow through the valve, and then, when the solenoid is not activated, a spring returns the valve to the closed position.

Figure 14.25 shows the sequential function chart program. When the start switch is activated, fill 1 and fill 2 occur simultaneously as a result of the actions of pumps 1 and 2 being switched on. When limit switch 1 is activated, fill 1 ceases; likewise, when limit switch 3 is activated, fill 2 ceases. We then have the containers for fluid 1 and fluid 2 full. The action that occurs when both limit switch 1 and 3 are activated is that the containers start to empty, the action being the opening of valves 1 and 2. When limit switches 2 and 4 are

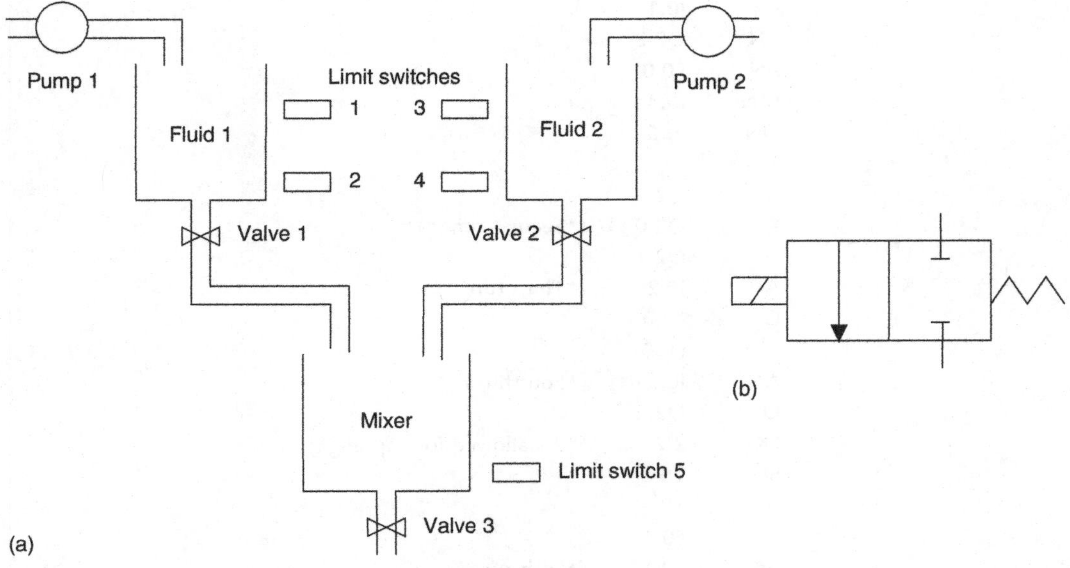

Figure 14.24: (a) The mixing operation, and (b) a valve.

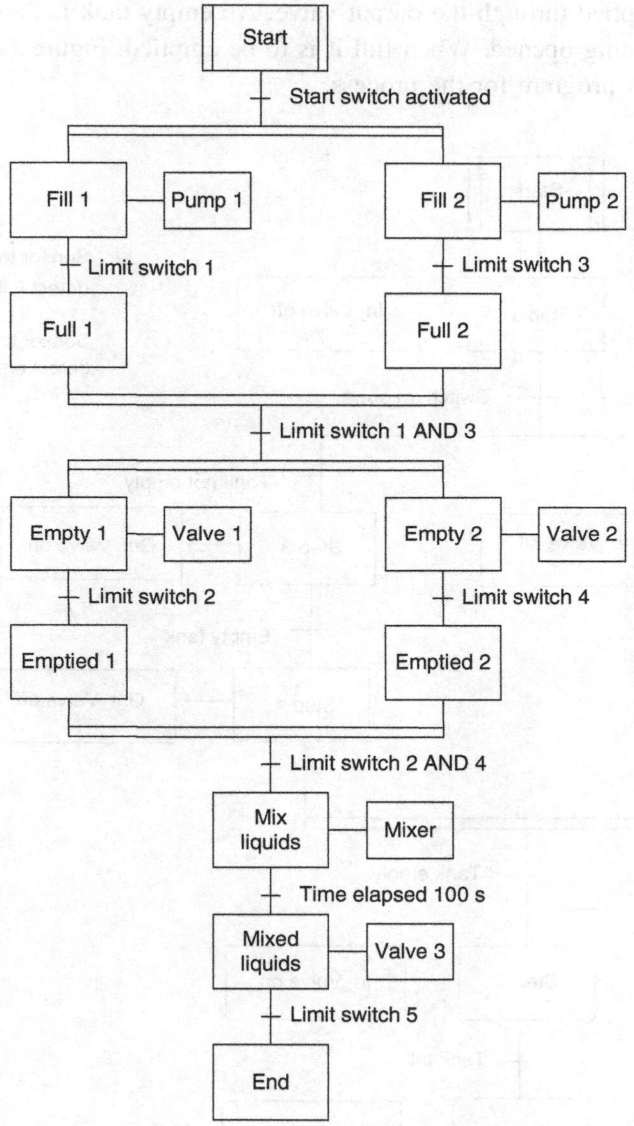

Figure 14.25: The mixing operation program.

activated, the containers are empty. The next stage, the mixing of the liquids, is then determined when limit switch 2 and limit switch 4 are both activated. After a time of 100 s, the mixing ceases and the mixed liquids empty through valve 3. When limit switch 5 is activated, the program reaches the end of its cycle and the entire sequence is then repeated.

As a further illustration, consider a process to control liquid in a storage tank that can be emptied by a valve. Before power is supplied, the valve on the pipe supplying the tank has to be closed. On power being supplied, the tank might be already full or empty.

If full it is to be emptied through the output valve. An empty tank is then to be filled by the input valve being opened. When full it is to be emptied. Figure 14.26 shows the sequential flow chart program for the process.

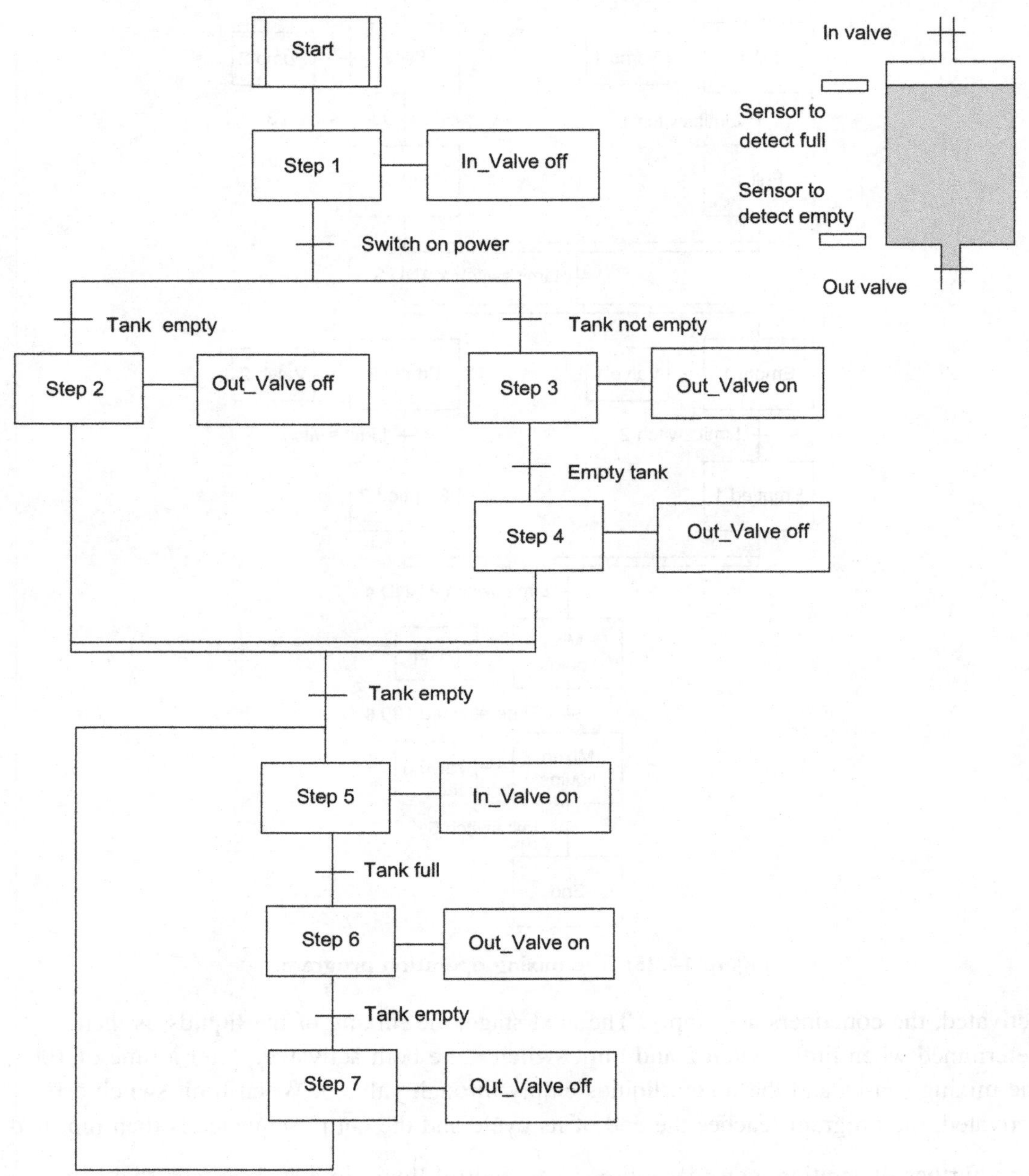

Figure 14.26: Filling of a storage tank.

14.5 A Selection Example: A Drinks Machine

Consider a drinks machine that allows the selection of tea or coffee, milk or no milk, sugar or no sugar, and will supply the required hot drink on the insertion of a coin. Figure 14.27 shows the function block diagram. There is an output from the first OR gate of a signal when tea or coffee is selected. The first AND gate gives an output when there has been a selection of tea or coffee and a coin has been inserted into the machine Finally, the output from this gate goes to the final AND gate which gives an output which combines hot water with the tea. Milk and sugar are optional additions which can occur after a coin has been inserted. Figure 14.28 shows a ladder program for the drinks machine.

14.6 A Data Comparison Example: A Fan Heater

Consider a simple example of the use of the data greater-than comparison. An electric heater is to heat up when switched on, and then after 10 s a fan starts and then after 30 s another fan starts. Figure 14.29 shows a possible ladder program and Figure 14.30 a functional block diagram.

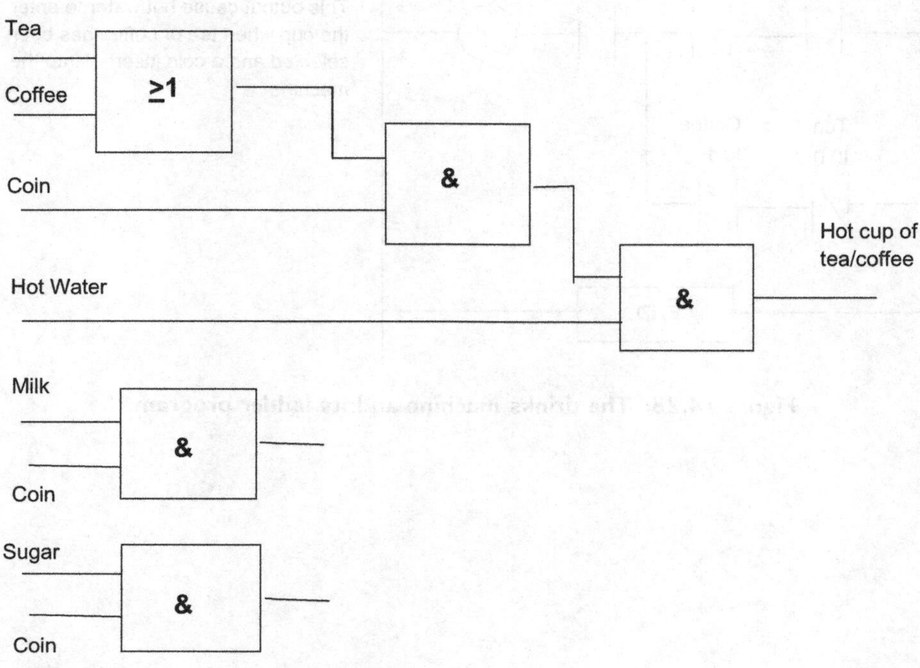

Figure 14.27: The drinks machine and its FBD program.

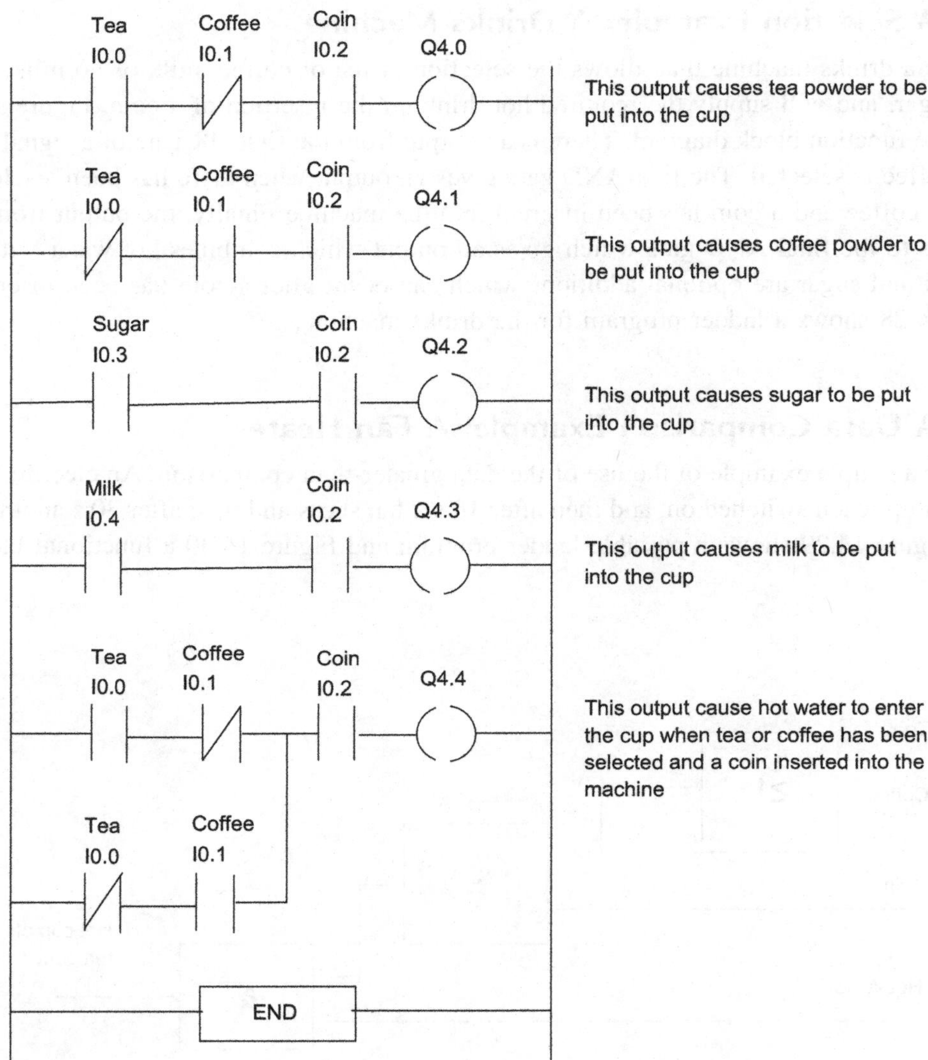

Figure 14.28: The drinks machine and its ladder program.

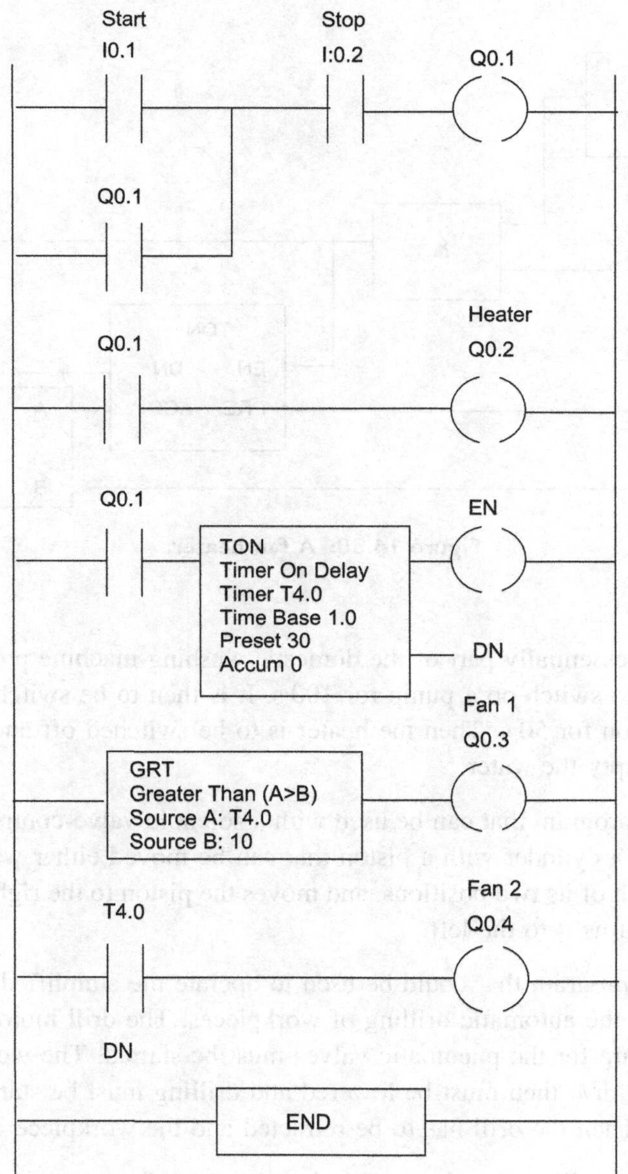

Figure 14.29: A fan heater.

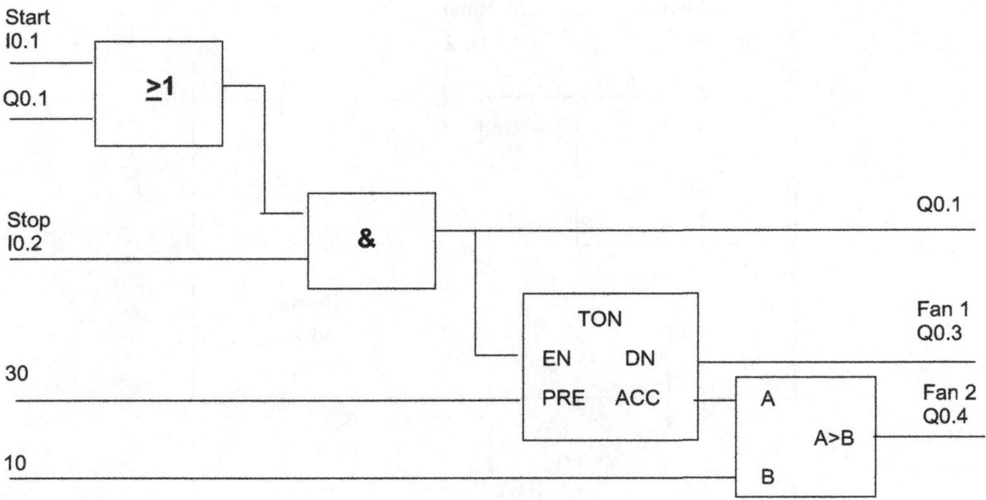

Figure 14.30: A fan heater.

Problems

1. This problem is essentially part of the domestic washing-machine program. Devise a ladder program to switch on a pump for 100 s. It is then to be switched off and a heater switched on for 50 s. Then the heater is to be switched off and another pump is to be used to empty the water.
2. Devise a ladder program that can be used with a solenoid valve-controlled double-acting cylinder, that is, a cylinder with a piston that can be moved either way by means of solenoids for each of its two positions, and moves the piston to the right, holds it there for 2 s, and then returns it to the left.
3. Devise a ladder program that could be used to operate the simplified task shown in Figure 14.31 for the automatic drilling of workpieces. The drill motor and the pump for the air pressure for the pneumatic valves must be started. The workpiece has to be clamped. The drill then must be lowered and drilling must be started to the required depth. Then the drill has to be retracted and the workpiece unclamped.
4. What are the principles to be observed in installing a safe emergency stop system with a PLC?
5. The inputs from the limit switches, the start switch, and the outputs to the solenoids of the valves shown in Figure 14.32a are connected to a PLC that has the ladder program shown in Figure 14.32b. What is the sequence of the cylinders?

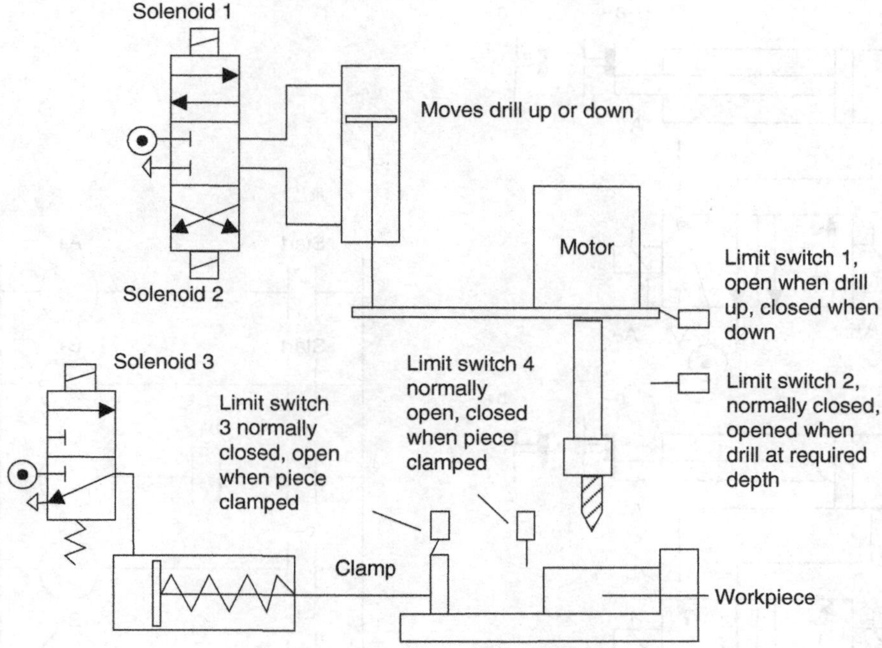

Figure 14.31: Diagram for Problem 3.

6. The inputs from the limit switches, the start switch, and the outputs to the solenoids of the valves shown in Figure 14.33a are connected to a PLC that has the ladder program shown in Figure 14.33b. What is the sequence of the cylinders?
7. Figure 14.34 shows a ladder program involving a counter C460, inputs X400 and X401, internal relays M100 and M101, and an output Y430. X400 is the start switch. Explain how the output Y430 is switched on.
8. Write a ladder program that will switch on two motors when the start switch is operated, then switch off one motor after 200 s and the other motor after a further 100 s. When both motors have been switched off, a third motor is to be switched on for 50 s. The cycle is then to repeat itself unless a stop switch has been activated.
9. Write a ladder program to switch on a motor when the start switch is momentarily activated, with the motor remaining on for 50 s. At the end of that time a second motor is to be switched on for a further 50 s. A third motor is to be switched on 10 s before the second motor switches off and is to remain on for 50 s. The cycle is then to repeat itself unless a stop switch has been activated.

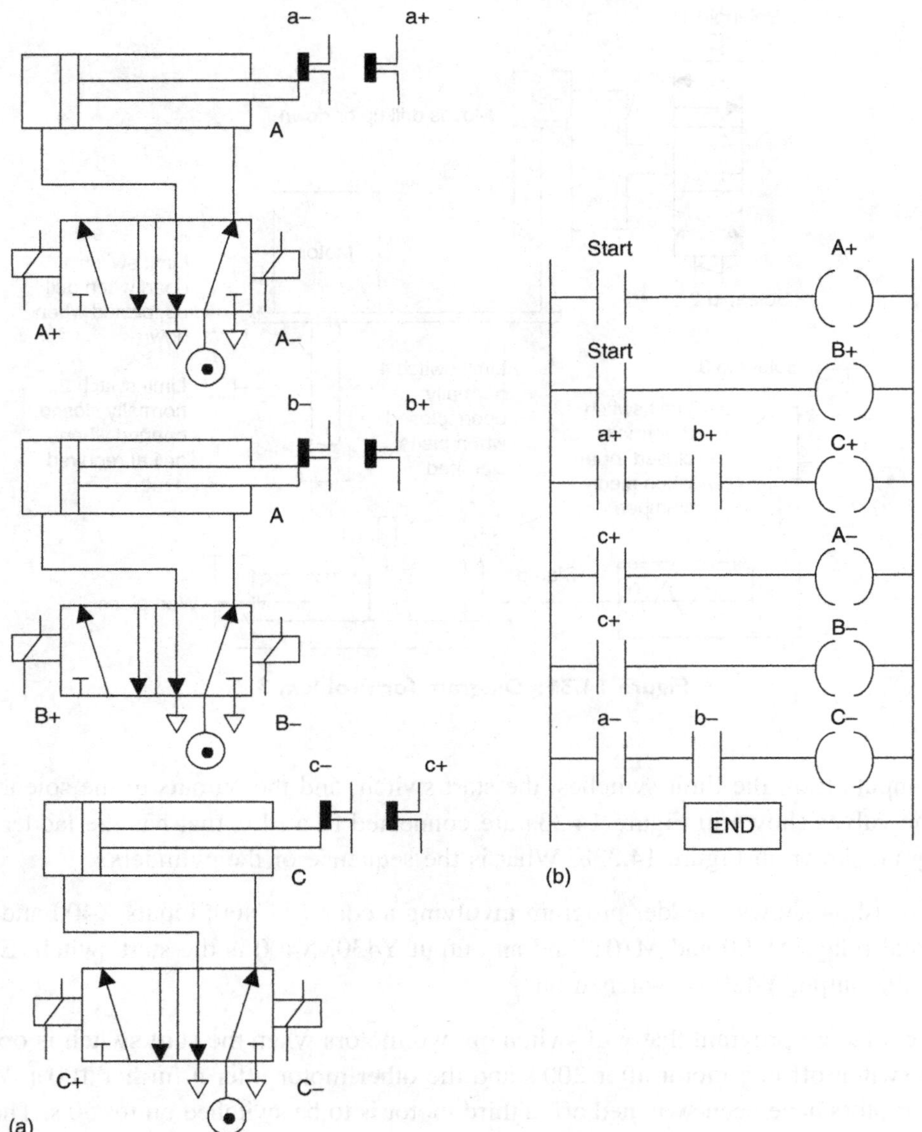

Figure 14.32: Diagram for Problem 5.

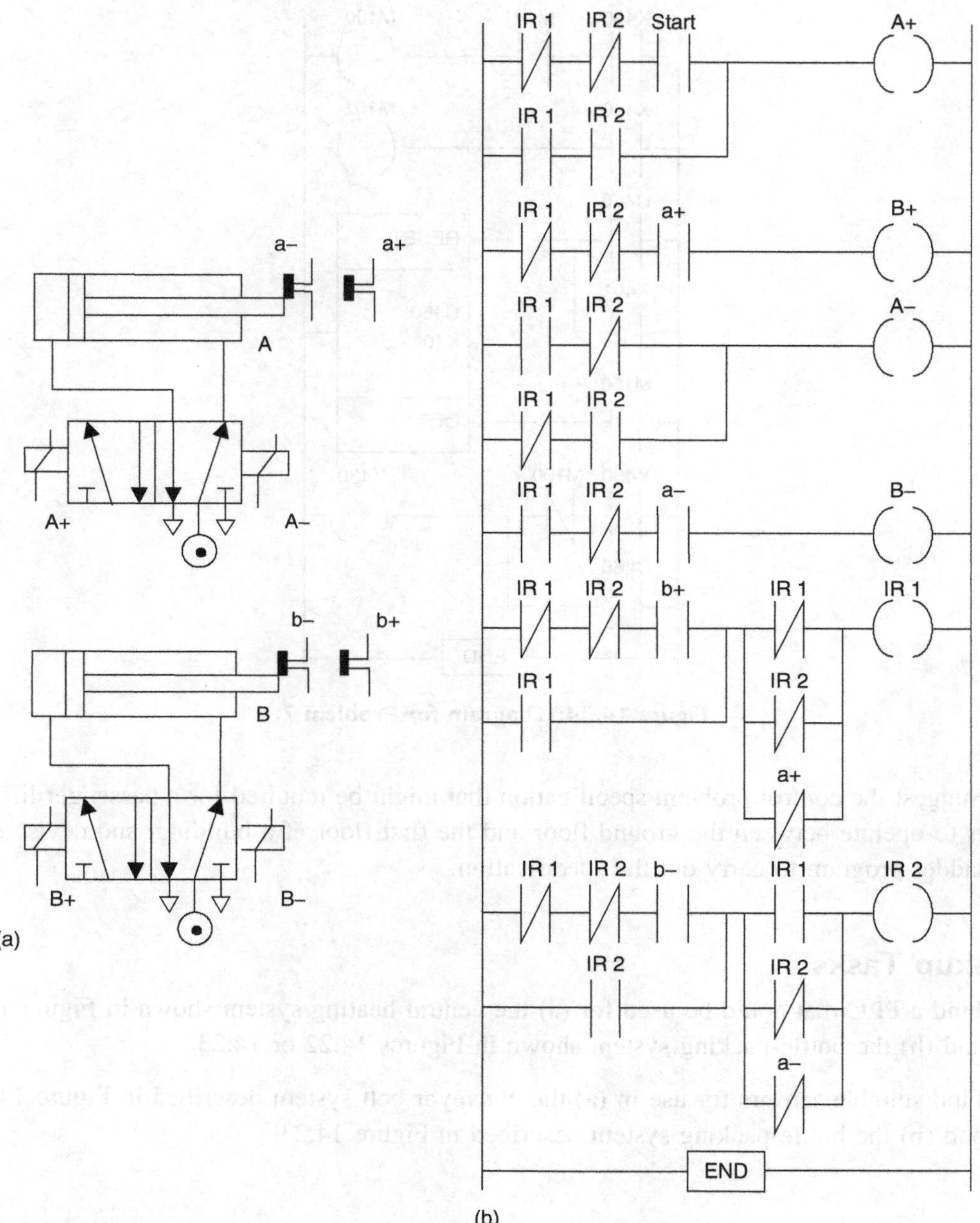

Figure 14.33: Diagram for Problem 6.

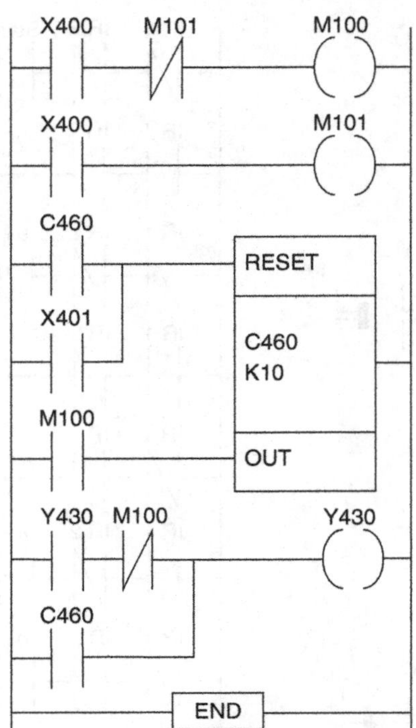

Figure 14.34: Diagram for Problem 7.

10. Suggest the control problem specification that might be required for a passenger lift that is to operate between the ground floor and the first floor of a building, and devise a ladder program to carry out the specification.

Lookup Tasks

11. Find a PLC that could be used for (a) the central heating system shown in Figure 14.7 and (b) the bottle-packing system shown in Figures 14.22 or 14.23.

12. Find suitable sensors for use in (a) the conveyor belt system described in Figure 14.20 and (b) the bottle-packing system described in Figure 14.23.

Appendix: Symbols

Ladder Programs

	Semi-graphic form	*Full graphic form*
A horizontal link along which power can flow	---------	─────────
Interconnection of horizontal and vertical power flows	---+---	──┤
Left-hand power connection of a ladder rung	+----	├────
Right-hand power connection of a ladder rung	----+	────┤
Normally open contact	---| |---	──| |──
Normally closed contact	---|/|---	──|/|──
Positive transition-sensing contact, power flow occurs when associated variable changes from 0 to 1.	---|P|---	──|P|──
Negative tranistion-sensing contact, power flow occurs when assoaciated variable changes from 1 to 0	---|N|---	──|N|──
Output coil: if the power flow to it is on then the coil state is on	---( )---	──( )──
Set coil	---(S)---	──(S)──
Reset coil	---(R)---	──(R)──
Retentive memory coil, the state of the associated variable is retained on PLC power fail	---(M)---	──(M)──

W. Bolton: Programmable Logic Controllers, Sixth Edition. http://dx.doi.org/10.1016/B978-0-12-802929-9.09993-3

Function Blocks

	Semi-graphic form	*Full graphic form*
Horizontal and vertical lines		
Interconnection of horizontal and vertical signal flows		
Crossing horizontal and vertical signal flow		
Blocks with connections		
Connectors	---> AV_WEIGHT > > AV_WEIGHT >---	⟶ AV_WEIGHT> > AV_WEIGHT ⟶

Commonly Encountered Blocks

BOOL is a Boolean signal, INT is an integer, REAL is a floating point number, ANY is any form of signal

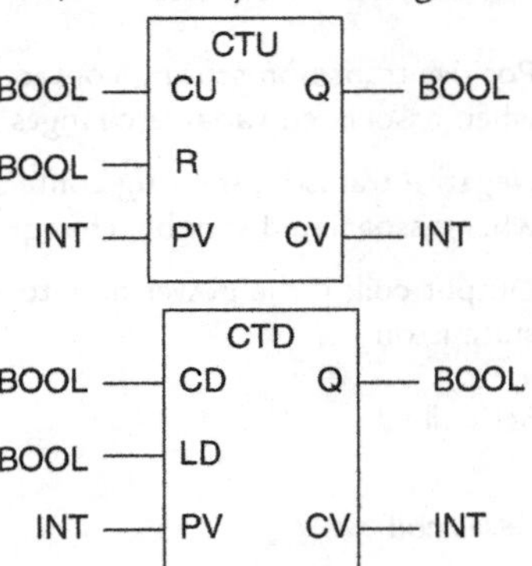

Up-counter counts the number of rising edges at input CU. PV defines the maximum value of the counter. Each new rising edge at CU increments CV by 1. Output Q occurs after set count. R is the reset.

Down-counter counts down the number of rising edges at input CU. PV defines the starting value of the counter. Each new rising edge at CU decrements CV by 1. Output Q occurs when count reaches zero.

Up-down counter. It can be used to count up on one input and down on the other.

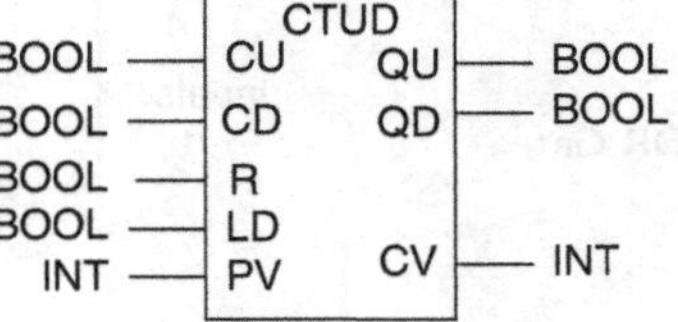

On-delay timer. When input IN goes true, the elapsed time at about ET starts to increase and when it reaches the set time, specified by input PT, the output Q goes true.

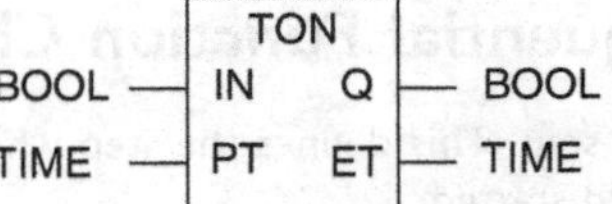

Off-delay timer. When input IN goes true, the output Q follows and remains true for the set time after which the input Q goes false.

Pulse timer. When input IN goes true, output Q follows and remains true for the pulse duration specified by input PT.

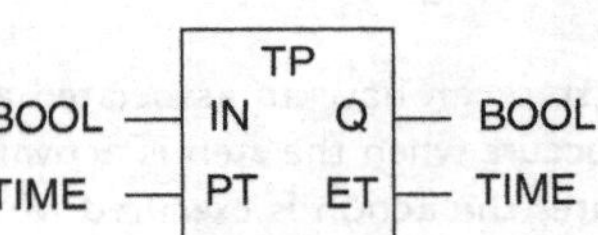

Logic Gates

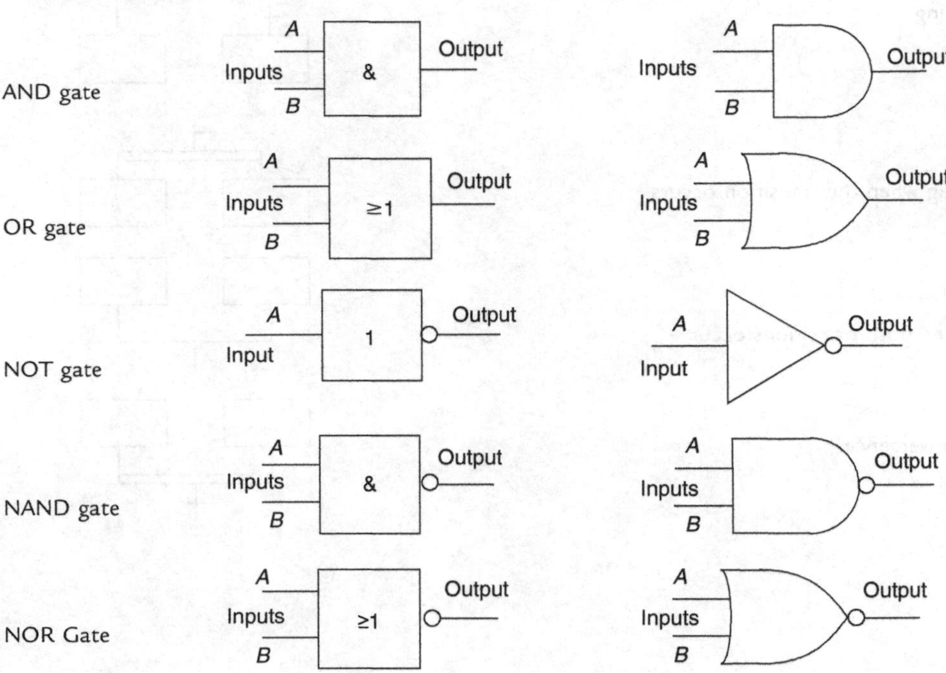

XOR Gate

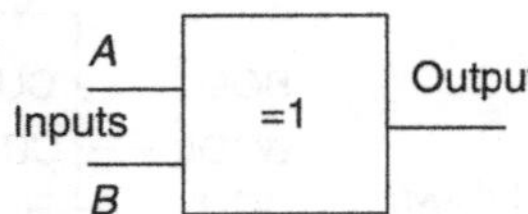

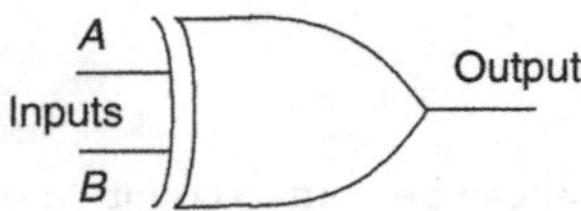

Sequential Function Charts

Start step. This defines the step which will be activated when the PLC is cold-started.

Transition condition. Every transition must have a condition. One that always occurs should be shown with the condition TRUE.

Step in a program

Every step can have an associated action. An action describes the behavior that occurs when the step is activated. Each action can have a qualifier: N indicates the action is executed while the step is active. If no qualifier is indicated it is taken to be N.

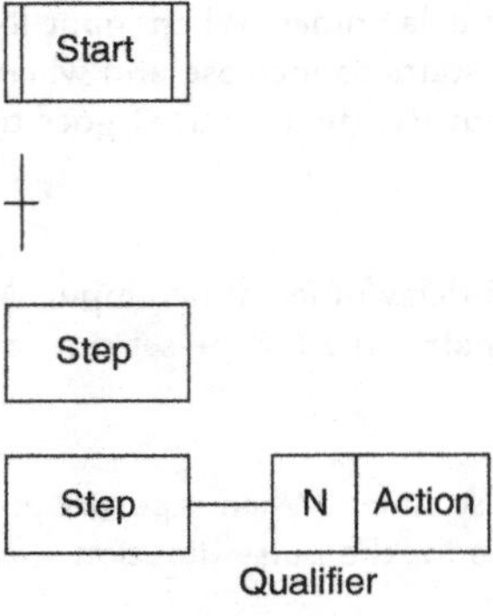

D: time-delayed action which starts after a given time.

Selective branching

Parallel branching when the transition occurs

Convergence when both transitions occur

Simultaneous convergence

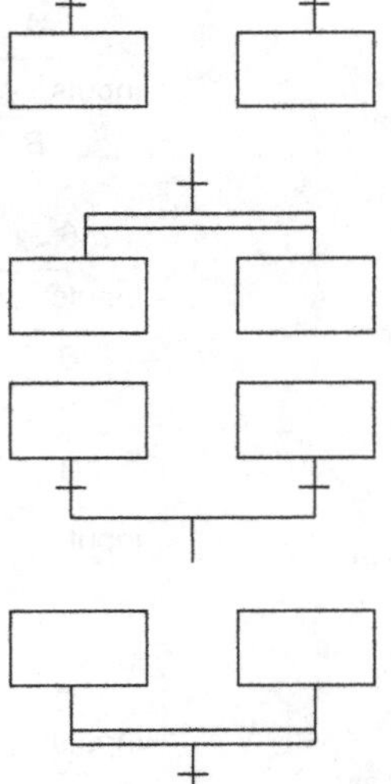

Instruction List (IEC 61131-3 Symbols)

LD Start a rung with an open contact

LDN Start a rung with a closed contact

ST An output

S Set true

R Reset false

AND Boolean AND

ANDN Boolean NAND

OR Boolean OR

ORN Boolean NOR

XOR Boolean XOR

NOT Boolean NOT

ADD Addition

SUB Subtraction

MUL Multiplication

DIV Division

Structured Text

X:= Y Y represents an expression that produces a new value for the variable X.

Operators

(. . .) Parenthesized (bracketed) expression

Function(. . .) List of parameters of a function

** Raising to a power

−, NOT Negation, Boolean NOT

*, /, MOD Multiplication, division, modulus operation

+, − Addition, subtraction

<, >, <=, >= Less than, greater than, less than or equal to, greater than or equal to

=, <> Equality, inequality

AND, & Boolean AND

XOR Boolean XOR

OR Boolean OR

Conditional and Iteration Statements

IF ... THEN ... ELSE is used when selected statements are to be executed when certain conditions occur.

The FOR ... DO iteration statement allows a set of statements to be repeated, depending on the value of the iteration integer variable.

The WHILE ... DO iteration statement allows one or more statements to be executed while a particular Boolean expression remains true.

The REPEAT ... UNTIL iteration statement allows one or more statements to be executed and repeated while a particular Boolean expression remains true.

Answers

Chapter 1

1. D.
2. A.
3. C.
4. A.
5. A.
6. C.
7. See Figure 1.4.
8. See Section 1.3.1.
9. 2 × 1024.
10. See Section 1.3.2 for an explanation of sourcing and sinking and 1.3.1 and Figure 1.15 for relay and transistor outputs.

Chapter 2

1. A.
2. A.
3. B.
4. D.
5. C.
6. A.
7. A.
8. A.

W. Bolton: Programmable Logic Controllers, Sixth Edition. http://dx.doi.org/10.1016/B978-0-12-802929-9.09994-5

9. B.
10. B.
11. C.
12. B.
13. C.
14. C.
15. See (a) Figure 2.6, (b) Section 2.1.4, (c) Section 2.1.3, (d) Section 2.1.8.
16. See Section 2.2.3.
17. See Section 2.2.4.
18. For example, (a) photoelectric transmissive system, (b) capacitive proximity sensor, (c) mechanical limit switch, (d) inductive proximity sensor.
19. See Section 2.2.4. Consider the behavior of *RL* circuits.
20. Stepper motor with 5° step.
21. (a) Photoelectric transmissive system, (b) direction control valve operated cylinder.

Chapter 3

1. (a) 3, (b) 63, (c) 13.
2. (a) 110 0100, (b) 1001 0010, (c) 1111 1111.
3. (a) 159, (b) 3411, (c) 1660.
4. (a) E, (b) 51, (c) A02.
5. (a) 1110, (b) 11101, (c) 1010 0110 0101.
6. (a) 250, (b) 12, (c) 1376.
7. (a) 24, (b) 411, (c) 620.
8. (a) 010 111 000, (b) 001 000 010, (c) 110 111 011.
9. (a) 0010 0000, (b) 0011 0101, (c) 1001 0010.
10. (a) 1111 1111, (b) 1101 1101, (c) 1000 0011.
11. (a) −16, (b) −55, (c) −40.

12. (a) 0.110010×2^{-3}, (b) 0.1100×2^{-4}, (c) $0.1000\ 0100 \times 2^{4}$.
13. See Sections 3.7 and 3.8.
14. (a) 1 AND 1, (b) 1 OR 1, (c) 1 AND NOT 1.
15. D 1, CLK 1.
16. It is a D latch as in Table 3.4.

Chapter 4

1. C.
2. D.
3. C.
4. B.
5. B.
6. A.
7. B.
8. A.
9. C.
10. C.
11. A.
12. C.
13. A.
14. D.
15. D.
16. (a) 0, (b) 1.
17. To detect message corruption.
18. See Section 4.5.
19. Input 1 kΩ, output 100 kΩ.
20. See Section 4.4.

21. See Sections (a) 4.5.5, (b) 4.5.2 and 4.5.3.
22. See Section 4.4.

Chapter 5

1. A.
2. D.
3. B.
4. B.
5. B.
6. B.
7. B.
8. D.
9. C.
10. A.
11. A.
12. B.
13. D.
14. C.
15. B.
16. A.
17. C.
18. C.
19. D.
20. C.
21. A.
22. C.
23. A.

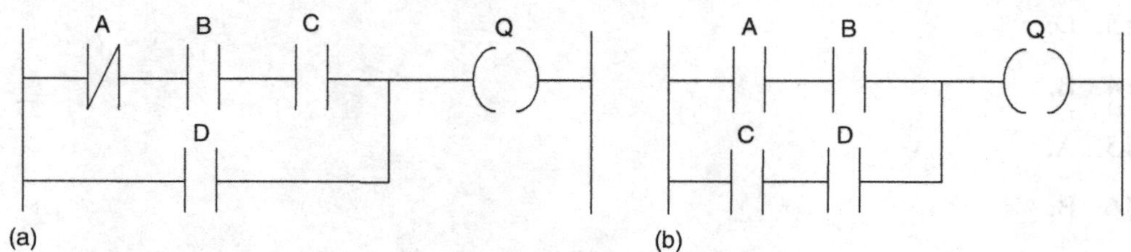

Figure A.1: Chapter 5, Problem 27.

24. D.
25. See (a) Figure 5.8, (b) Figure 5.10, (c) Figure 5.19, (d) Figure 5.10, (e) Figure 5.11, (f) Figure 5.5(a), (g) an AND system as in Figure 5.8.
26. (a) An OR gate as in Figure 5.28, (b) as in Figure 5.30, (c) an OR gate as in Figure 5.28.
27. (a) As in Figure 5.33, (b) see Figure A.1(a), (c) see Figure A.1(b).
28. (a) $Q = A + B$, (b) $Q = A \cdot B \cdot \bar{C}$, (c) $Q = A \cdot \bar{B}$.

Chapter 6

1. C.
2. A.
3. B.
4. D.
5. A.
6. C.
7. A.
8. B.
9. D.
10. A.
11. A.
12. C.

13. D.
14. B.
15. A.
16. B.
17. A.
18. C.
19. A.
20. D.
21. C.
22. A.
23. C.
24. B.
25. See Figure A.2.
26.
```
WHILE NOT (Level_switch1 AND Drain_valve)
Valve1 :=1
END_WHILE
```

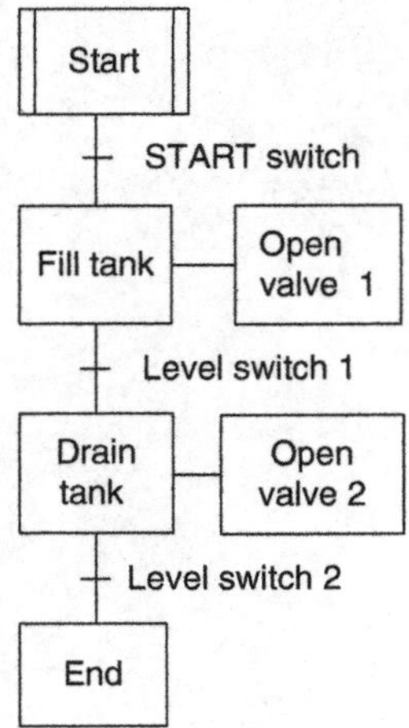

Figure A.2: Chapter 6, Problem 25.

27.
```
CASE temperature_setting OF
Furnace_switch :=1;
1 : temp :=40;
2 : temp :=50;
3 ; temp :=60; fan1 :=1;
4 : temp :=70; fan2 :=1;
ELSE
Furnace_switch :=0;
END_Case
```

Chapter 7

1. D.
2. B.
3. C.
4. A.
5. C.
6. C.
7. C.
8. B.
9. A.
10. A.
11. C.
12. D.
13. B.
14. B.
15. B.
16. C.
17. A.
18. A.

19. A.
20. B.
21. A.
22. A.
23. A.
23. B.
24. See (a) Figure 7.8, (b) Figure 7.9 or 7.10, (c) Figure 7.22.

Chapter 8

1. C
2. B
3. A
4. B
5. B
6. A
7. Call and return subroutines, which are blocks of program code; see Section 8.2.

Chapter 9

1. C.
2. A.
3. D.
4. D.
5. D.
6. D.
7. C.
8. C.
9. B.

10. C.
11. A.
12. A.
13. A.
14. B.
15. D.
16. B.
17. A.
18. D.
19. C.
20. D.
21. B.
22. See (a) Figure 9.4, (b) Figure 9.10, (c) Figure 9.12, (d) see Figure A.3.

Chapter 10

1. C.
2. A.
3. C.

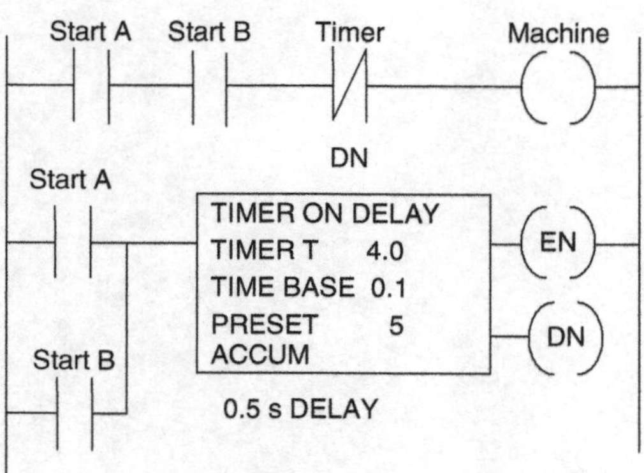

Figure A.3: Chapter 9, Problem 22d.

4. B.
5. B.
6. B.
7. B.
8. D.
9. C.
10. A.
11. A.
12. B.
13. B.
14. C.
15. B.
16. D.
17. C.
18. C.
19. A.
20. See (a) Figure 10.4, (b) Figure 10.10, (c) see Figure A.4(a), (d) see Figure A.4(b), (e) similar to Figure 10.10 or 10.11.

Chapter 11

1. D.
2. C.
3. C.
4. D.
5. A.
6. A.
7. C.

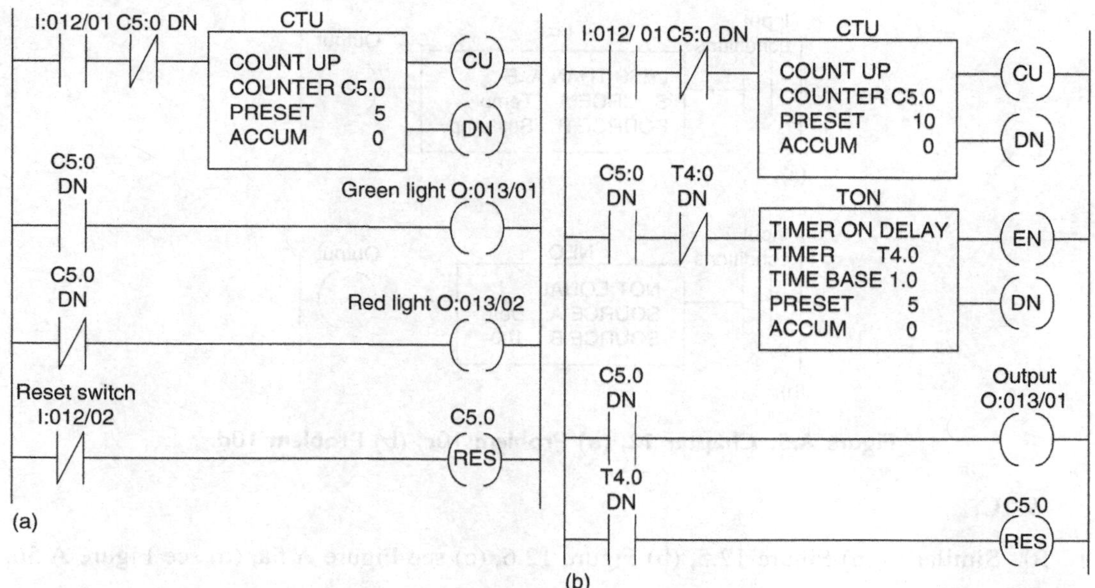

Figure A.4: Chapter 10, (a) Problem 20c, (b) Problem 20d.

8. C.
9. D.
10. (a) As Figure 11.1/11.2 with a constant input to In 1/X400, so entering a 1 at each shift, (b) as in Figure 11.3 but instead of a faulty item, a hook with an item, and instead of a good item, hooks with no items.

Chapter 12

1. C.
2. C.
3. B.
4. B.
5. A.
6. A.
7. B.
8. B.

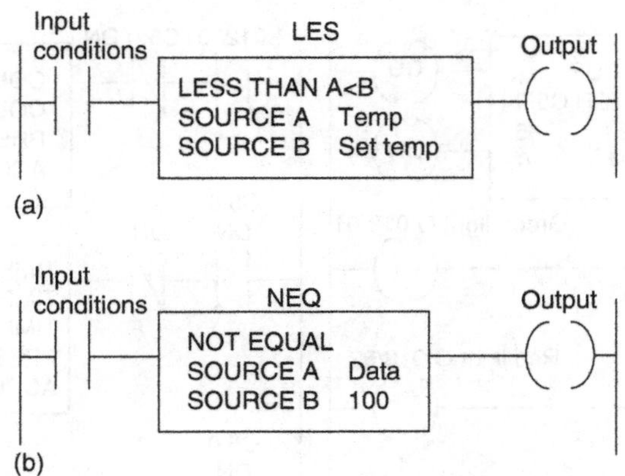

Figure A.5: Chapter 12, (a) Problem 10c, (b) Problem 10d.

9. C.
10. Similar to (a) Figure 12.5, (b) Figure 12.6, (c) see Figure A.5a, (d) see Figure A.5b.
11. See Section 12.4.
12. See Section 12.4.

Chapter 13

1. B.
2. B.
3. C.
4. B.
5. C.
6. A.
7. A.
8. See Section 13.3.1.
9. Power failure, supply off, power tripped.
10. Wiring fault, device fault.
11. See Figure A.6.
12. See Figure A.7.

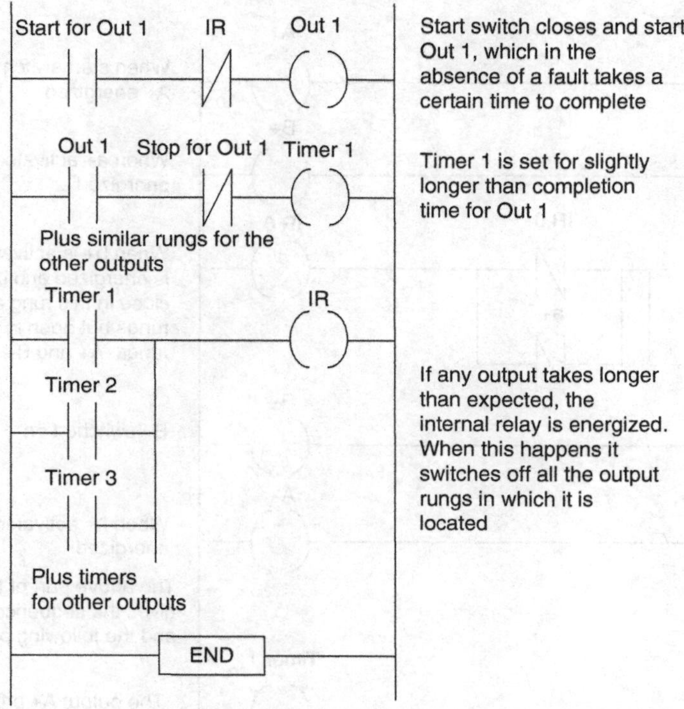

Figure A.6: Chapter 13, Problem 11.

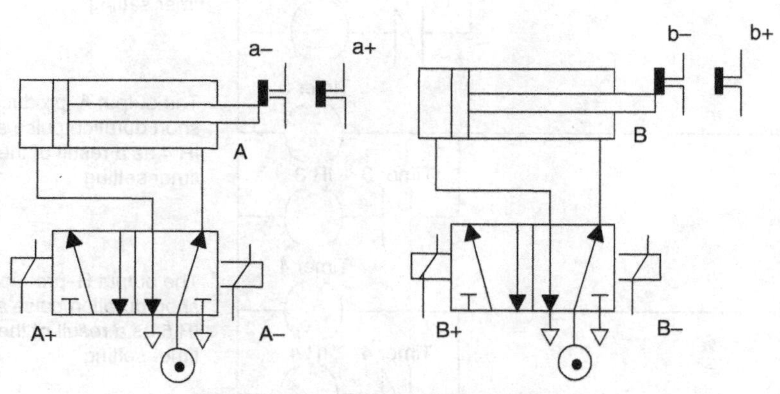

Figure A.7: Chapter 13, Problem 12.

(Continued)

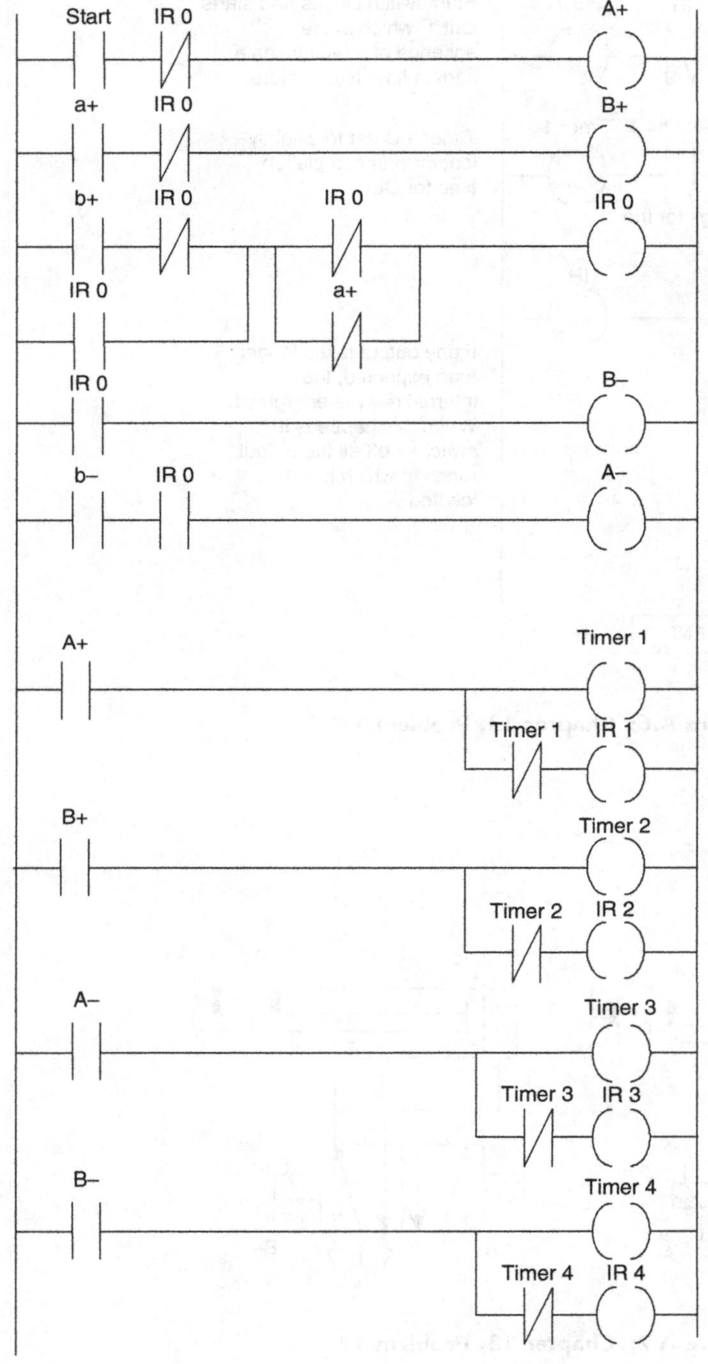

When start switch closed, A+ energized

When a+ activated, B+ energized

When b+ is activated, IR 0 is energized and its contacts close in this rung and later rungs but open in earlier rungs. A+ and B+ switched off

B– switched on

When b– activated, A– energized

The above part of the program gives the sequence A+, B+, B–, A– and the following part the diagnostics

The output A+ produces a short duration pulse at IR 1 as a result of the timer setting

The output B+ produces a short duration pulse at IR 2 as a result of the timer setting

The output A–produces a short duration pulse at IR 4 as a result of the timer setting

The output B–produces a short duration pulse at IR 5 as a result of the timer setting

Figure A.7: Cont'd

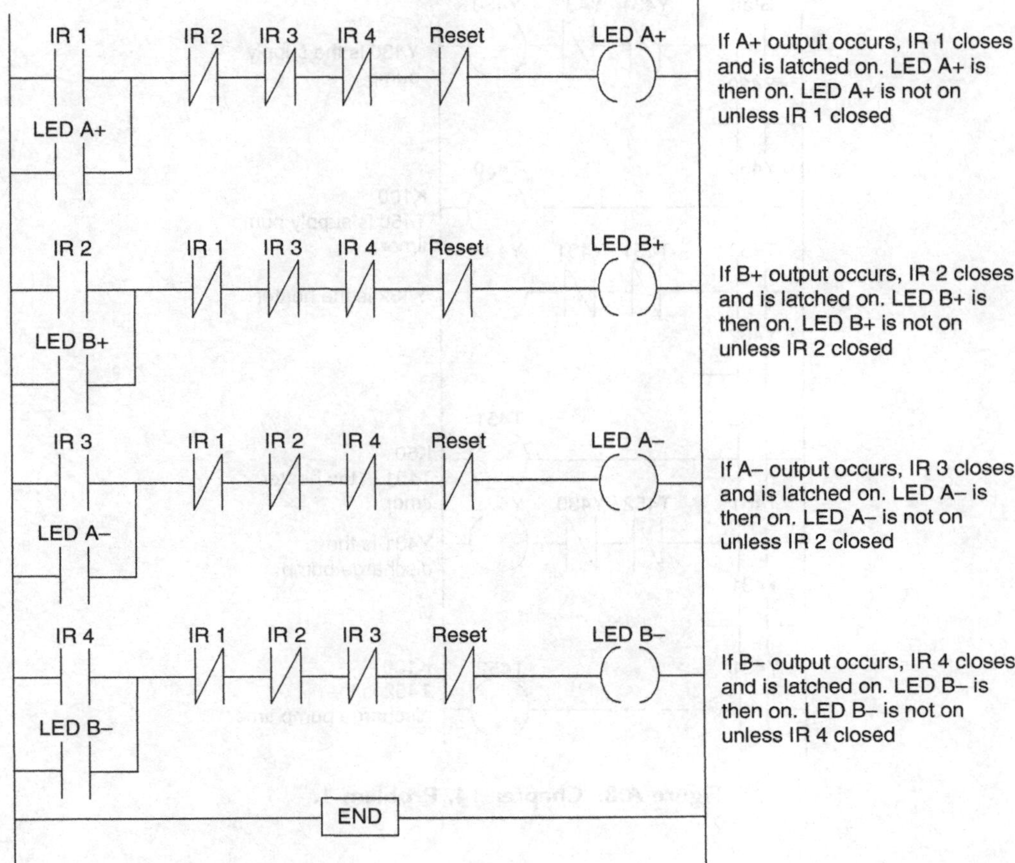

Figure A.7: Cont'd

Chapter 14

1. See Figure A.8.
2. See Figure A.9.
3. See Figure A.10 for a basic answer.
4. Hardwired emergency stop button, not dependent on software.
5. A+ and B+, C+, A– and B–, C–.
6. A+, B+, A–, B–, A+, A–.
7. M100 and M101 activated. Ten pulses on X401 counted. Then output.
8. See Figure A.11.

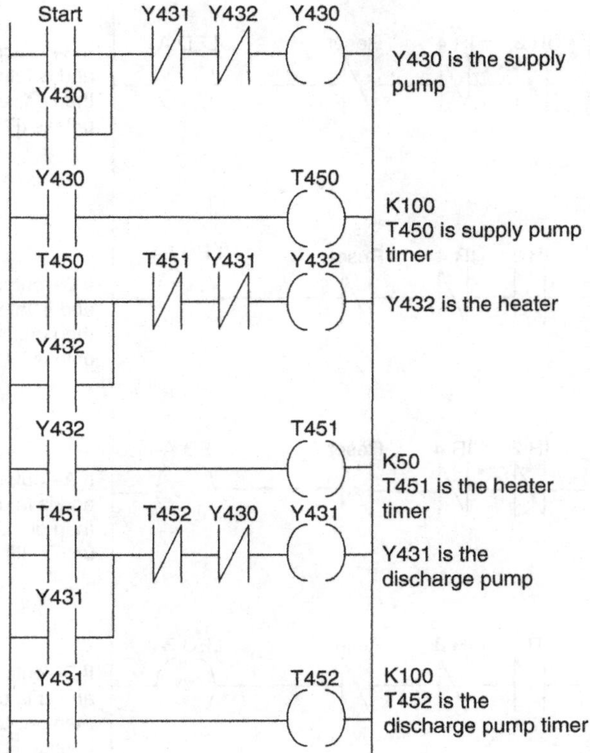

Figure A.8: Chapter 14, Problem 1.

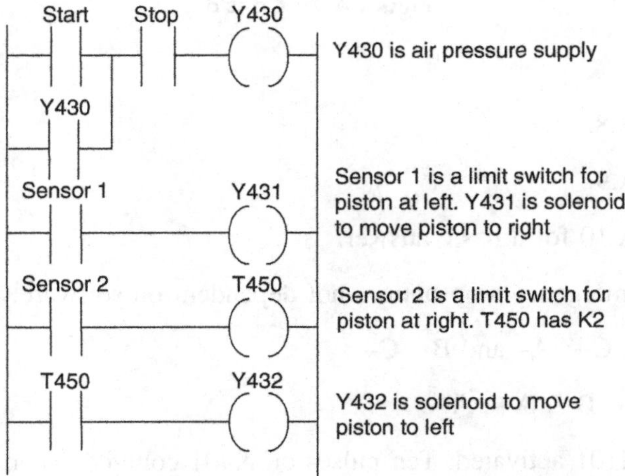

Figure A.9: Chapter 14, Problem 2.

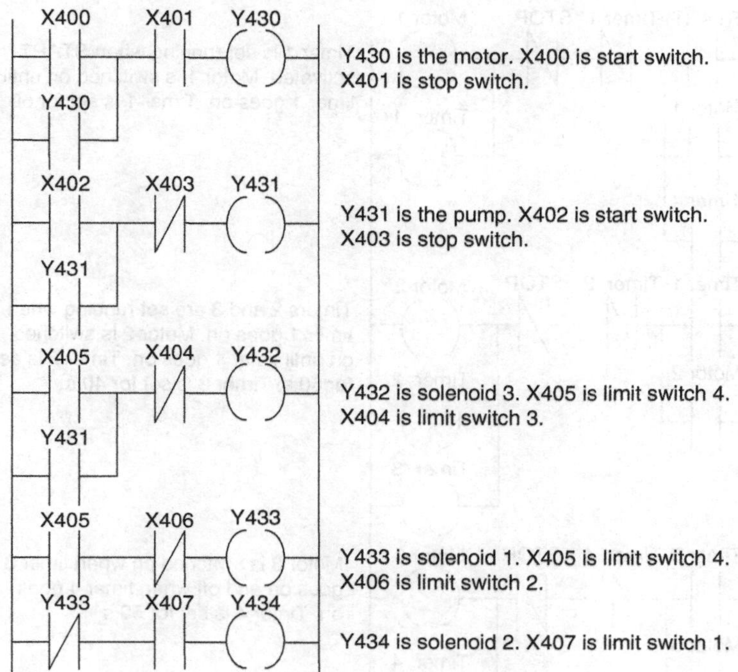
X400
X401
Y430
Y430 is the motor. X400 is start switch.
X401 is stop switch.
Y430
X402
X403
Y431
Y431 is the pump. X402 is start switch.
X403 is stop switch.
Y431
X405
X404
Y432
Y432 is solenoid 3. X405 is limit switch 4.
X404 is limit switch 3.
Y431
X405
X406
Y433
Y433 is solenoid 1. X405 is limit switch 4.
X406 is limit switch 2.
Y433
X407
Y434
Y434 is solenoid 2. X407 is limit switch 1.

Figure A.10: Chapter 14, Problem 3.

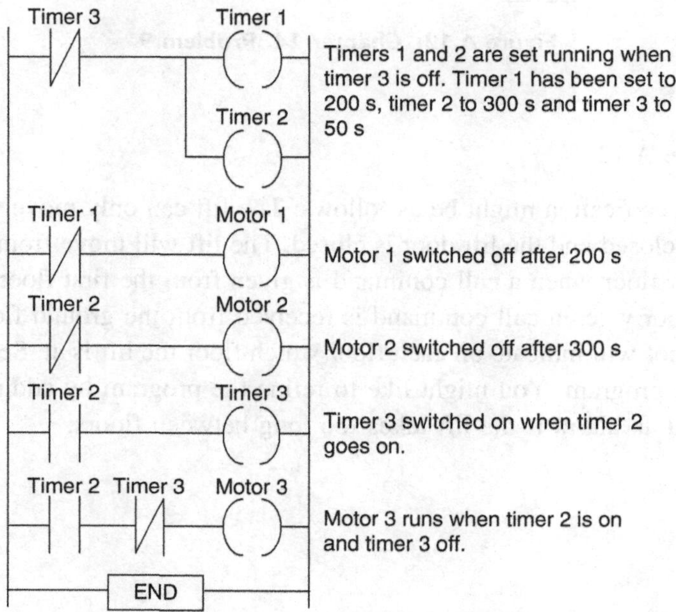
Timer 3
Timer 1
Timers 1 and 2 are set running when timer 3 is off. Timer 1 has been set to 200 s, timer 2 to 300 s and timer 3 to 50 s
Timer 2
Timer 1
Motor 1
Motor 1 switched off after 200 s
Timer 2
Motor 2
Motor 2 switched off after 300 s
Timer 2
Timer 3
Timer 3 switched on when timer 2 goes on.
Timer 2
Timer 3
Motor 3
Motor 3 runs when timer 2 is on and timer 3 off.
END

Figure A.11: Chapter 14, Problem 8.

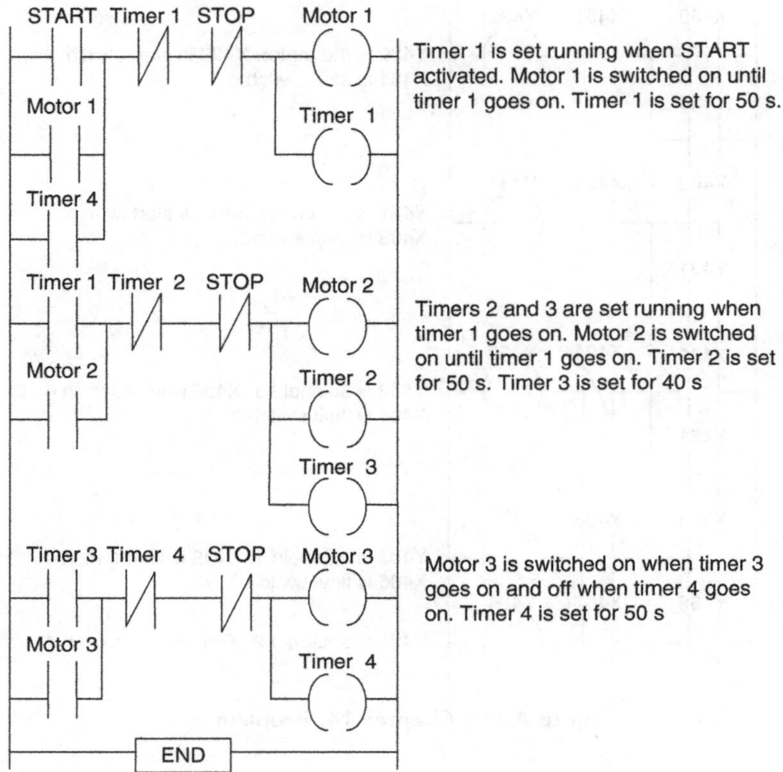

Figure A.12: Chapter 14, Problem 9.

9. See Figure A.12.

10. A basic specification might be as follows: The lift can only move when both access doors are closed and the lift door is closed. The lift will move from the ground floor to the first floor when a call command is given from the first floor and move to the ground floor when a call command is received from the ground floor. Signal lamps at each floor will indicate on each floor which floor the lift is at. See Figure A.13 for a possible program. You might like to refine the program by adding a timer which will sound an alarm if the lift takes too long between floors.

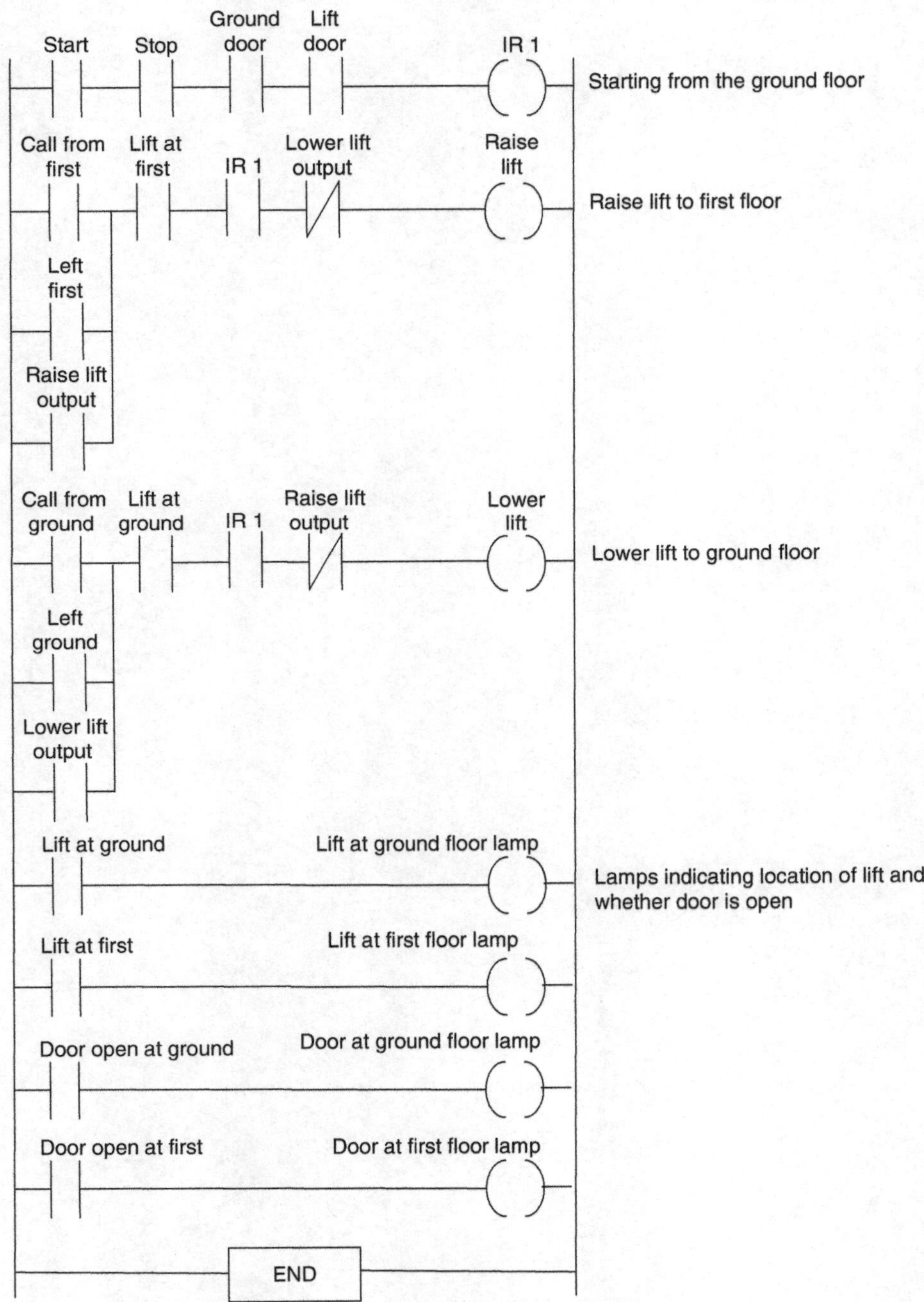

Figure A.13: Chapter 14, Problem 10.

Index

Note: Page numbers followed by *f* indicate figures and *t* indicate tables.

Q

R

S

Printed in the United States
By Bookmasters